AF726534

ACOUSTIC SENSING AND PROBING

Fourth Course of the International School on Physical Acoustics

ACOUSTIC SENSING AND PROBING

3 – 10 October 1991 Erice, Italy

Editor

A. Alipi

Coordination by

A. Petri

World Scientific

Singapore • New Jersey • London • Hong Kong

Published by

World Scientific Publishing Co Pte Ltd
5 Toh Tuck Link, Singapore 596224
USA office: 27 Warren Street, Suite 401-402, Hackensack, NJ 07601
UK office: 57 Shelton Street, Covent Garden, London WC2H 9HE

British Library Cataloguing-in-Publication Data
A catalogue record for this book is available from the British Library.

ACOUSTIC SENSING AND PROBING
Proceedings of the 4th Course of the International School on Physical Acoustics

Copyright © 1992 by World Scientific Publishing Co Pte Ltd

All rights reserved. This book, or parts thereof, may not be reproduced in any form or by any means, electronic or mechanical, including photocopying, recording or any information storage and retrieval system now known or to be invented, without written permission from the publisher.

For photocopying of material in this volume, please pay a copying fee through the Copyright Clearance Center, Inc., 222 Rosewood Drive, Danvers, MA 01923, USA. In this case permission to photocopy is not required from the publisher.

ISBN-13 978-981-02-0912-4
ISBN-10 981-02-0912-6

The organization of the Course has been made possible through the contribution of

Consiglio Nazionale delle Ricerche
Elettronica San Giorgio-Elsag
Ente Nazionale per l'Energia Elettrica

and the administrative support of

Istituto di Acustica "O.M. Corbino"

which all we wish to acknowledge.

International School of Physical Acoustics
Fourth course on
ACOUSTIC SENSING AND PROBING
Erice, Italy, October 3-10, 1991

DIRECTORS

Prof. A. Alippi
Institute of Acoustics, CNR
Rome, ITALY

Prof. W.G. Mayer
Georgetown University
Washington DC, USA

LECTURERS

L. Adler
Ohio State University
Columbus OH, USA

W. Arnold
Fraunhofer Institute IFZP
Saarbrücken, GERMANY

M.A. Breazeale
The University of Mississippi
MS, USA

E.H. Brown
University of Colorado
Boulder CO, USA

H. Engan
University of Trondheim
Trondheim, NORWAY

Yu. V. Gulyaev
Institute of Radio Engin. & Electronics
Moscow, USSR

P. Khuri-Yakub
Stanford University
CA, USA

W. Lauterborn
Institut Angewandte Physik
Darmstadt, GERMANY

W.G. Mayer
Georgetown University
Washington DC, USA

G. Quentin
Université Paris VII
Paris, FRANCE

R. Reibold
Physikalish-Technische Bundesanstalt
Braunschweig, GERMANY

A. Sliwinski
University of Gdansk
Gdansk, POLAND

FOREWORD

Here to follow are the Proceedings of the 4th Course of the International School of Physical Acoustics, which was held in Erice, Italy, from Oct 3 through Oct 10, 1991, at the International Center for Scientific Culture "Ettore Majorana". We are proud to announce that the School is now well and lively active among all the other Schools that for years have been holding courses at the E. Majorana Center. We began some ten years ago, with a topic, "Fundamentals of ultrasonic wave propagation", that was chosen on purpose to be as general as possible, in order to test the needs of the acoustic community and possibility of continuing with topics of a more specific nature. It appeared through the years that the seed was sown in a fertile field: participation increased steadily, and the School kept on growing through the comments and works of its participants.

Ultrasonics sensing and probing was the subject selected for the present course, where the scholars were intended to be inticed into exchanging their ideas and experiences with those involved in the many fields of applied science where sound waves can contribute to give information about material and structure defects.

The interaction of sound waves with the medium of propagation results in effects that can be used either to study the ultrasonic field itself, or to give information on the medium. The sensing and/or probing of the ultrasound field is studied together with the sensing and/or probing by the ultrasound field, unified by the effect of the interaction of the field with the medium.

We called in papers from the two sides, and hoped to have cross fertilization of interests. That really stands as one of the main scopes of the School: exchanging experience and comparing results among people that have a common denominator in their research work.

We do hope that the target has been partially fulfilled; at least, we presume we did it in the past from the support the participants gave us to continue the School.

We also hope the Proceedings can enrich the readers and stimulate their interest in ultrasonic sensing and probing.

Finally, we wish to express our gratitude to all those who contributed to the success of the School: the sponsoring Institutions named on the front page; the people at the E. Majorana Center in Erice for their warm support and hospitality; the staff at the Corbino Institute of Acoustics in Rome for the organization, particularly Lucia Covi, for her constant and intelligent catalizing action, and Gisella Menichelli for her care in editing the manuscripts.

Adriano Alippi

TABLE OF CONTENTS

I. INVITED PAPERS

MATERIAL CHARACTERIZATION BY ULTRASONIC SPECTROSCOPY

LASZLO ADLER
Department of Welding Engineering
The Ohio State University
190 W. 19th Avenue
Columbus, OH 43210

Abstract

Ultrasonic Spectroscopy is the study of ultrasonic waves resolved into their Fourier frequency components. Since most ultrasonic wave interaction with materials are frequency dependent, ultrasonic spectroscopy is a viable tool for material characterizations. Several examples will be included in this paper, such as flaw and porosity characterization, where the frequency dependent scattering mechanism is used for inversion.

1. Introduction

Ultrasonic Spectroscopy is the frequency domain study of ultrasonic wave interactions with a material. Ultrasonic spectroscopy proves valuables because many material properties are manifested as amplitude or phase changes in the ultrasonic waves interrogating the specimen.

Initial material characterization studies were carried out utilizing narrowband ultrasonic pulses. Gericke[1] pioneered the use of broadband pulses for material property investigation. Spectrum analysis of the pulse yielded information over a range of frequencies. He also appears to have been the first to use the term "ultrasonic spectroscopy". In an analogy with light, the specimen was insonified with 'white sound'. Defects or material structure would alter the 'color' of the scattered waves.

Other investigators refined ultrasonic spectroscopic instrumentation and broadened the list of applications[2]. In addition to the use of spectroscopy for flaw characterization, it has proven useful for determining surface properties to evaluate the volume fraction of porous materials, investigating microstructure, assessment of adhesive and solid state bonds, etc. In this paper, after describing a basic ultrasonic spectroscopy system, a few application of material characterization will be described.

2. Ultrasonic Spectroscopic Systems

All ultrasonic spectrum analysis systems contain provision s for (1) generating ultrasound, (2) receiving a portion of the ultrasound which has interacted with the material under study, and (3) analyzing the received wave to determine the magnitude and phase of the signal over a range of frequencies. In a common system configuration (Figure 1), an electrical signal generated by the transmitter is applied to the transmitting transducer. Electrical energy is converted to mechanical energy by the transducer, producing an ultrasonic wave. Interactions of the wave with the material alter its amplitude, phase and direction. A receiving transducer intercepts part of the energy and converts the ultrasonic energy back to an electrical signal. Because the signal is usually small, an amplifier is used to boost its amplitude. An analysis system sorts out the amplitude and timing of the ultrasonic interactions to present magnitude and phase spectra.

Since experimental artifacts must be removed from the displayed spectra, it has been proposed[3] that the spectroscopic instrument be modelled as a linear, time-invariant (LTI) system. Components can be analyzed in detail and their interaction to form the instrument may be determined. The behavior of a LTI system is described by its impulse response (in the time domain) or its frequency response (in the frequency domain). The two descriptions of system response are equivalent and are related by the Fourier transform pair.

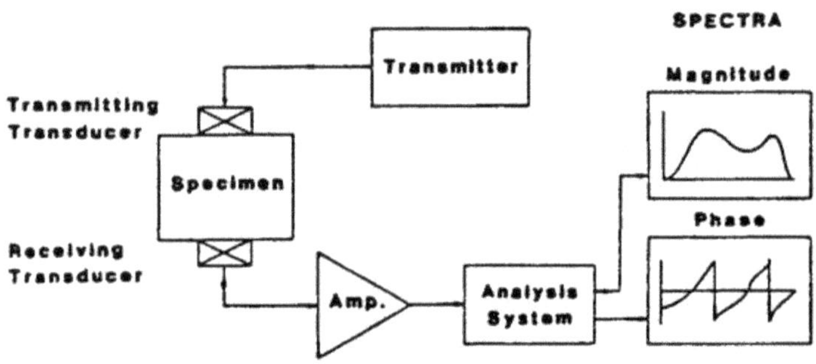

Figure 1 Generalized ultrasonic spectroscopy system.

Importantly, if the input [I(f)] and output [0(f)] from the specimen can be determined, then the ultrasonic transfer function of the material h(t) [or H(f)] may be determined by deconvolution.

Figure 2 illustrates an ultrasonic spectroscope modelled as a LTI system. Notice that for most of the components, the signal of interest is an electrical waveform. However, for an important section of the system the relevant signal is an ultrasonic wave propagating through the coupling media and material under study. to evaluate the material transfer function, the characteristics of each of the system components must be determined, and its effect removed by deconvolution.

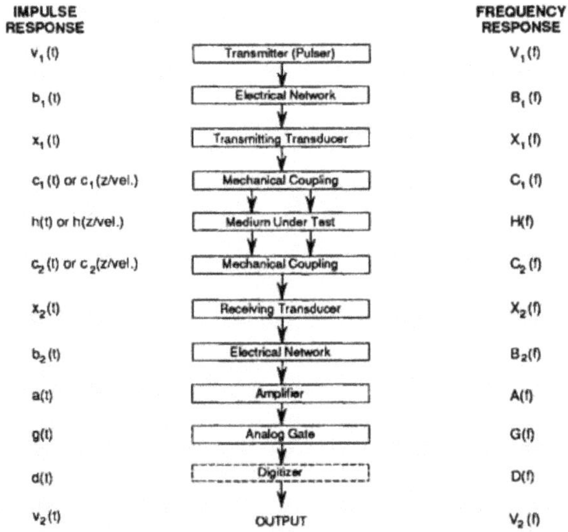

Figure 2 Elements of an Ultrasonic Spectroscopic system Modelled as a Linear, Time-Invariant System.

Ideally, components should exhibit an amplitude and phase response which is uniform for a wide range of frequencies. Piezoelectric transmitting and receiving transducers and often the mechanical coupling layers have transfer functions which are notoriously modulated. Although the nonuniformities may be compensated for in the deconvolution processing step, signal-to-noise ratio varies across the spectrum.

Assuming we have been able to characterize all components in the spectroscopic system, we move on to the investigation of frequency-dependent ultrasonic wave interactions with a material.

3. Flaw Characterization

Recent interest in quantitative flaw characterization in metals has stimulated several developments in the study of ultrasonic scattering from discontinuities in solids. The characterization of a discontinuity (size, orientation, shape, etc.) from the information obtained from scattered ultrasonic waves is a problem of great interest.

A new approach started by Gericke[1] was the development of a contact ultrasonic spectrum analysis system to characterize the geometry of hidden flaws in solids. He used a broadband rather than a single frequency ultrasonic pulse and he found that the spectral components of the echoes were strongly affected by the geometry of the discontinuity. Whaley and Cook[4] developed an immersion spectrum analysis system. Adler and Whaley[5] carried out a systematic study of the spectral response of scattered waves from circular reflectors in water by using this system. They developed a model to measure the size and orientation of circular reflectors. In this model the edges of disks are assumed to be Huygens sources of the scattered field in analogy to the Sommerfeld approach to diffraction of light waves by aperture in a screen. In the course of these investigations it became evident that quantitative flaw characterization will be possible only if elastic wave interaction with known discontinuities can be described in a simple but accurate manner. The so-called inverse problem can be attacked only if the direct problem is understood, i.e., if it is known how the parameters of the scattered ultrasonic field are affected by known discontinuities. The need to analyze ultrasonic scattering data on the basis of simple theoretical predictions became obvious.

Mathematically, ultrasonic wave scattering in solids is treated by using the equations of elastodynamic theory, with boundary conditions appropriate to the problem. Exact solutions exist only for a few idealized configurations. Several approximate solutions have been developed recently. The problem is classified as to whether the discontinuities are bounded by smooth closed surfaces such as for spheroids, or by planar surfaces having sharper edges, as for crack-like discontinuities. Another classification of the theory is based on the applicability in the region of ka where k is the wave number and a is some dimension of the discontinuity. For $ka < 1$ and for spheroidal discontinuities a theory based on the Born approximation was developed by Gubernatis, Domany, and Krumhans[6]. Other approximate theories in this region were also developed.

For crack-like planar defects in the region $ka > 1$, Adler and Lewis[7] used Keller's geometrical diffraction theory to analyze their experimental data. Achenbach, Gautesen, and McMaken[8] extended geometrical diffraction theory to the three-dimensional diffraction by cracks in solids, where wave motions are governed by a scalar and a vector waves equation, whose solutions are coupled by the boundary conditions on the diffracting obstacles. In this paper theoretical and experimental results will be presented in both regions of $ka \geq 1$ and $ka \leq 1$.

3.1 Theoretical Results

The scattering of ultrasonic waves by flaws in a solid is schematically illustrated on Figure 3. An incident wave is scattered by a flaw and the scattered wave is received at a point which is described by polar angle θ and azimuthal angle ϕ. It is convenient to present theoretical results in three-dimensional display. The choices of parameter are the scattered amplitudes vs. frequency and scattering polar angle for fixed azimuthal angle of amplitude vs. polar angles and azimuthal angles for a fixed frequency.

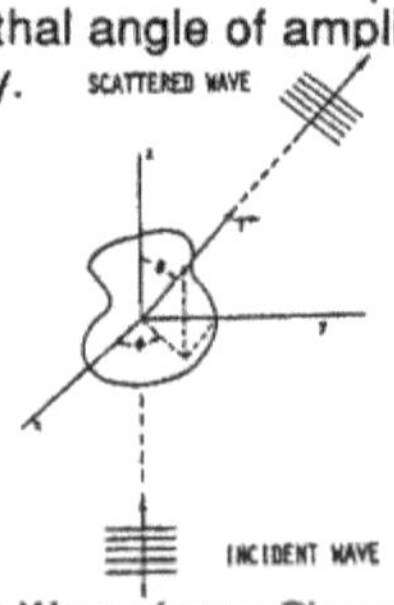

Figure 3 Scattering of Ultrasonic Waves from a Discontinuity.

3.1.1 Ellipsoidal Cavities. ka ≤ 1

For the region of ka ≤ 1 and for ellipsoidal cavities the Born approximation calculation was used[6]. For an incident L wave the scattered L and S wave amplitudes are plotted on Figure 4. The flaw is an oblate spheroidal cavity in titanium with dimensions of 200μ x 400μ. For the frequency range from 3 to 7 MHz the corresponding ka is from 0.6 to 1.4.

On Figure 5 the calculated amplitude field is plotted which is scattered from the same oblate spheroidal cavity at the frequency of 5.1 MHz. The polar angle changes from 120° to 30° and the azimuthal angle changes from 0° to 360°.

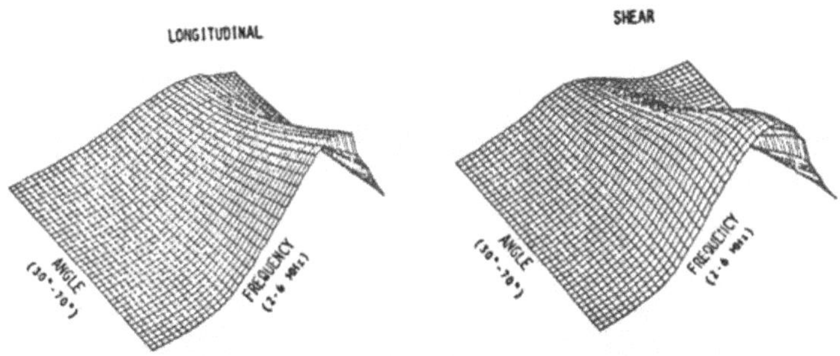

Figure 4 Calculated Intensity Distribution of L and S Waves Scattered from 200 x 800μ Oblate Spheroidal Cavity.

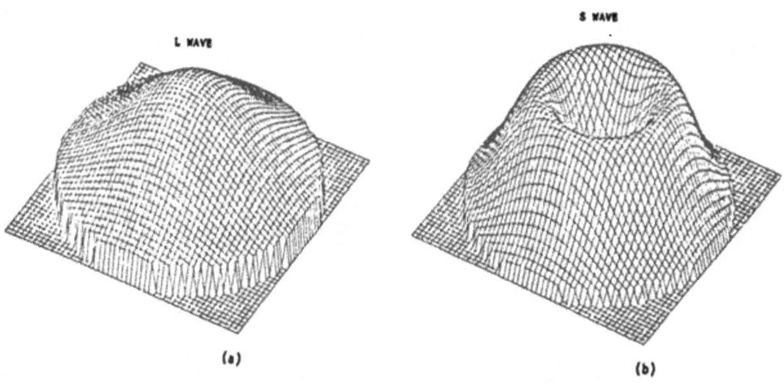

Figure 5 Calculated Scattering Field Profile for Incident L Wave of 5.1 MHz from a 200 x 800 Oblate Spheroidal Cavity. (a) Scattered L Wave; (b) Scattered S Wave.

3.1.2. Elliptical Cracks. ka ≥ 1

This region of scattering is commonly called the diffraction region. The diffraction of L waves by two-dimensional crack-like flaws was calculated using elastodynamic theory[8]. On Figure 7 the scattered amplitude from an elliptical crack (in titanium) of major axis b and minor axis a are plotted vs. frequency and azimuthal orientation as the receiver's position changed from along the major to the minor axis. The polar angle is 60°. The four different plots correspond to aspect ratios (b/a) of 1, 2, 4, and 8. The major axis b in each case is 2500μ. The spectra along the minor axis shows periodicity which is inversely proportional to the aspect ratio. The other theoretical prediction is that the energy scattered along the major axis is always less than the energy along the minor axis. The difference is increasing with increasing aspect ratios.

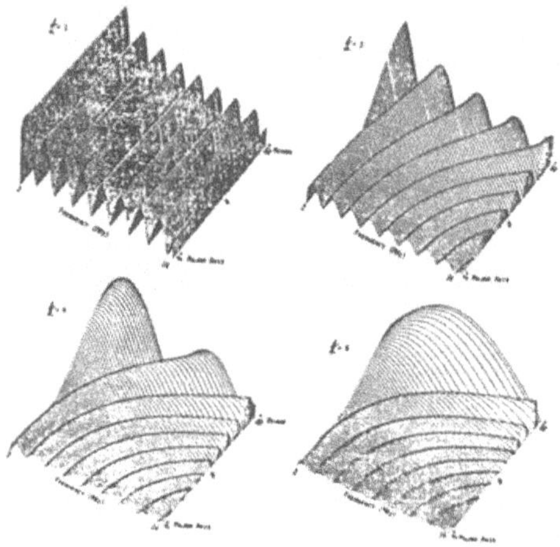

Figure 6 Theoretical Amplitude Spectra of Normal Incidence L Waves Scattered from Elliptical Cracks. Polar Angle is 60°.

3.2. Experimental Technique and Procedure

The technique used is shown on Figure 7. The sample, which is a titanium alloy disk (2.5 x 10 cm) with a flat surface, is immersed in water. The transmitter launches a longitudinal wave toward the liquid-solid interface at an angle. For nonnormal incidence both L and T waves are produced in the metal. The cavity can be insonified either by the L wave or by the T wave with incident angle α. At the cavity, the waves are scattered and mode conversion occurs. The scattered waves are received and can be analyzed separately due to their separation in time. The defects studied were cracks (circular and elliptical) and cavities (oblate and prolate spheroidal) in diffusion-bonded titanium. Defect dimensions ranged from 200μ to 2500μ.

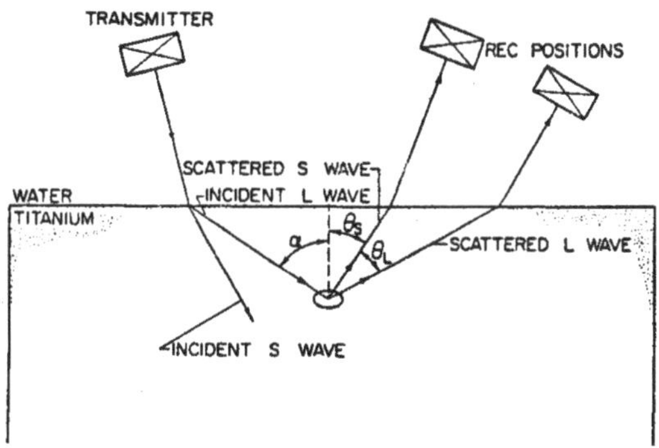

Figure 7 Experimental Technique

3.3 Results

3.3.1. Cavities

Various key parameters of the scattered field were identified throughout the experiments as characteristics of a given cavity.

3.3.2. Frequency Dependence

The spectra of the scattered L and S waves from an L wave normally incident on various spheroidal inclusions was recorded from 30° to 60° in two degree increments, and compared to the spectra predicted by the Born approximation theory. As mentioned before, the theory should apply for $ka < 1$. Considering the radius along the direction of incidence to be a, the L wave values for the 200 x 800 μm oblate spheroid range from 0.2 to 0.8 for frequencies of 2 to 8 MHz for normal incidence. Figure 8 compares theory with experiment for this oblate spheroid for two angles of both L and S scattered waves. The theory has been corrected by the transfer function of the ultrasonic spectroscopic system. The agreement for both L and S wave scattering is reasonably good.

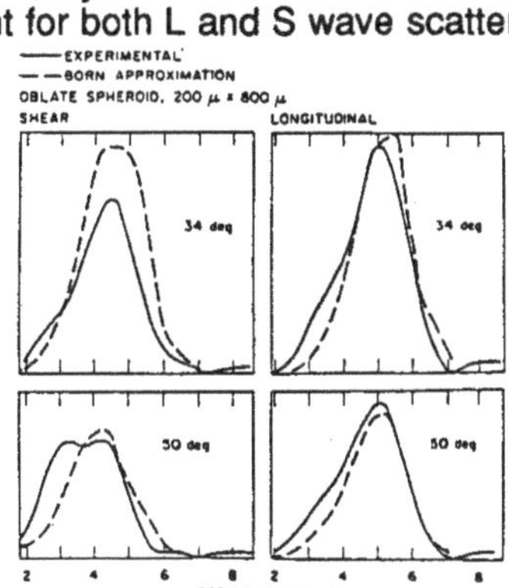

Figure 8 Power Spectra for L-S and L-L Scattering

3.3.3. Planar Cracks

For normally incident L waves the amplitude of the diffracted L wave is calculated as a function of frequency for a 2500μ circular crack in titanium. Typical data are shown on Fig. 9 at 45° scattering angle. The dota are experimental data points. The amplitude of the first few cycles agrees well at higher frequencies. The experimental results are lower than predicted by the theory. The position of maxima and minima, however, agree well between experiment and theory. The locations of the maxima are significant for the inversion process as will be shown in the next section. On Fig. 10, both theoretical and experimental amplitude spectra are shown for different azimuthal angles. The agreement between theory and experiment is good for the larger angles and poor for the smaller angles, i.e., near the major axis. A possible explanation of this phenomenon is the available scattered energy is higher (as predicted by Fig. 4) as one approaches the minor axis and therefore the analysis is more meaningful.

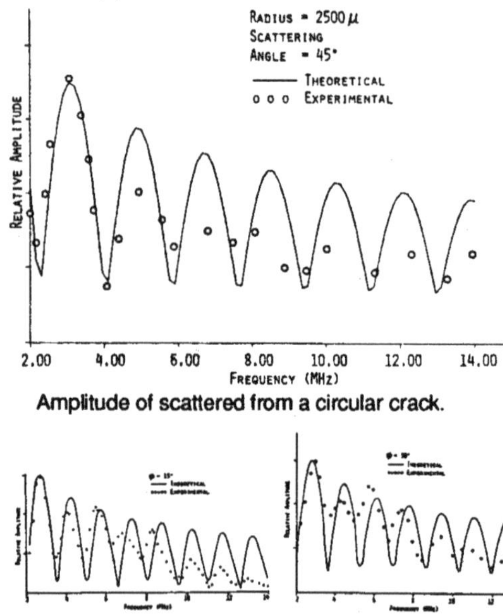

Figure 9 Amplitude of scattered from a circular crack.

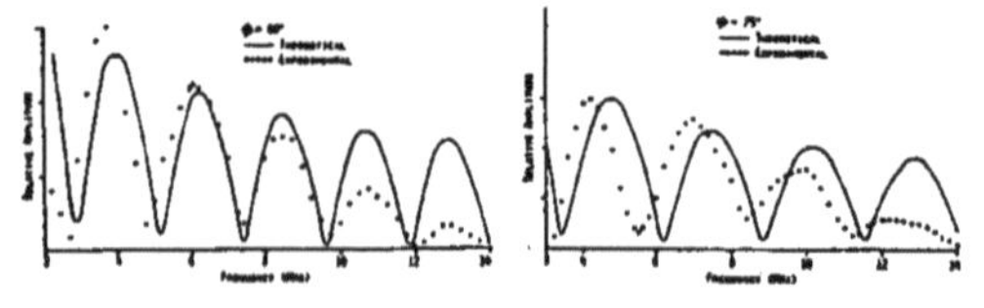

Figure 10 Amplitude spectra of scattered L waves from a 2500μ x 1250μ elliptical crack in titanium along different azimuthal directions. Polar angle 60°.

3.4. Determination of Crack Size from the Diffraction Spectra

It has been shown by Achenbach, Adler, Lewis, and McMaken[9] that one can carry out a simple inversion procedure on characteristics of the amplitude spectra. The dimension of the crack

$$a = (C_L) / [2 \sin \theta \, (\Delta f_{max})_{av}]$$

where θ is the scattering angle, C_L is the longitudinal velocity in the solid, and $(\Delta f_{max})_{av}$ is the average frequency spacing between two consecutive frequency maxima. Table 1 illustrates the results collected from several measurements for four different size elliptical cracks. The measured and actual dimensions are in very good agreement.

Table 1

Determination of Major (B) and Minor Axis (A) of Elliptical Cracks in Titanium from Scattered Amplitude Spectra

(B) In Microns		(A) In Microns	
Actual	Measured	Actual	Measured
2500	2498	2500	2498
2500	2587	1250	1291
2500	2387	625	830
2500	--	312	285

4. Characterization of Porous Material

Porosity in structural materials limits their ultimate strength and hence their utility. Consequently nondestructive methods of characterizing the amount and nature of the porosity are important for this and other reasons. The number of pores as well as their distribution in size, shape and orientation are quantities which an ideal characterization method would determine. Ultrasonic methods are particularly good candidates since they are sensitive to these parameters. One of the fundamental measurements in ultrasonics is the attenuation of an incident beam or pulse. As will be discussed the spectral behavior of the ultrasonic attenuation is sensitive to the volume fraction and average pore size.

4.1 Theory[10]

Scattering of ultrasound from porosity will be described in terms of the following model. Consider an isotropic elastic medium which is uniform and everywhere homogeneous except between two parallel planes. Between the two planes (the first at $z = 0$ and the second at $z = z_0$) there is assumed to be a spatially uniform distribution of spherical voids (pores). All of the pores are assumed to have the same radius, a_p.

An idealized scattering experiment is modeled as follows. A longitudinally polarized plane wave is incident along the z-axis which is normal to the planes bounding the porous region. A portion of the incident wave is transmitted coherently. Fundamental measurements are made to determine the reflection and transmission. Clearly both depend on the frequency dependent attenuation of the incident pulse as it propagates through the porous region. Figure 11 shows a schematic of the idealized experimental geometry.

The description of the scattering and attenuation is greatly simplified in the limit that the volume fraction of pores, c, is small[11]. This allows multiple scattering to be ignored. The scattering amplitude becomes just the coherent sum of the amplitude generated at each pore. The volume fractions measured in our experiments varied from 0% - 6% and, therefore, $c \ll 1$. It should be noted that this porosity range encompasses all premium aluminum castings used in aerospace applications.

The attenuation $\alpha(k)$ describes the power scattered out of the main beam as a function of the frequency. On the other hand, the power scattered by each pore is described by its total cross-section $\delta(k)$. In the small volume fraction limit, one may thus expect the attenuation to be proportional to the scattering cross-section. Here k is the magnitude of the longitudinal wave vector where $k = 2\pi f/V_L$; f is the frequency and V_L denotes the longitudinal velocity.

The attenuation, reflection, and transmission can be understood qualitatively by considering the low and high frequency limits for long wavelengths, $ka_p \ll 1$, the power scattered per pore decreases as frequency to the fourth power $[\alpha(k) \;\; k^4]$. Consequently, the sample is relatively transparent and little ultrasound is reflected. At high frequencies ($ka \gg 1$) scattering can be described by ray concepts an the geometric cross-section of the flaw. For $ka_p \gg 1$, the attenuation, reflection, and transmission become relatively independent of frequency, and the sample becomes more opaque compared to lower frequencies.

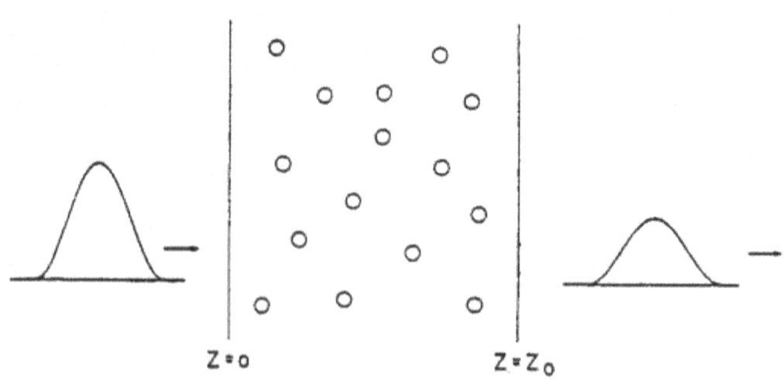

Figure 11 Schematic representation of the idealized experimental situation.

The transition from long - to short-wavelength behavior typically occurs when $ka_p \approx 1$. Since the transition is relatively abrupt, measurements of the transmission or reflection can be used to determine the pore size. Further, the overall strength of the signal can then be used to find the volume fraction of pres. For the aluminum alloy used in this study the pore size and porosity (pore concentrations) are given by the following equations.

Pore size: $$a_p = (V_L/2\pi)\,(1/f_p) \qquad (1)$$

Porosity: $$c = (V_L/2\pi x K)\,(\alpha_p/f_p) \qquad (2)$$

where V_L is the longitudinal velocity of each sample, K is a constant which depends on the material used, f_p and α_p are the frequency and the attenuation coefficient corresponding to the location of the turning point.

4.2 Attenuation Measurements

The attenuation of the ultrasonic waves in a porous sample is given as

$$\alpha(k) = \frac{1}{z - z_1} \qquad \ln \frac{A(k,z)}{A(k,z_1)} \qquad (3)$$

Here A(k,z) represents the amplitude at the front surface of the sample (z), and $A(k,z_1)$ represents the amplitude at the back surface of the sample (z_1). These amplitudes are corrected for beam divergence using diffraction corrections for A(k,z) and $A(k,z_1)$. Corrections are also made for the liquid-solid interface losses. The attenuation coefficient $\alpha(k)$ is then obtained by dividing the front surface spectrum by the backwall echo spectrum as shown in Figure 12. The result attenuation coefficient versus frequency curve is also shown in Figure 12.

Determination of the pore radius was made by dividing the frequency dependent attenuation coefficient by its frequency, thus approximating a single peak near $ka_p = 1$. The value of f_p, determined from the $\alpha(k)/f_p$ graph, was then substituted into the following equation

$$a_p = 1.05 \times V_L/(2\alpha \times f_p) \qquad (4)$$

where V_L is the longitudinal velocity of each sample. The volume fraction of porosity was calculated by substituting the value of a_p obtained from Eq. 4 and $\alpha(k_p)$, the value of the attenuation coefficient located at f_p, into Eq. 5.

$$c = 1.22 \times \alpha(k_p) \times a_p \qquad (5)$$

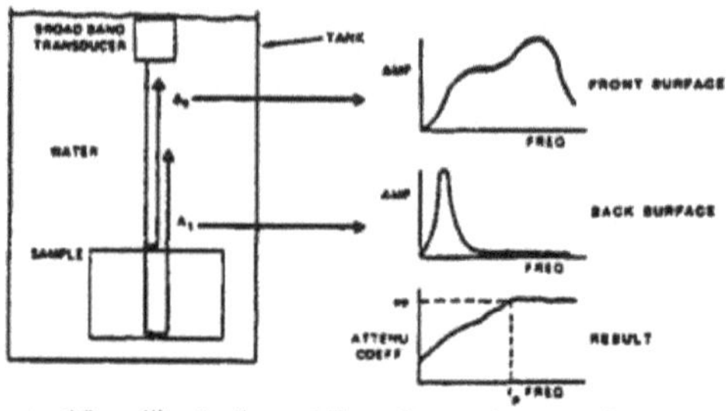

Figure 12 Illustration of the ultrasonic porosity assessment technique.

A typical result of the attenuation coefficient frequency dependence for high porosity material is shown in Figure 13. The sample is 4.6% porosity. The attenuation coefficient increases with frequency up to about 10 MHz at which point it levels off to a value of about 2.8 Np/cm. The behavior of this curve below 2 MHz should be ignored because the transducer puts out very little energy in this region. For a lower porosity material such as 2.18% porosity), the attenuation coefficient is shown in Figure 14. From these figures one may conclude that at least two qualitative features of the spectral dependence of the porosity induced attenuation may be extracted: (i) the frequency at which the attenuation curve reaches a relatively constant plateau value, and (ii) the value of the attenuation coefficient in the plateau region. These quantities may be used to evaluate the average pore size and the volume fraction of porosity c.

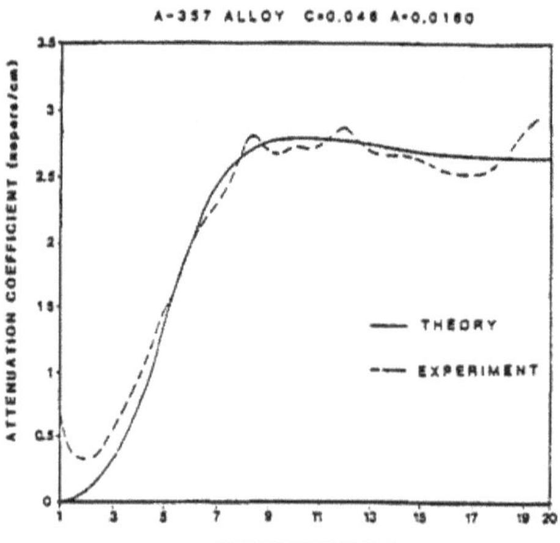

Figure 13 Sample 1920

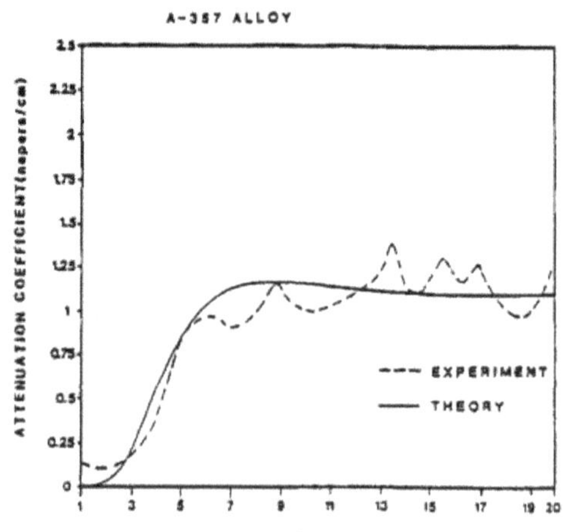

Figure 14 Sample 1510

5. Characterization of Solid-State Bond of Layered Substrate[12]

In nondestructive evaluation the inspection of bond quality between dissimilar materials is of primary importance and is one of the most difficult problems to solve. Most previous work dealt with the frequency analysis of the normal incidence reflected signal and showed that is has very poor sensitivity to bond quality. Instead of using bulk waves, one can investigate bond quality in a layer-substrate structure (aluminum on steel or steel on aluminum) using generalized Lamb waves. The two materials have shear velocities very close to one another and interesting behavior of the lowest velocity mode is expected theoretically at low values of the frequency-thickness parameter fh. Depending on the ratio of the longitudinal velocities in the layer and the substrate, the initial slope of the dispersion curve of this mode is either positive or negative and then the curve exhibits a turning point with slope reversal.

In Figure 15 we plotted the dispersion curves of the various modes for a steel layer on an aluminum substrate immersed in water. The behavior of the slowest acoustical mode is very interesting. The phase velocity of the surface wave first decreases with frequency then reaches a minimum value at fh ~0.5 MHz x mm and then increases. Both the initial slope and the minimum position depend on the longitudinal wave velocities.

In order to calculate the two curves plotted in Figure 15, rigid boundary conditions are used between the layer and the substrate. Such conditions are only valid for good bonding. The program that was used can be modified in order to study the "weak bond" and "no bond" cases (results are shown on Figure 16).

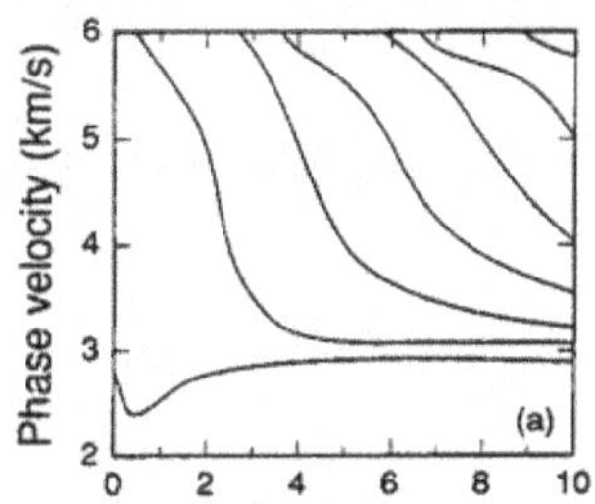

Figure 15 Theoretical dispersion curves for various guided modes. Steel layer on an aluminum substrate.

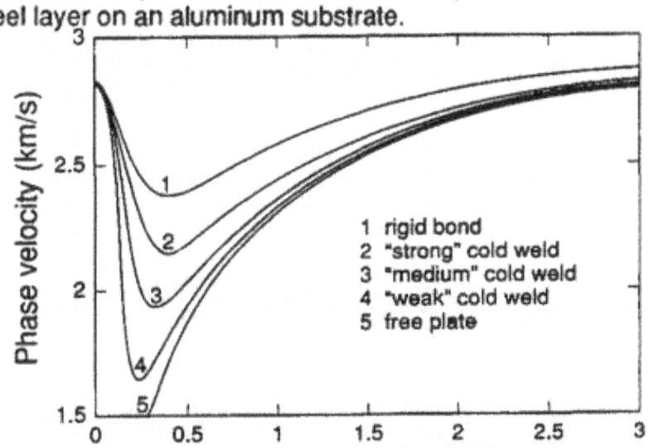

Figure 16 Theoretical dispersion curve of the lowest-velocity mode for various bond qualities.

5.1 Experimental Techniques

As usual for Lamb waves, the dispersion curves are plotted versus the product fh. Two techniques have been used in order to measure the variations of the phase velocity versus frequency. The first one (Figure 17a) is based on the existence of a minimum in the reflection coefficient when the Snell's law is satisfied for one of the waves generated in the sample, that is when

$$\sin \theta = V_w / v(f) \tag{6}$$

where v_w is the velocity of sound in the surrounding water.

An ultrasonic tone burst is reflected from the sample. The incidence θ_i and reflection θ_R angles are varied simultaneously in such a way that they remain equal to one another and for each center frequency of the tone burst one measures the angles θ_i where a minimum reflection occurs (Figure 17b). From the angles, using relation (6), one obtains the velocities v_i of waves generated at this frequency.

The second experimental setup is shown in Figure 18a. It uses a single transducer emitting a broadband pulse. A signal is received by the same transducer if the frequency and incident angle θ_i are such that a leaky wave is generated in the sample. The wave is reflected by the right-hand corner, comes back to the left, and because of the leaking property of this wave, part of its energy goes back towards the transducer. A typical example of the signal received by the transducer is plotted in Figure 18b. In this signal we select, using a gate, the part corresponding to the reflection at the corner and then carry out a Fast Fourier Transform analysis. The obtained frequency spectrum (Figure 18) exhibits maxima at frequencies f_i. The velocity v_i of the generated waves is given by Eq. 2, and we plot the dispersion curves as the points corresponding to frequencies f_i on the horizontal line v_i. Then we change the angle of incidence θ_i and do the same for various angles of incidence.

Both techniques were used in this study to obtain the dispersion curves for bonded layer-substrate structures. Frequencies used in the experiments ranged from 0.4 MHz to 15 MHz.

5.2. Results

In Figure 19, we compare experimental results with theoretical predictions for a steel layer on aluminum when the bond is a good one. Agreement between theory and experiment is good, especially for the slowest mode. The characteristic features of this mode are well verified by experiment (initial decrease towards a minimum and then smooth increase toward an asymptotic value when frequency increases). For the faster modes, the experimental points correspond, in general, to velocities slightly lower than predicted theoretically.

On the other hand, in the case of "very weak" bond (Figure 20) between the steel layer and the aluminum substrate, the experimental points fit the classical Lamb mode curve quite well for a layer of the same thickness in water.

That is, the velocity of the slowest mode (A_0 mode) increases monotonically towards an asymptotic value equal to the Rayleigh velocity on steel when frequency increases. The two other Lamb modes observed experimentally are the S_0 mode and the A_1 mode. The experimental results confirm that, in this case, the layer vibrations are the same as for a free layer. Experimental results are also shown together with the two extreme case (good and poor bonds) for intermediate "weak" quality bond. For this case, the data points are scattered between their limits and this can be related to the theoretical results presented on Figure 21.

The fact that the data points for a "weakly" bonded sample are scattered instead of being placed on a given dispersion curve is because weakly bonded inertia-friction welded samples exhibit non-uniform quality of the bond along the welded interface.

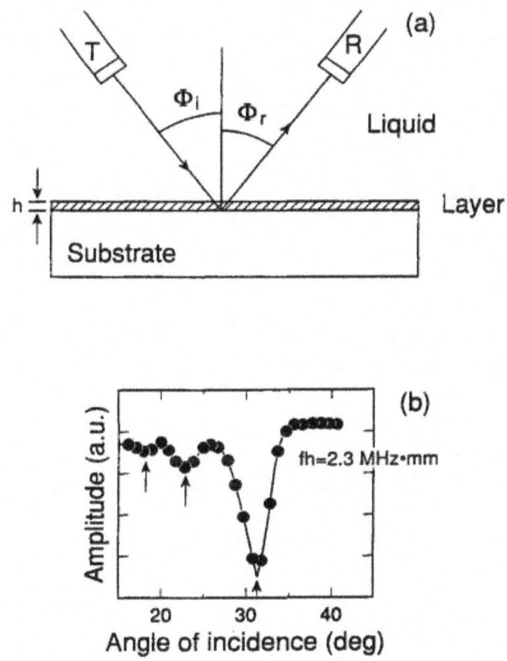

Figure 17 a and b (a) Experimental technique using the minimum of the reflection coefficient. (b) Angular dependence of the reflected amplitude.

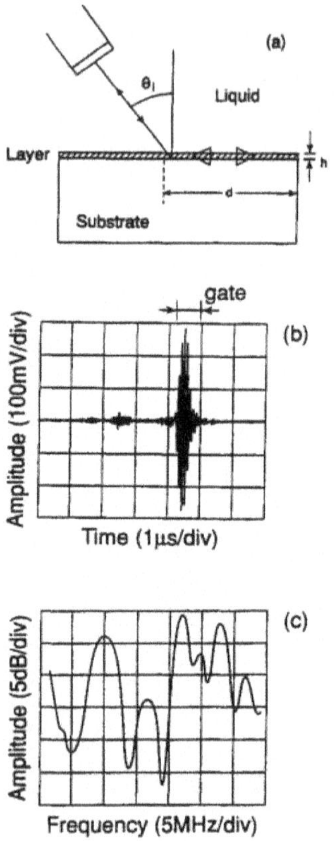

(a) Experimental technique using reflection at the sample's edge.

(b) Typical time signal obtained with the configuration of (a).

(c) Frequency spectrum of the previous signal [(b)].

Figure 18 a, b, c

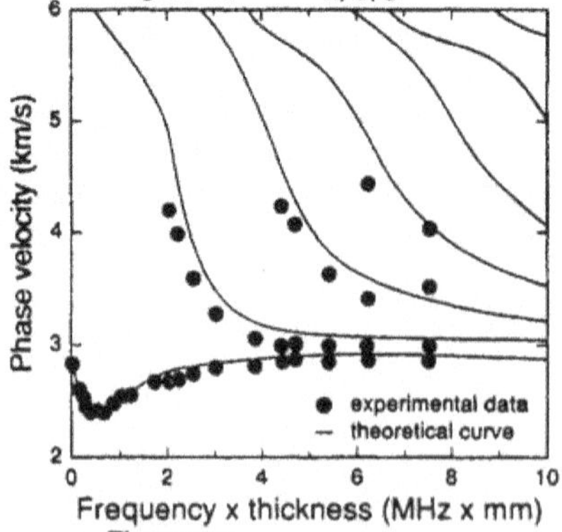

Figure 19

Theoretical/experimental comparison of dispersion curves for a "good" bond between a steel layer andan aluminum substrate.

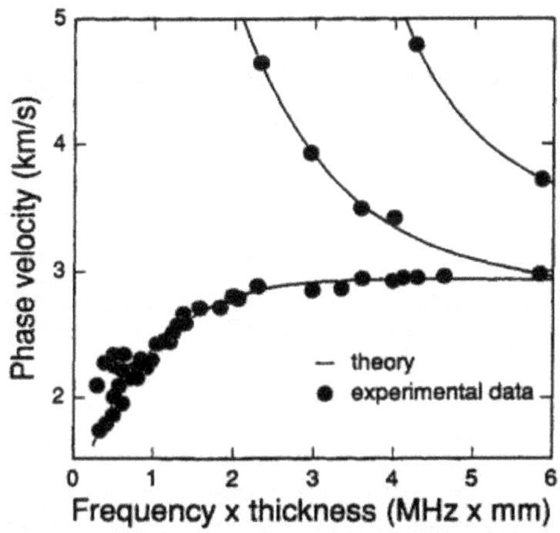

Figure 20 Theoretical/experimental comparison of dispersion curves for a "weak bond between a steel layer and an aluminum substrate.

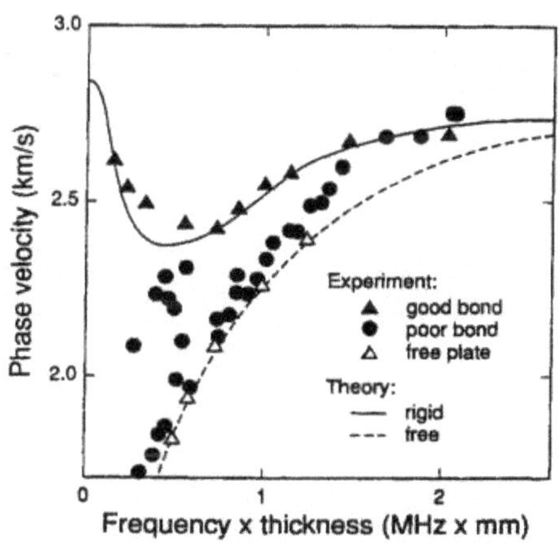

Figure 21 Experimental dispersion curve of the lowest-velocity mode for various bond qualities.

6. Acknowledgements

This work is supported by the Office of Naval Research under contract number N00014-88-KO452; Scientific officer: Dr. Y. Rajapakse.

7. References

1. O. R. Gericke, *J. Acoust. Soc. Am.* **35**, 364 (1963).

2. Fitting, D. W. and Adler, L., *Ultrasonic Spectrum Analysis for Nondestructive Evaluation*. Plenum Pub. Co. (New York), (1981).

3. Frederick, J. R. and Seydel, J. A., *Bulletin Number 185 of the Welding Research Council*. (July, 1973).

4. H. L. Whaley and K. V. Cook, *Mat. Eval.* **28**, 61 (1970).

5. L. Adler and H. L. Whaley, *J. Acoust. Soc. Am.* **51**, 881 (1972).

6. J. E. Gubernatis et all., *J. Appl. Phys.* **48** (1977); **48**, 2812 (1977).

7. L. Adler and D. K. Lewis, *J. Acoust. Soc. Am.* **57A**, Supp. No. 1 (1975).

8. J. D. Achenbach and A. K. Gautesen, *J. Acoust. Soc. Am.* ***61***, 413 (1976).

9. Achenbach, J. D., Adler, L., Lewis, D. K. and McMaken, H., *J. Acoust. Soc. Am.* **66**, 1848 (1979).

10. L. Adler, J. H. Rose and C. Mobley, *J. Appl. Phys.* **59**, 336 (1986).

11. J. E. Gubernatis and E. Donany, *Rev. Progress in QNDE* **2A** (1983).

12. L. Adler et al., *J. Appl. Phys.* **68** 6072 (1990).

NON-DESTRUCTIVE CHARACTERIZATION OF MATERIALS BY HIGH-FREQUENCY ACOUSTICS

W. ARNOLD
Fraunhofer-Institute, Bldg. 37, University
D-6600 Saarbrücken 11,
Fed. Rep. Germany

ABSTRACT

This paper describes in short some new ultrasonic techniques developed for non-descructive characterization of materials.

1. High-Frequency Acoustics

Non-destructive characterization (NDC) of components made of advanced materials requires special efforts. Since these materials are used for high-strength applications, NDC-techniques are necessary which allow one to detect small defects. This results from fracture mechanical considerations. Often it is required to detect defects with a linear dimension smaller than 100 µm. We have developed high-frequency ultrasonic techniques which fulfill these requirements and which are described in detail elsewhere /1/.

Our system comprises two rf-transmitters, and one wide-band receiver. Depending on the ndt-task to be performed, narrowband (burst-signals with 20 W peak power from 10-200 MHz center frequency, 1-99 oscillations) or exponentially decaying voltage spikes (300 V, 3 ns rise-time, 10 ns decay-time) may excite the PVDF-transducer (Fig. 1). Careful theoretical and experimental analysis of the signals obtained from various defects show that it is possible to detect spherical inclusions of 30 µm diameter with a signal-to-noise (SNR) ratio of 6 dB in materials with an ultrasonic absorption less than 1 dB/cm at 50 MHz /1/. There are quite a few systems on the market described as high-frequency ultrasonic inspection systems /2/. Most of them use only spike-excitation of the probe which is easy to realize electronically. If one needs to perform measurements at elevated frequencies, however, it is advantageous to work with narrow-band excitation of the probe in order to obtain sufficient SNR, since the total attenuation in technical materials increases as a function of frequency rapidly. For example in ceramics, the attenuation starts to increase rapidly beyond 1 db/cm due to Rayleigh-scattering above 50 MHz, as published previously /3/ and verified by us independently.

We display the results of the scanning of a component with ultrasonic waves in the form of conventional C-scans. The ultrasonic data can be also displayed after digitization in a "three-dimensional way" and assembling these data in a computer /4/. By selecting any

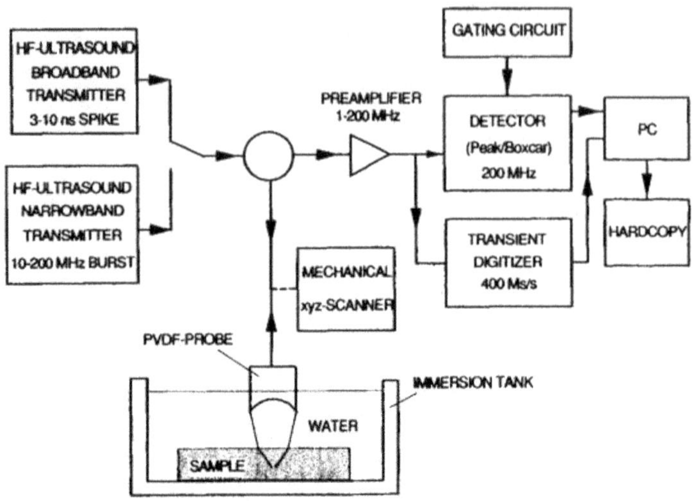

Fig. 1: Block diagram of the system used for high-frequency ultrasonic C-scan imaging.

desired cross section planes inside the volume, one is able to "travel" through the volume of the sample and to extract very rapidly all conventional C-scan and B-scan images, allowing one to obtain a clear picture of the distribution of defects in the sample and its homogeneity (Fig. 2). In addition, using representations of the data in the frequency domain, one often gets a clear image of the distribution of the second phases possibly present, because they scatter ultrasound more strongly than the matrix (Fig. 3).

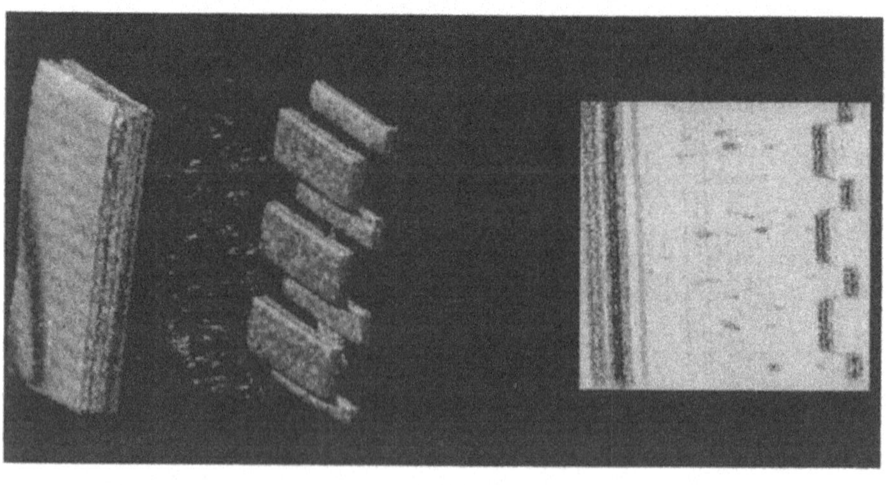

infiltrated SiC ceramic with three 0.5 mm deep vertical groves on the rear side. Three-dimensional viewing (left side), and slices perpendicular or parallel to the time axis (right side), corresponding to conventional C- and B-scans, can be chosen. The dots visible indicate the location of defects (from /4/).

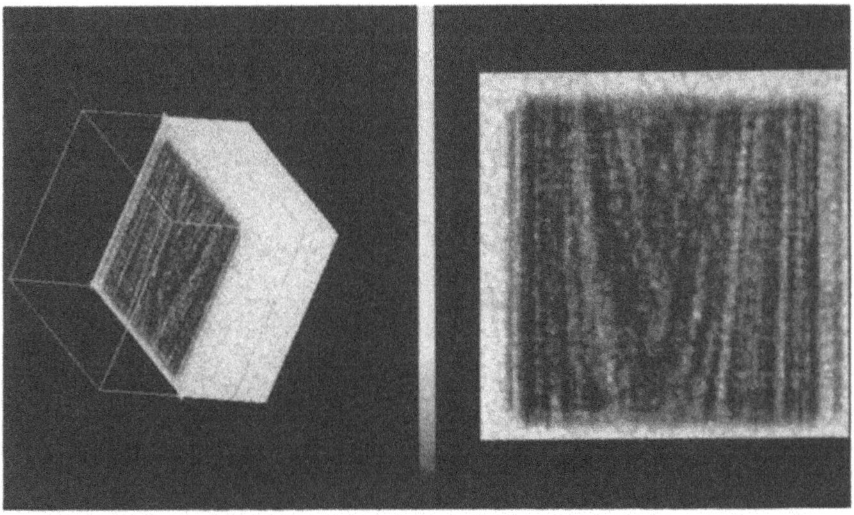

Fig. 3: Frequency-domain image of a powder metallurgical aluminum alloy component containing 2% of 15 µm Cu particles. The flow pattern developed during the production process is visible owing to the particles enhanced ultrasonic scattering ($\propto f^4$). The depth axis represents frequency. Here, an image is obtained at 25 MHz. The other axis represents x- and y- coordinate (30 x 30 mm^2) (from /4/).

2. Acoustic Microscopy

If the frequency of the ultrasonic system is further increased, one eventually obtains a resolution comparable to the one of optical microscopy /5/. These systems all work with rf-bursts from 100 MHz up to 2 GHz, and are called "Scanning Acoustic Microscopes". At this frequency the penetration into the material is only a few microns. Acoustic Microscopy allows one to study elastic material properties with high spatial resolution. This was pointed out previously, also at presentations at former "Summer School of Physical Acoustics" /6/. At present our efforts concentrate on the measurements of elastic effects related to the so-called R-curve behavior in Al_2O_3- and SiC ceramics (Fig. 4), on the evaluation of creep damage /7/, and on the stress distribution along grain boundaries as measured by the V(z)-technique. Additionally, SAM allows one to monitor surface damage caused by grinding, polishing and other surface processing techniques of ceramics. In this case, the number of indications describing deviations from surface homogeneity are directly correlated with the bending strength of the material under consideration (Fig. 5).

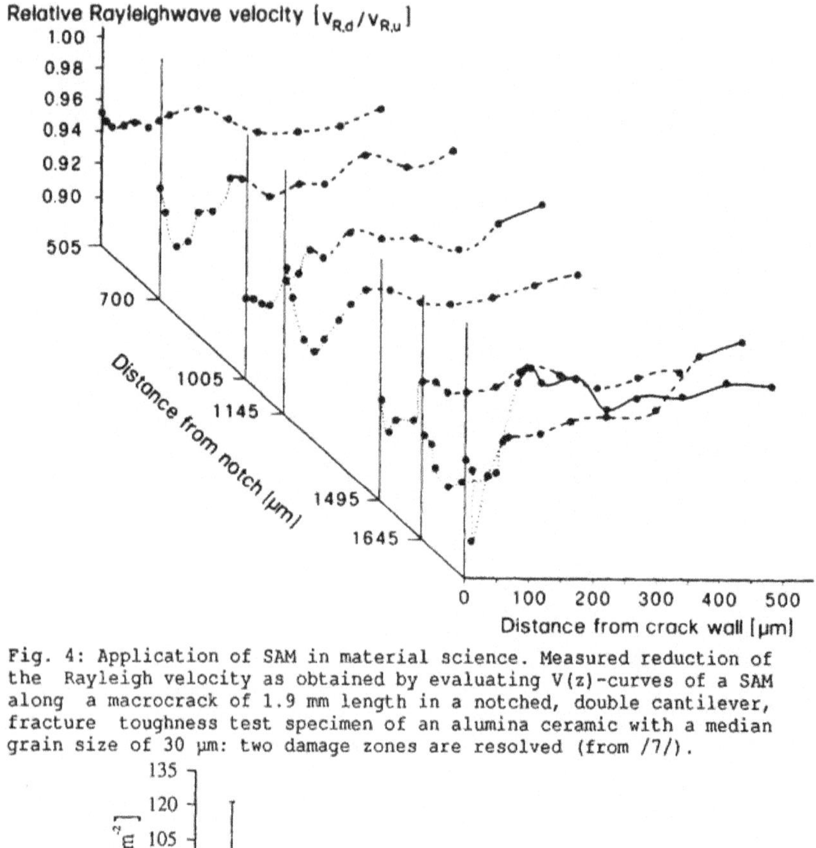

Fig. 4: Application of SAM in material science. Measured reduction of the Rayleigh velocity as obtained by evaluating V(z)-curves of a SAM along a macrocrack of 1.9 mm length in a notched, double cantilever, fracture toughness test specimen of an alumina ceramic with a median grain size of 30 μm: two damage zones are resolved (from /7/).

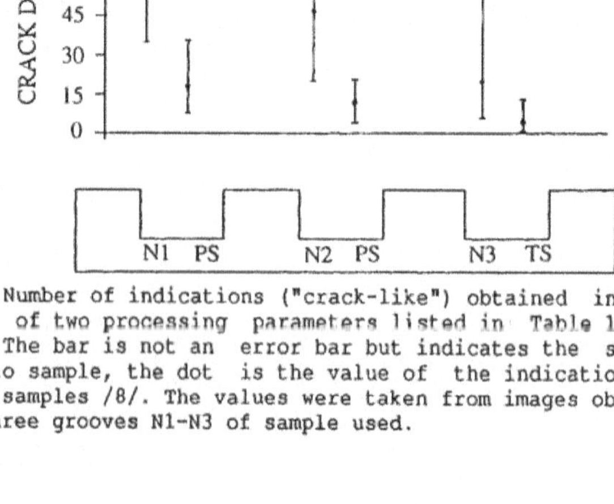

Fig. 5: Number of indications ("crack-like") obtained in a SAM as a function of two processing parameters listed in Table 1 on a ZrO_2 sample. The bar is not an error bar but indicates the scatter from sample to sample, the dot is the value of the indications averaged over all samples /8/. The values were taken from images obtained at 1 GHz at three grooves N1-N3 of sample used.

	IN-FEED a_e (mm)	FEED SPEED v_{ft} (mm/s)	CUTTING SPEED v_c (m/s)	MATERIAL REMOVAL RATE PER UNIT WHEEL WIDTH Q'_w (mm^2/s)	
N1	0.03	167	25	5	PS
N2	0.03	84	25	2.5	PS
N3	0.5	10	25	5	TS

PS : ALTERNATING GRINDING
TS : CREEP FEED GRINDING

Table 1: Processing parameters related to data shown in Fig. 5.

3. Depth-Hardening Measurements

Ultrasonic scattering can be used for the characterization of the microstructure, in particular for the evaluation of sintering processes, creep damages, etc.. Ultrasonic backscattering exploiting Rayleigh-scattering has been studied previously in order to monitor grain-sizes in polycristalline materials non-destructively /9/. As it turns out, the same technique can be used to measure depth-hardening /10, 11/ (Fig. 5) for industrial applications /12/.

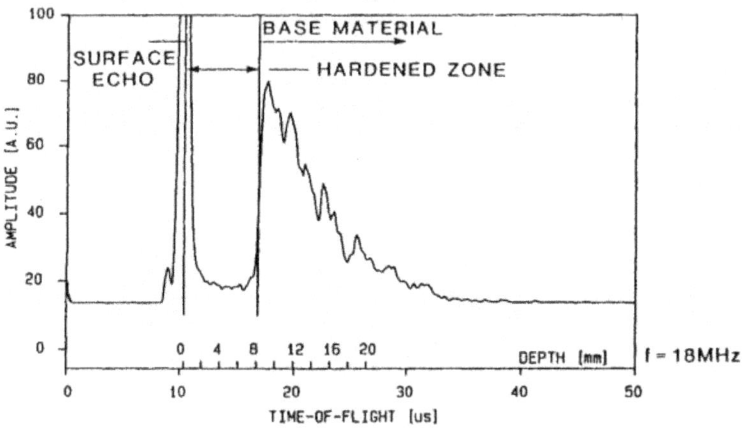

Fig. 5: Averaged backscattered ultrasonic signal from a hardened component. Due to its fine-grained structure, almost no scattering is observed during the passage of the pulse through the hardened layer. Only in the transition zone adjacent to the structure of the base material, strong scattering occurs. Together with the known sound velocity, the hardening depth t can be determined (from /10/).

4. Laser Ultrasonics

Short laser pulses can be used to generate ultrasonic pulses in polycristalline materials, and with interferometric techniques it is possible to detect the ultrasonic signals in a non-contact way. The damping of the spectral components of the ultrasonic pulses can be exploited to measure the total attenuation $\alpha(f)$ as a function of frequency f. Knowing $\alpha(f)$, it is then straightforward to deduce the grainsize by using the Rayleigh-approximation. This technique has been applied to polycristalline metal plates with a thickness from 0.5 to a few mm's, and grain sizes between 20 µm and 80 µm /13/.

Laser ultrasonics allows one to deduce in polycristalline materials the absorptive, i.e. the inelastic part α_a. The processes which determine αa, such as dislocation movements, thermoelastic losses, Snoek efect are all known, however not their overall effect in polycristalline materials (at lest to my knowledge). The experimental set-up is displayed in Fig. 6. A short laser pulse generates broadband ultrasonic signals which are partially mode-converted at each reflection at the surface and scattered at each grain boundary. Eventually, most of the ultrasonic energy is transformed into shear modes. The overall decay of the energy can be measured by an interferometer at any given point of the surface of the sample. The sample must be small enough so that its volume is equally filled with ultrasonic energy. The technique is similar to the one published previously using piezoelectric transducers /14/. Because the decaying ultrasonic energy is measured by an interferoemter in the technique described here, no additional damping of the reverberations occur due to coupling. From Fig. 7, it can be seen clearly that the decay time due to absorption α_a is much longer than the one due to scattering α_s. There are typically 1-3 orders of difference between the amplitudes of the two parameters. At present we can say that this absorptive part is caused mostly by dislocation movement, although well-known theories describing dislocation absorption in single-crystals cannot be used without proper refinement for polycristalline materials. Further details will be published elsewhere.

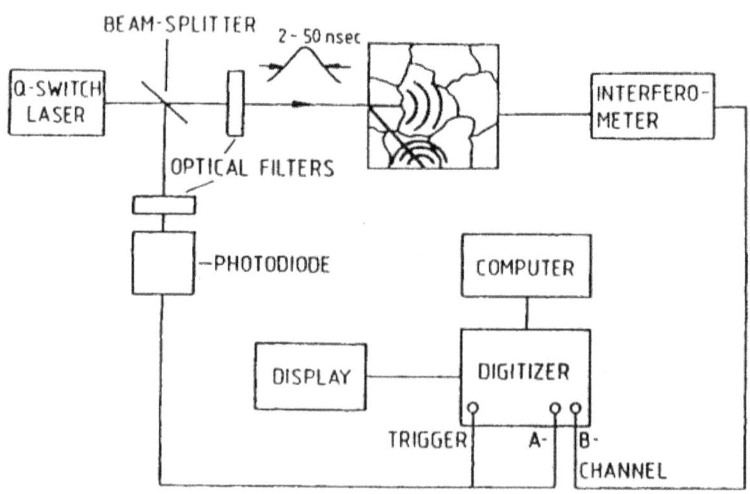

Fig. 6: Principle of measurement of the ultrasonic absorption in polycristalline samples by laser ultrasonics /14/.

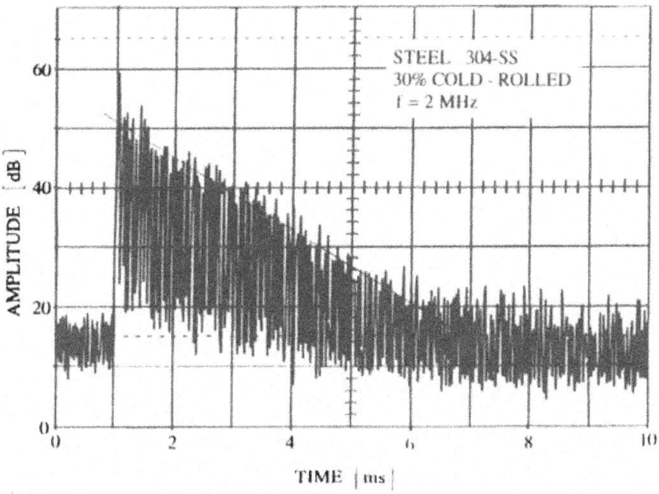

Fig. 7: Decay pattern of the ultrasonic signals as obtained from the interferometer output in an steel sample with a grain size of a few tenth of µm. The frequency of 3 MHz was selected by a filter after broadband amplification. The interferometer detects only vertical displacements. But because most of the ultrasonic signals with transverse polarization are reflected under an oblique angle to the surface, they are detected by the interferometer /14/.

5. Acknowledgement

I should like to thank W. Bernauer, S. Faßbender, B. Haberer, R. Herzer, U. Netzelmann, M. Paul, S. Pangraz, H. Stolz, U. Rabe, and H. Willems for close collaboration. The financial support for some of the research described here by the German Ministry of Research and Technology, the European Commission, and the European Commission on Coal and Steel is gratefully acknowlegded.

6. References

/1/ S. Pangraz, H. Simon, R. Herzer, and W. Arnold, in "Acoustical Imaging" Eds. H. Lee and G. Wade, 18, Plenum Press (1991), p. 189

/2/ see for example colecetd papers in "Non-Destructive Testing" Proc. 12th World. Conf. NDT, Eds. J. Boogard and G.M. van Djik, Elsevier Science Publishers and catalogue of Conf. Exhibition

/3/ A.G. Evans, B.R. Tittmann, L. Ahlberg, B.T. Khuri-Yakub, and G.S. Kino, J. Appl. Phys. 49, 2669 (1978)

/4/ U. Netzelmann, R. Herzer, H. Stolz, and W. Arnold, in "Acoustical Imaging" Eds. H. Ermert and H.P. Harjes, Plenum Press, to be published (1992).

/5/ C. F. Quate, A. Atalar, and H.K. Wickramasinghe, Proc. IEEE, 67, 1092 (1979)

/6/ J. Attal, in "Ultrasonic Signal Processing" Proc. 3rd Int. School of Phys. Acoustics, Ed. A. Alippi, World Scientific (1989), p. 69
/7/ S. Pangraz and W. Arnold, subm. for publ. in Am. J. Ceram. (1992)
/8/ S. Pangraz, unpublished
/9/ K. Goebbels in "Research Techniques in Non-Destructive Testing" Ed. R.S. Sharpe, 4, Academic Press (1980), p. 87
/10/H. Willems, in "Rev. Progr. in Quant. Non-Destr. Eval.", Eds. D.O. Thompson and D.E. Chimenti, 10B, Plenum Press (1991), 1707
/11/D. Bruche, R. Herzer, and E. Schneider, in "Moderne Analysen und Verfahren in der zfP" Proc. DGzfP-Jahrestagung, (1990) 18, 434
/12/R. Neumann und H. Willems, in "DACH-Jahrestagung" Proc. "DGzfP-Jahrestag." 28.2 (1991), p. 637
/13/S. Faßbender, M. Kulakov, B. Hoffmann, M. Paul, H. Peukert, and W. Arnold, in "Non-Destructive Material Characterization" Proc. 5th Int. Symp. Non-Destr. Charact. Mat., Eds. P. Höller, V. Hauk, G. Dobmann, C. Ruud, and R. Green, Springer Berlin (1989), p. 337
/14/B. Haberer, M. Paul, unpublished

THE CAPACITATIVE MICROPHONE AS A SENSOR IN MEASUREMENT OF ACOUSTICAL NONLINEARITY

M. A. BREAZEALE
National Center for Physical Acoustics
Coliseum Drive
University, MS 38677

ABSTRACT

A capacitive receiver is described for use in measurement of the absolute amplitude of ultrasonic waves in solids between 10 and 60 MHz. A sensitivity of 10^{-4} angstroms means that this capacitive receiver is ideally suited for measurement of nonlinear distortion of ultrasonic waves in solids. Measured results are given for nonlinearity parameters in a number of solids between 4° K and room temperature.

INTRODUCTION

The capacitive receiver has been used as a detector of acoustical signals for many years. It has characteristics which may present real advantages on occasion: it usually is plane; it is very sensitive; it has a flat frequency response; it is capable of absolute calibration; when it is used as a transmitter one finds that there is a frequency doubling without bias, so bias often is used in the capacitive transmitter; and the capacitive system is very simple to use. By describing the system we have been using I felt that I might suggest additional applications.

Our first system used a capacitive receiver at 30 MHz for measuring the absolute amplitude of ultrasonic waves in crystalline solids[1]. Later we adapted the system for measuring absolute amplitudes in crystalline solids as a function of temperature from 4° K up to room temperature[2]. In other experiments we have measured to temperatures as high as 350° K[3]. Since we subsequently have used the capacitive receiver at 60 and 90 MHz, there is not an obvious frequency limitation to the system either. We have observed that the sensitivity is great enough that we can measure a displacement amplitude as small as 10-4 angstroms[4]. By measuring the amplitudes of the fundamental and the second harmonics, we have been able to determine the nonlinearity parameters and the third order elastic constants of a large number of solids[5].

EXPERIMENTAL ARRANGEMENT

The arrangement for using the capacitive receiver is shown in Fig. 1, where each of the components is labelled. Typically, the capacitive receiver has a diameter of 1 cm and a spacing of 5 microns. Sample size is preferably 2.5 cm in diameter and 2.5 cm in length. The system for using the capacitive receiver is shown in block diagram form in Fig. 2. Absolute measurement of amplitudes is possible because the funciton generator produces a substitutional signal whose amplitude and frequency are accurately known. This signal can be compared directly with the unknown signal passing through the sample. An equivalent circuit of the capacitive receiver is shown in Fig. 3. An alternating signal V is applied across C, passes through C_2, and R_2, a known resistance. Sample vibration causes the spacing to change by an amount

$$S = S_0 + 2A \sin \omega t \qquad (1)$$

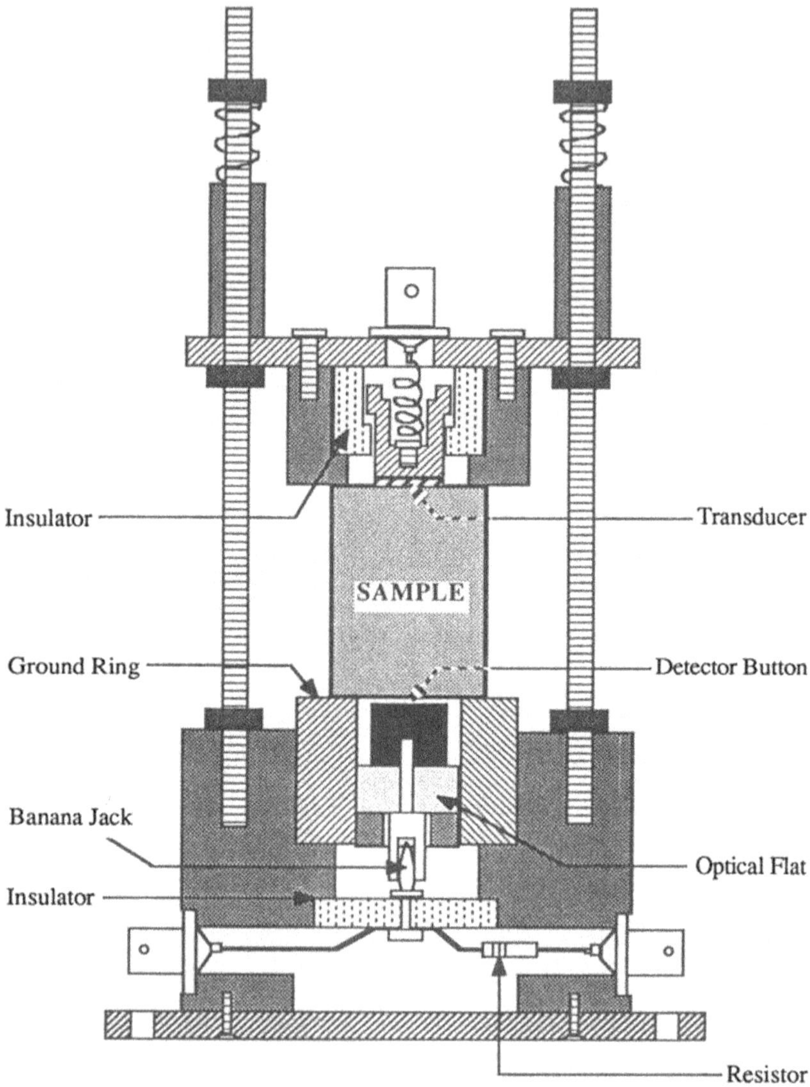

Figure 1. Capacitive detector for the measurement at room temperature

where A is the wave amplitude to be measured in the sample. This spacing change causes the capacitance to change by

$$\frac{1}{C} = \frac{1}{C_0}\left(1 + \frac{2A}{S_0}\sin \omega t\right) \tag{2}$$

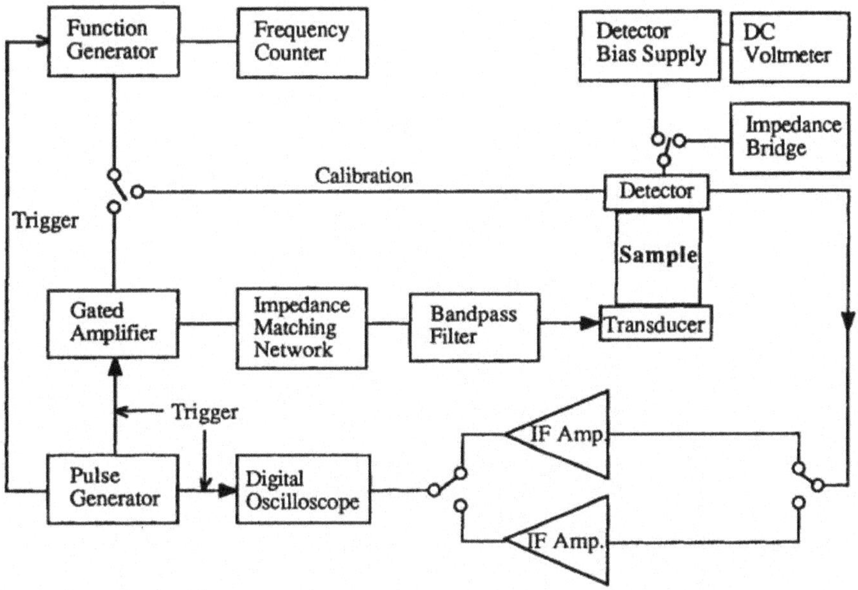

Figure 2. Block diagram of the room temperature experimental setup

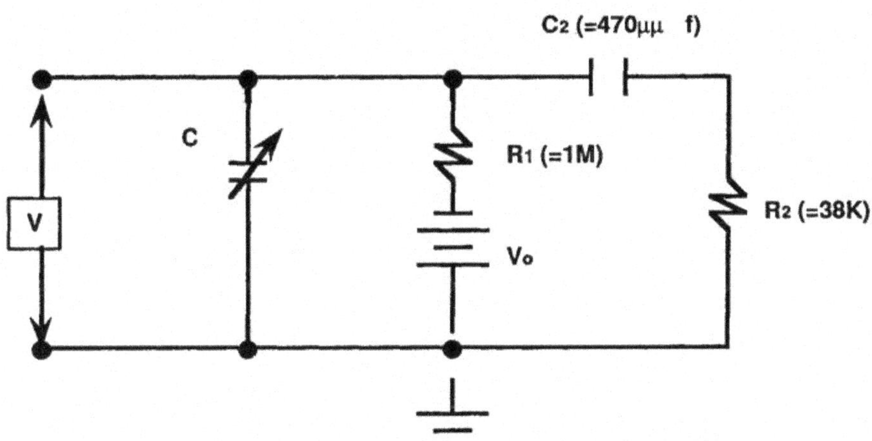

Figure 3. Equivalent circuit of the capacitive receiver

where

$$C_0 = \frac{\varepsilon\alpha}{S_0} \tag{3}$$

and ε is the permitivity and α is the area of the receiver button. For a bias on the capacitive receiver of magnitude V_b, the signal is

$$V = \frac{2AV_b}{S_0} \tag{4}$$

Since $i_D = V\omega C_D$, the current generated by the wave amplitude A is

$$i_D = \frac{2AV_b\omega C_D}{S_0} \tag{5}$$

USE OF THE CAPACITIVE RECEIVER

The capacitive receiver described can be used to advantage in many experiments in which one is determining the physical properties of crystalline solids. This is in addition to the many applications of the principles described at lower frequencies in gases. In this section a few examples will be given. The first example is its use to demonstrate the effect of diffraction and the use of diffraction to measure wedge angle in the sample. An echo pattern was obtained at a frequency of 10 MHz in a copper single crystal with propagation in the [100] direction. A transducer of diameter 1.27 cm generated an ultrasonic tone burst in a copper single crystal having a length of 1.717 cm. The resulting echo train received by a capacitive receiver at the opposite end of the copper sample is shown in Fig. 4 (b). It is clear that the llth echo corresponds to the second maximum of the envelope plotted in Fig. 4 (a). At the second maximum the ordinate $2kan\theta$ has a magnitude 5. Thus, the wedge angle of the sample $\theta = 5/(2kan) = 2.4 \times 10^{-3}$ radians, or 0.14 degrees (since $2a = d = 1.27$cm, $k = 1.5 \times 10^4$ m^{-1} and $n = 11$). This is a wedge angle of $\theta = 8.4$ minutes. Subsequent measurement of the wedge angle of the sample with an autocollimator gave a wedge angle of $\theta = 8.1$ minutes, a difference of less than 3% from that measured ultrasonically. If the objective were to measure attenuation in this sample, one could do so[6]. However, the usefulness of the capacitive receiver in this case already has been demonstrated.

The echo train in a much more parallel sample of PZT is shown in Fig. 5. The attenuation in this sample can be measured directly since the effect of nonparallelism of the two surfaces is negligible. The maximum wedge angle in this sample is 2.1 seconds of arc, so the effect of diffraction in this sample is barely noticeable. One prefers to have samples as good as this for most ultrasonics experiments; however, such accuracy may not be practical in all cases.

The capacitive receiver is capable of producing well-defined echo trains even when one is using it to study the nonlinear behavior of solids. Figure 6 shows an echo train in a copper single crystal in the [100] direction. The upper trace shows the 30 MHz signal applied to a quartz transducer at one end of the sample. The middle trace shows the echo train observed at the fundamental frequency by the capacitive receiver at the other end. The bottom trace shows the second harmonic echo train. Higher harmonics than the second can be measured as well.

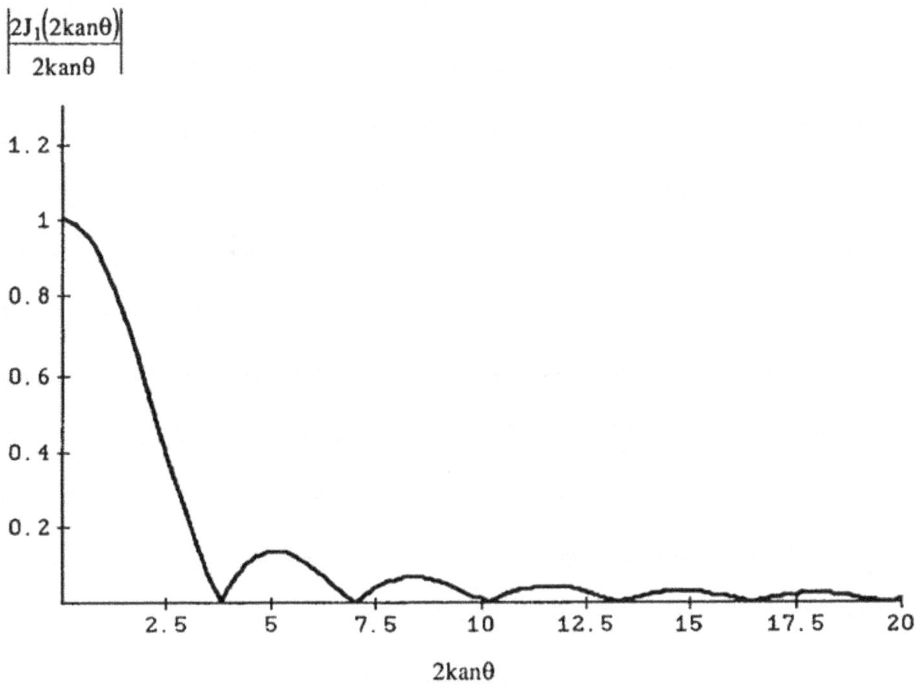

Figure 4(a). Plot of [$2J_1(2ka n\theta)/2ka n\theta$]

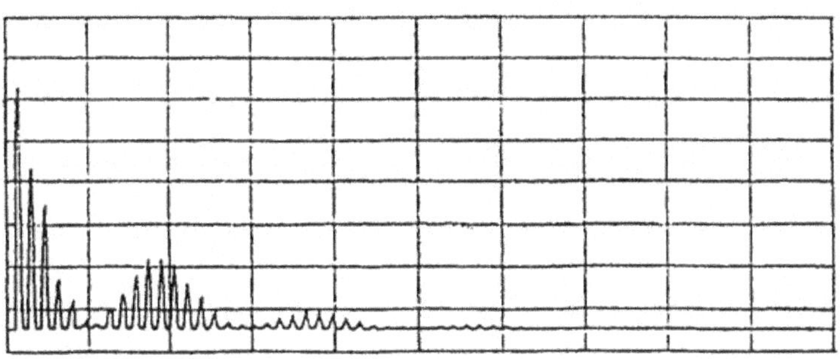

Figure 4(b). Plot of echo pattern obtained at frequency 10 MHz with 1.727 cm copper single crystal [100].

MEASUREMENT OF NONLINEARITY PARAMETERS

Since it is an absolute measurement device, and since it has a flat frequency response, the capacitive receiver is ideally suited to the measurement of the nonlinearity parameters of solids. Generally speaking, the nonlinearity parameters of solids are linear

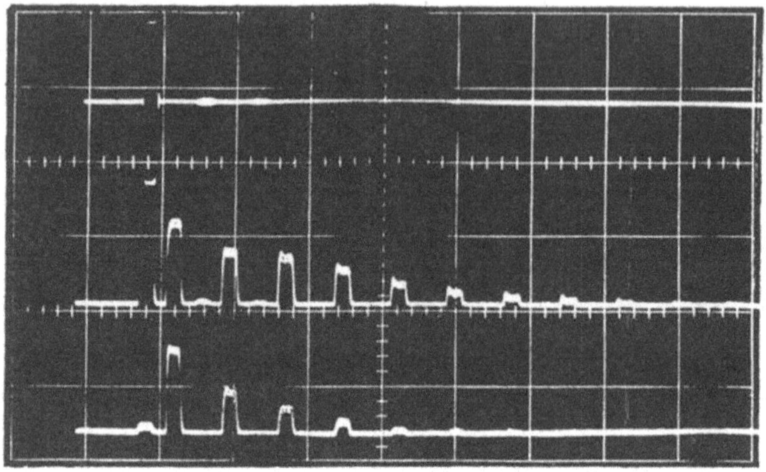

Top: Pulse applied to generating transducer.
Middle: Echo pattern at fundamental frequency.
Bottom: Echo pattern at second harmonic.

Figure 5. Echo trains in a [100] copper sample

combinations of second order elastic constants and third order elastic constants. Second order elastic constants are readily available from measurement of the velocity of ultrasonic waves in various directions in a solid. The third order elastic constants, however, are not so readily available. Various means of measuring third order elastic constants have been the subject of intensive investigation over the past quarter-century because of the fundamental role played by third order elastic constants in theories describing the nonlinear behavior of solids. In addition, the fundamental role played by the nonlinearity parameter, especially in technology, is realized more and more. This section describes the measurement of the nonlinearity parameter by use of the capacitive receiver and gives results for a number of solids. Evaluation of the third order elastic constants from similar data previously has been described[5].

The propagation of an ultrasonic wave in a cubic lattice can be described by a nonlinear equation which takes the form

$$\rho_0 \frac{\partial^2 U}{\partial t^2} = K2 \frac{\partial^2 U}{\partial a^2} + (3K_2 + K_3) \frac{\partial U}{\partial a} \frac{\partial^2 U}{\partial a^2} \qquad (6)$$

where the expressions for K_2 in terms of second order elastic constants, and K_3 in terms of third order elastic constants, are given in Table I. If one generates a sinusoidal wave at the end of the sample a = 0, then the solution after propagation through the sample a distance a is given by

$$U = A_1 \sin(ka - \omega t) + A_1^2 k^2 a \beta \cos 2(ka - \omega t) \ldots \qquad (7)$$

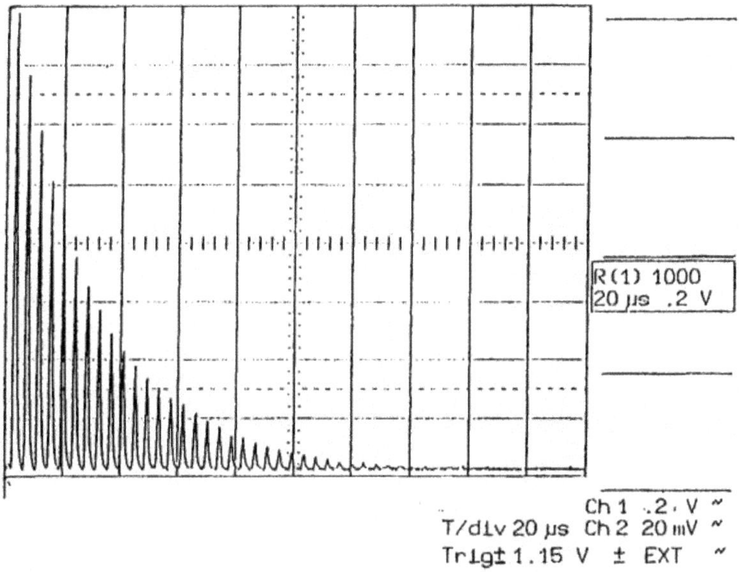

Figure 6. Ultrasonic echo pattern of S1-Unpolarized ceramics at frequency 15 MHz

Table 1 K_2 and K_3 for [100], [110], and [111] directions in a cubic lattice.

Direction	K_2	K_3
[100]	C_{11}	C_{111}
[110]	$\frac{C_{11} + C_{12} + 2C_{44}}{2}$	$\frac{C_{111} + 3C_{112} + 12C_{166}}{4}$
[111]	$\frac{C_{11} + 2C_{12} + 4C_{44}}{3}$	$\frac{C_{111} + 6C_{112} + 12C_{144} + 24C_{166} + 2C_{123} + 16C_{456}}{9}$

where A_1 is the amplitude of the fundamental. The amplitude of the second harmonic A_2 also contains the propagation constant $k = 2\pi/\lambda$, the sample length a, and the nonlinearity parameter

$$\beta = -\left[\frac{3K_2 + K_3}{K_2}\right] \tag{8}$$

One can use the capacitive receiver to measure the amplitudes of the fundamental and the second harmonic absolutely. Once this is done, one can use the data and

$$\beta = \frac{A_2}{A_1^2 k^2 a} \tag{9}$$

evaluated from the amplitude of the second harmonic in Eq. 7 to determine the nonlinearity parameter β. The nonlinearity parameter depends on the Born-Mayer hardness parameter, and thus in crystalline solids the magnitude of the nonlinearity parameter is determined by the shape of the interatomic potential function. Measurement has been made with the capacitive receiver of the nonlinearity parameters of a number of solids, both crystalline and isotropic. Some results of measurements between room temperature and liquid helium temperatures are given in Figs. 7, 8, and 9. Interpretation of these data in terms of models of the solid state is in progress. As more data are accumulated for different crystalline symmetries it is assumed that more detailed theories of the solid state will be possible, and that mathematical description of such phase transitions as the Martensite transition, the transition at the Curie temperature, or the transition in high T_c superconductors will be possible.

REFERENCES

1. W.B. Gauster and M.A. Breazeale, Rev. Sci. Instrum. 37, 1544-1548 (1966).

2. R.D. Peters, M.A. Breazeale and V.K. Paré, Rev. Sci. Instrum. 39, 1505-1506 (1968).

3. W. Cao, G.R. Barsch, W. Jiang and M.A. Breazeale, Phys. Rev. B 38, 10244-10255 (1988).

4. R.D. Peters and M.A. Breazeale, Appl. Phys. Letters 12, 106-108 (1968).

5. M.A. Breazeale and Jacob Philip, in Physical Acoustics, W.P. Mason and R.N. Thurston, eds. Vol. XVII, Academic Press, New York, pp. 1-60 (1984).

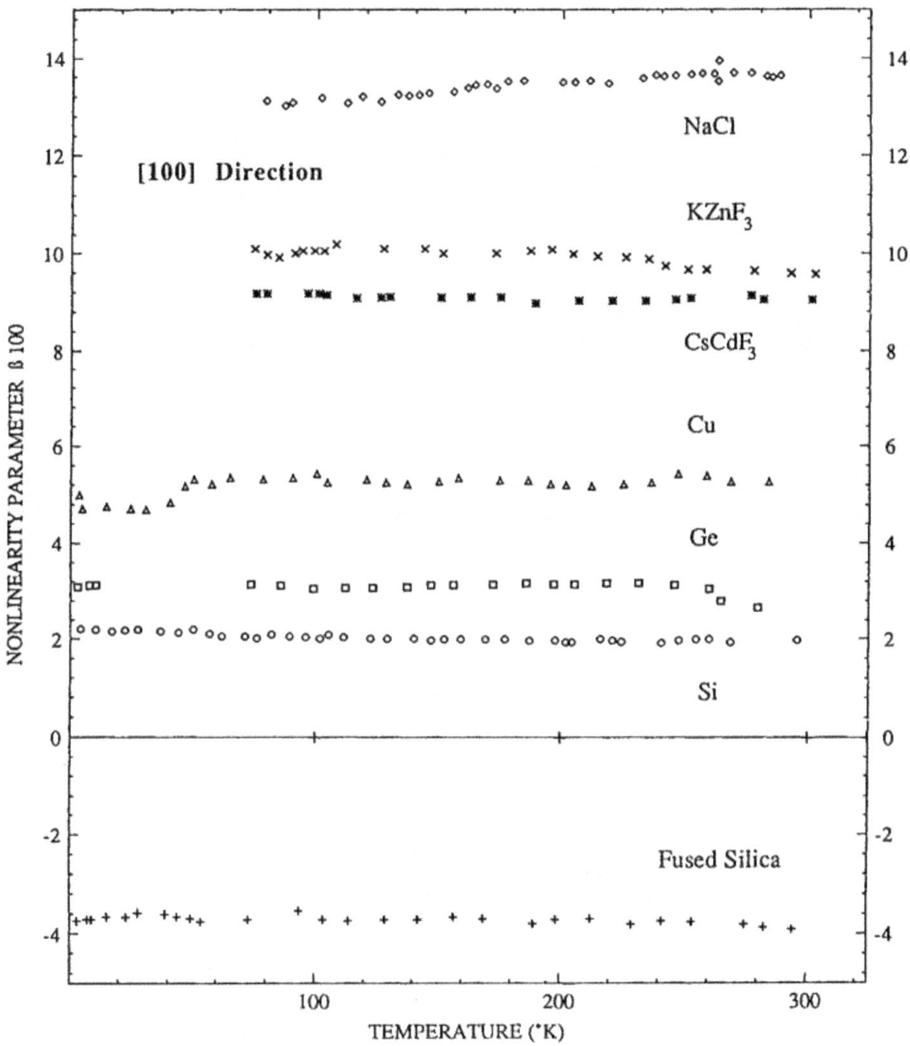

Figure 7. Temperature variation of nonlinearity parameters in the [100] direction

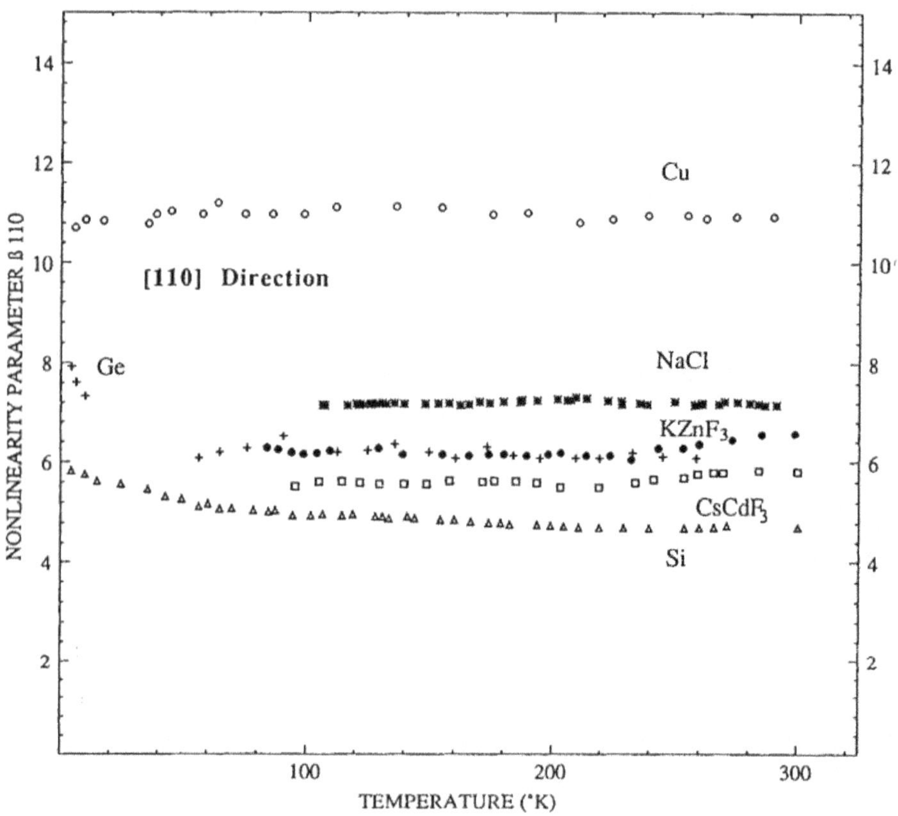

Figure 8. Temperature variation of nonlinearity parameters in the [110] direction

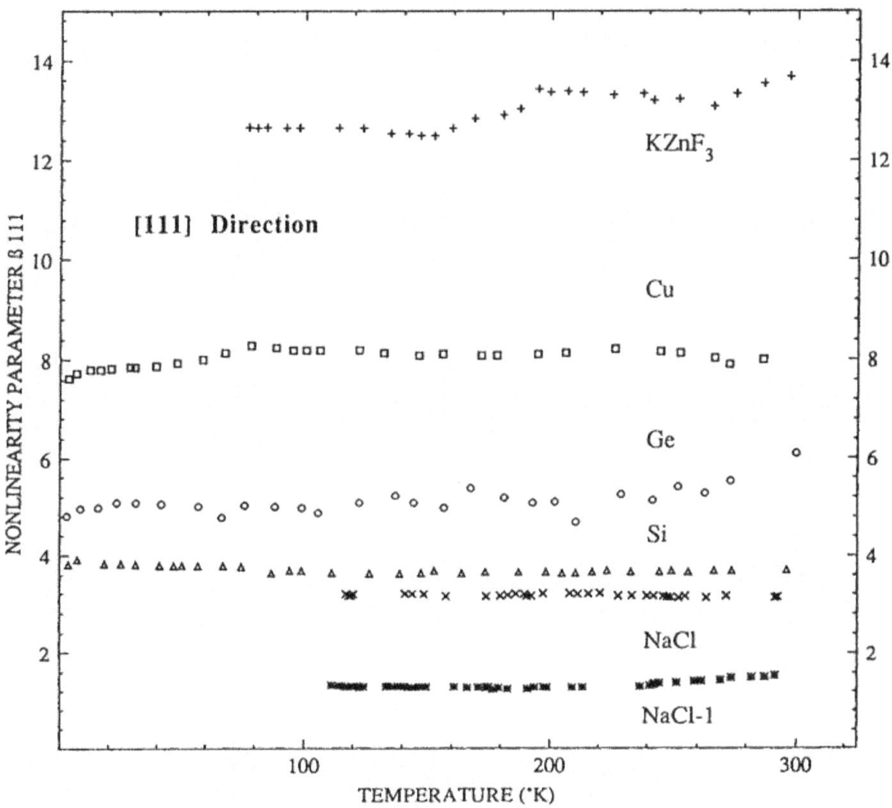

Figure 9. Temperature variation of nonlinearity parameters in the [111] direction

ACOUSTIC REMOTE SENSING IN THE AIR AND THE SEA

EDMUND H. BROWN
CIRES, University of Colorado, Campus Box 449
Boulder, Colorado 80309-0449, USA*

ABSTRACT

After a definition and brief history of active acoustic remote sensing, the paper derives the acoustic equation and gives examples of its application for appropriate frequency ranges to turbulence in the atmosphere and oceans and to clouds of particulates or plankton.

1. Introduction

1.1 *Areas of Research*

Acoustic remote sensing is one of the most useful methods in our technological kit of tools for research and modification of our environment. In the ocean *sonars* provide information on depths and bottom profiles, locate schools of fish, trace motions of pollutants in the benthic boundary layer, and can measure the spectrum of wind-driven waves in the oceanic surface layer. In the atmospheric boundary layer, where we - other than "frequent fliers" and astronauts - live most of the time, *Echosondes* (or *Sodars*) provide data on winds, temperatures (by means of the hybrid Sodar/Radar or RASS), wind shears, turbulence parameters, inversion heights, and pollution levels. The vast range of useful acoustic frequencies (and, correspondingly the types of instruments) for atmospheric and, even more, oceanic observations suggests, for a short paper, a restriction to measurements in our immediate surroundings, say, to about a few kilometers in the atmosphere and a few hundred meters in the ocean. Thus, the final section of this paper will derive an equation - the *Acoustic Equation* - suitable for these regions, and apply it to obtain *Scattering Cross-Sections*, the *Echosonde Equation* (which provides the link between experiment and theory), and the *Spatial Fourier Decomposition* ("eddies") of fields of turbulence or internal waves. Some of the results in this paper appeared in the earlier work of Brown and Hall[1] and Brown[2] .

---

* Address for Correspondence: E.H.Brown, 905 15th Street, Boulder, CO 80302, USA.

[1] E.H.Brown and F.F.Hall, Jr., Advances in atmospheric acoustics, **Rev. Geophys. and Space Phys., 16,** 47-110 (1978).

[2] E.H.Brown, Acoustic remote sensing in the air and the sea, **Proc. Internat. Symp.on Acoustic Remote Sensing of the Atmosphere and Oceans,** (Available from the University of Calgary, ISBN: 0-88953-031-9) (1982).

1.2 . *Early History*

The earliest form of acoustic remote sensing - *range finding* - antedates the use by, or even the existence of humans. Various species of bats emit sound pulses with frequencies up to 200 kHz, and determine range to ground, cave walls, or prey from the time delay of the echoes. Similarly, some dolphin species determine ranges to shores, colleagues, or schools of fish in the same way. For humans, the use of sound for remote sensing could not begin until the speed of sound was known. The first measurement of the *speed of sound in air* by Mersenne[3] used rather primitive methods - eyes and ears for detection of a musket shot at night, with the delay between the two timed by his own pulse or, later, by the swings of a pendulum. A few decades after Mersenne's work, William Derham[4] invented an improved pendulum timer that, together with longer test paths, allowed more accurate measurements of sound speed. These results, presented by Derham at the Royal Society, induced the president of the Society, Isaac Newton, to try to improve his incorrect (because based on the assumption that sound propagation was an isothermal process) prediction by unconvincing mathematics. The discrepancy remained for almost a hundred years until Laplace[5] obtained the correct value by assuming an adiabatic process.

With a reasonable estimate of ambient sound speed in hand, Derham was able to pursue answers to a number of questions about the nature of sound, such as whether sound travelled at the same speed uphill as down, whether the sound of a musket shot proceeded at the same rate with the musket pointed toward or away from the observer, or whether the sound speed depended on various states of the weather, questions that were not as trivial then as they appear today (at the time "authorities" such as Gassendi claimed that sound consisted of a stream of particles and could not be affected by changes in wind or temperature!). On the question of the effect of fog or mist, Derham contradicted his own observations and claimed that they impeded the propagation and intensity of sound.

---

3 F.M.Mersenne, **Harmonicorum Liber.**, Paris (1636).

4 W. Derham, Experimenta et observationes de soni motu, **Phil.Trans.Roy.Soc. London, 26,** 2-32, (1708).

5 P.S.Laplace, **Méchanique Céleste,** Paris (1799).

However, on the question of the effect of wind, Derham obtained the correct answer, that wind along the sound path linearly increased or decreased the measured apparent speed, depending on whether the wind was blowing toward or away from the observer. Values of wind were obtained by another of Derham's inventions, throwing up chaff (from grain - not radar chaff) and measuring the distance it travelled before hitting the ground. Recently, Delaney[6] plotted Derham's data on sound speed *vs.* wind, clearly showing the linear dependence as claimed. Finally, Derham advanced and explained the echo-delay method for range finding. For example, using a building on the other bank as a reflector he obtained reasonably accurate measurements of the width of the Thames near London. Plans to make a similar measurement of the width of the English Channel were aborted by the outbreak of war between the Dutch and the French. All in all we should consider Derham as the first pioneer in acoustic remote sensing, that is, the application of sound to measurement of environmental properties.

In the second half of the 19th century, financed by a society for retired seamen, John Tyndall conducted research on sound propagation over the ocean and through fogs, with the hope of improving warning systems, such as fog horns. His results, which were reported in a paper[7] at the Royal Society, showed that sound propagated with smaller loss through fog than through clear air, and that sound projected horizontally over the ocean through clear air returned echoes even though no ships were present to cause reflections. He attributed the echoes to backscatter from "flocculence" in air temperatures and supported this explanation by creating a laboratory model of the process. Later, he turned his apparatus to project pulses vertically upward into the atmosphere and again obtained returning echoes. We could consider this vertical system to be the first *monostatic echosonde* (*sodar*).

For the *speed of sound in water*, although several sporadic attempts to measure it occurred earlier, according to Hersey[8] the first reliable value, 1500 m/s , was obtained by François Beudant for sea water near Marseilles in 1820. Later, reporting their own measurements in Lake Geneva in 1826, Daniel Colladon and Charles

---

[6] M.E.Delaney, Sound propagation in the atmosphere: an historical review, **Acustica, 38** , 201-223 (1977).

[7] J.Tyndall, On the atmosphere as a vehicle of sound, **Trans.Roy.Soc.London,** Ser.A, **164,** 183-244 (1874).

[8] J.B.Hersey, A chronical of man's use of ocean acoustics, **Oceanus,20,** 8-21(1977).

Sturm[9] credit Beudant and his experimental value. Using the observed value, throughout the remainder of the 19th century researchers attempted to exploit echo ranging in the ocean for depth sounding, with varying degrees of success. By World War I, however, the use of depth sounders to provide single measurements became more common. In 1919 M. Marti patented the facsimile recorder, an instrument that - combined with constant ship motion - provided complete bottom profiles. Range finding depends on the simplest propagation process - *reflection*. In 1957 George Batchelor[10] derived the equation for the *scattering* of sound by turbulence, and by fluctuations in an arbitrary "passive additive," such as temperature or salinity. Apparently, Batchelor's paper had little or no impact on the naval or oceanographic communities, although his results, later on, assisted in correction of an error that appeared in the equivalent theory of scattering in the atmosphere.

1.3 *Recent Developments.*

The gradual introduction of turbulence theory into studies of the small-scale dynamic structure of the atmosphere by Taylor[11] and others revived interest in the scattering of sound by what Tyndall had called "flocculence." In 1941, the same year that Kolmogorov[12] and Obukhov[13] independently derived the "two-third's law" for inertial range turbulence, Obukhov[14] found a relation between the acoustic scattering cross section and the spectral density of turbulent kinetic energy. Unfortunately Obukhov neglected a factor and, as a consequence, the theoretical prediction given by Tatarskii[15] (based on Obukhov's work) in the first publication of his book diverged widely from the experimental values found by

---

9 J.-D.Colladon and J.C.F.Sturm, Memoire on the compression of liquids, **Ann.Chim. Phys. ,36,** 225-257 (1827).

10 G.K.Batchelor, Wave scattering due to turbulence, in **Symposium on Naval Hydraulics**, edited by F.S.Sherman, 409-423 (National Academy of Sciences, Washington, D.C.) (1957).

11 G.I.Taylor, Eddy motion in the atmosphere, **Phil.Trans.Roy.Soc.London, Ser.A, 215,** 1-26 (1915).

12 A.H.Kolmogorov, The local structure of turbulence in an incompressible viscous fluid for very large Reynolds numbers (in Russian), **Dokl.Akad.Nauk SSSR,39,**301 (1941).

13 A.M.Obukhov, Energy distribution in the spectrum of turbulent flow (in Russian), **Dokl.Akad.Nauk SSSR, 32,** 22 (1941).

14 A.M.Obukhov, Scattering of sound in turbulent flow (in Russian), **Dokl.Akad.Nauk SSSR,30,** 611 (1941).

15 V.I.Tatarskii, **Wave Propagation in a Turbulent Medium** (translated from the Russian by R.A.Silverman), Dover, New York (1962).

Kallistratova[16] for scattering angles near 90°. However, Tatarski's revised book[17] contained the correct scattering cross-sections found by Monin[18] (and earlier, Batchelor).

For either the atmosphere or the ocean, sharing common arenas places acoustic sensing in close contact with several other remote sensing methods, such as tropospheric radio propagation, optical propagation (*e.g.* lidar), or combined systems such as *radio-acoustic sounding systems* (RASS). In pursuit of the cause of signal fading on "troposcatter" transhorizon micro-wave paths, Gilman, Coxhead, and Willis[19] devised an early form of the echosonde. Their instrument used a loudspeaker to transmit pulses upward into the atmosphere, a microphone to receive scattered or reflected waves, and a motion picture camera to record the received (and filtered) signal that was displayed on an oscilloscope. The poor antenna beam pattern, required operation at the relatively high carrier frequency of 4 kHz. Since sound attenuation in the atmosphere increases approximately as the square of the frequency, the range of their instrument was limited to about 100 m. However, despite the minimal performance of their system, they were able to observe low level inversions and correlate them with radio signal fading.

Research on the same problem of troposcatter radio propagation induced McAllister[20,21] to return to the echosonde concept, but with three major changes: a vastly improved transmitting antenna consisting of a parabolic dish with a pressure transducer mounted at the focus; use of the same antenna to receive (with a transmit/receive switching circuit) to produce a much narrower beam and thereby greatly increase the signal-to-noise ratio (more than making up for the increase in sensitivity achieved with a microphone); and, perhaps most important, the display of the results on a facsimile recorder in the same way as used in ocean

---

16 M.A.Kallistratova, Procedure for investigating sound scattering in the atmosphere (English translation),**Sov.Phys.Acoust.**, **5**, 512-514 (1959).

17 V.I.Tatarskii, **The Effects of the Turbulent Atmosphere on Wave Propagation** (available from the National Technical Information Service, Springfield, VA)(1971).

18 A.S.Monin, Characteristics of the scattering of sound in a turbulent atmosphere, (English translation) **Sov.Phys.Acoust.**, **7**, 370-373 (1962).

19 G.W.Gilman, H.B.Coxhead, and F.H.Willis, Reflection of sound signals in the troposphere, **J.Acoust.Soc.Amer.**,**18**, 274-283 (1946).

21 L.G.McAllister, J.R.Pollard, A.R.Mahoney, and P.J.R.Shaw, Acoustic sounding - a new approach to the study of atmospheric structure, **Proc.IEEE**, **57**, 579-587 (1969).

sonar measurements. In the facsimile recorder, each pulse produces a vertical trace whose darkness varies with the received signal strength from each altitude, with a height on the ordinate of the chart proportional to the altitude in the atmosphere. Since the traces from succesive pulses are closely spaced along the abscissa, the eye can easily integrate corresponding features and detect the evolution of atmospheric structure with time. With this instrument beautiful pictures of inversions, internal gravity waves, wind shears, and thermal plumes were obtained.

Also in 1969, Little[22] used Monin's correct scattering cross-section to make a systems analysis of the echosonde and predict feasibility and limitations in a variety of situations. Paraphrasing some of his conclusions as given in a later talk: Remote sensing offers extraordinarily important opportunities. *In situ* measurements, which sample some particular physical property of the medium at a point, or over some very small area or volume immediately surrounding and in contact with the sensor, are perhaps practicable in cases where one is dealing with small numbers of sensors located near the surface of the earth in accessible geophysical regions. But our modern commercial and military activities increasingly depend on environmental information, often from remote parts of the world. Under such circumstances, we rapidly become information-limited, and the costs of procuring an adequate grid of such information by point measurements become prohibitively expensive.

Acoustic remote sensing is particularly useful, first, in the ocean below the surface, since the electrical conductivity of sea water produces high absorption of electromagnetic waves for wavelengths longer than those in the infrared window, and second, in the ABL (atmospheric boundary layer), since the power of acoustic waves scattered by turbulence fluctuations may be 10^6 times greater than that for electromagnetic waves. At the same time we should recognize that use of acoustic remote sensing complements other remote sensing methods such as radar, lidar, RASS, and satellite-borne instruments. Among others, the deployment of every useful system that can be brought to bear on an environmental problem has been championed by Weill[23] in several field campaigns.

---

22 C.G.Little, Acoustic methods for the remote probing of the lower atmosphere, **Proc.IEEE,57,** 571-578 (1969).

23 A.Weill, *et al.*, The Mesogers 84 experiment: a report, **Boundary Layer Meteor., 42** ,251-264 (1988).

2. The Acoustic Equation.

2.1 *Derivation.*

Derivation of a *useful* theory for sound propagation in the air or the sea and its application to acoustic remote sensing depends on a number of approximations, each of which must withstand scrutiny in regard to validity in a particular situation. **The goal of the derivation is to obtain an equation for a *single* unknown function that describes the interaction of sound with a random fluid flow.** A number of books , such as that of Richard von Mises[24], the 2-volume treatise of Monin and Yaglom[25], (mainly concerned with the atmosphere), and the article of John DeSanto[26] (stressing ocean acoustics), analyze more general situations. This paper includes a brief derivation to emphasize aspects peculiar to active acoustic remote sensing.

The derivation depends on the following assumptions:

1. Assume that every particle $X = X_i$, $i = 1,2,3,$ moving with the fluid lies within an (imaginary) closed surface with an interior large enough to have negligible molecular or quantum fluctuations but small enough to have fluid velocities and other properties constant and distinct within each such "particle" or "parcel." (Clearly, such a construct becomes difficult in the high atmosphere, such as the mesosphere!).
2. Assume that the velocities and properties of the *particles* X, together with their time and space first derivatives, form continuous fields over the *points* $x = x_i$ of space and the time t.
3. Assume the separability of the particle velocities $v = v_i$ and thermodynamic properties: pressure p, density ρ, temperature T, entropy density s, and sound speed, c , into parts related to a compressible field called "sound," and parts related to an incompressible field, either "turbulence" or "internal waves."
4. Assume the capability to turn the sound on or off (*active* remote sensing), which implies that turbulence or internal waves remain too weak to produce observable sound (obviously not true for jet engines!).

---

[24] R.von Mises, **Mathematical Theory of Compressible Fluid Flow,** Academic Press, New York (1958).

[25] A.S.Monin and A.M.Yaglom, **Statistical Fluid Mechanics:Mechanics of Turbulence,** The MIT Press, Cambridge, Mass. (1971).

[26] J.A.DeSanto, Theoretical Methods in Ocean Acoustics, in **Ocean Acoustics,** Edited by J.A.DeSanto, Springer Verlag, Berlin-Heidelberg (1979).

5. Assume that the sound also remains too weak to affect the turbulence or internal waves in any way. (Studies have recently shown that sending very intense sound beams through turbulent jets can either increase or decrease the strength of the turbulence, depending on the choice of frequency ranges!).

6. Assume that appropriate constants exist that normalize and non-dimensionalize all velocities and properties. This assumption permits comparisons of orders of magnitude and the neglect or omission of some of the quantities appearing in the equations. In particular, if **A** represents any normalized quantity of the compressible field (that is, of the sound), and **T** represents any normalized quantity of the incompressible field, then assume that $A = |A| << 1$, $T = |T| << 1$. Thus, neglect terms in the equations of the type A^n , $n = 2,3,4,\cdots$, (that is, restrict to *linear acoustics*), and neglect terms of the type AT^n , $n = 2,3,4,\cdots$, (that is, restrict to *linear* acoustic/turbulent *interactions*). Keep terms of the type T^n , $n = 1,2,3,\cdots$, unrestricted, since they will eventually fall out of the acoustic equation.

7. Assume a gradient in potential temperature, density, or suspended particulates so that the incompressible field can mix different values of these quantities and create random fluctuations throughout the scattering volume. For the particulate case, (*e.g.* clouds of particulates from dumping of pollutants in the ocean, from scattering layers of plankton, or from solid particles or water droplets in the atmosphere) the same conditions must hold that permit treating mists or fogs as a mixture of three "gases" - air, water vapor, and clouds of particles - (see Brown and Hall[1] for a description of some of these studies).

8. Assume that both sound propagation and incompressible flows satisfy the *adiabatic* condition, and that the fluctuations of properties associated with the incompressible field, such as sound speed, behave as "passive additives."

9. Finally, the derivation needs *Gauss' Theorem* :

$$\int_V dV\, \nabla \cdot \mathbf{f} = \int_S d\mathbf{S} \cdot \mathbf{f} \quad , \tag{1}$$

which holds with **f** and its first derivatives continuous and the surface S closed and orientable (no Klein bottles allowed!). Depending on convenience either vector or index notation may appear and in particular $\nabla \mathbf{f} = \partial_i f_i$, $d\mathbf{S} \cdot \mathbf{f} = dS_i f_i$, with repeated indices implying summation and the abbreviations $\partial_i = \partial/\partial_i = \{\partial/\partial x, \partial/\partial y, \partial/\partial z\}$, $\partial_t = \partial/\partial t$.

In accordance with assumption 3. , define the particle velocities $w = w_i$ for the incompressible field, $u = u_i$ for the sound, and the total particle velocity, $\bar{w} + v = \bar{w}_i + v_i$. Here the overbar indicates an average. A number of ways exist to define an average, such as, *e.g.* producing the sound in a set of pulses and forming the ensemble average (also indicated by < >) over it. Also define $w_i = \bar{w}_i + w'_i$, $u_i = \bar{u}_i + u'_i$, and $v_i = w'_i + u_i$, where a "prime" on a quantity indicates a fluctuation from the mean. The peculiar "break-out" of the mean of w reflects the fact that, unlike for other velocities, the Mach number or ratio $\bar{m} = |\bar{w}|/c_0$ must have special treatment since it can reach values near 0.1 (or even exceed this value in hurricanes, tornadoes , or chinooks), whereas the rms values of the Mach vector velocity fluctuations , $m'_{rms} = <|w'|^2>^{1/2}/c_0$, rarely reach values near 0.01. The derivation here will avoid the complications of a more realistic vertical profile of wind or current, and assume $\bar{w}$ constant in time and space. One should also note that $\bar{u} \neq$ constant.

With the above preliminaries out of the way, the actual derivation can begin. Conservation of *mass* leads to the equation

$$\partial_t \int_V dV\, \rho = - \int_S dS_i\, \rho v_i \ , \tag{2}$$

that is, the rate of change of mass in the volume V (LHS of Eq.2) must equal the rate of flow of mass in, or out through the boundary surface S (RHS of Eq.2). Conservation of momentum plus Newton's 2nd Law gives

$$\partial_t \int_V dV\, \rho v_i = -\int_S dS_j\, \rho v_i v_j - \int_S dS_i\, p - \int_V dV\, \rho' g\, \delta_{3i} \ , \tag{3}$$

where g is the acceleration of gravity, and δ_{ik} is the "Kronecker delta" : = 1 , for i = k , = 0, for i $\neq$ k. The LHS and RHS, again have obvious interpretations.

Now use Eq. 1. to convert the surface integrals in Eq. 2. and Eq. 3. into volume integrals and see that, since the volumes are arbitrary, the same equalities must hold for the integrands. Thus arrive at the first two of the following equations which, together with the three equations expressing the adiabatic condition and the behavior of passive additives, leads to the set :

$$\partial_t \rho + \partial_j(\rho v_j) = 0 \tag{4}$$

$$\partial_t (\rho v_i) + \partial_j(\rho v_i v_j) = - \partial_i p - \rho'_\omega g \,\delta_{i3} \tag{5}$$

$$\partial_t s + v_j \partial_j s = 0 \tag{6}$$

$$\partial_t \rho_\omega + v_j \partial_j \rho_\omega = 0 \tag{7}$$

$$\partial_t c + v_j \partial_j c = 0 . \tag{8}$$

The derivation to follow uses the partition of properties between sound quantities (subscript α) and turbulence quantities (subscript ω - a reminder of non-vanishing vorticity): $p = p_\omega + p_\alpha$, $p_\omega = p_0 + p'_\omega$, p_0 constant, $\rho = \rho_\omega + \rho_\alpha$, $\rho_\omega = \rho_0 + \rho'_\omega$, ρ_0 constant. But, as a quantity of second order in m' , one can neglect p'_ω and set $p_\omega \approx p_0$ in any terms containing acoustic properties. Eq. 6. represents the adiabatic condition, Eq. 7. states that in incompressible but inhomogeneous turbulence the density at a particle remains constant along a flow line, and similarly in Eq. 8. for the sound speed.

Taking the derivative of a product in the first LHS term in Eq. 5. and then substituting for $\partial_t \rho$ from Eq. 4. leads to the Euler equation

$$\underset{1}{\partial_t v_i} + \underset{2}{v_j \partial_j v_i} = - \underset{3}{(1/\rho)\, \partial_i p} - \underset{4}{(\rho'_\omega/\rho) g \,\delta_{i3}} . \tag{9}$$

The numbers under each term will help in the term-by-term reduction of the equations. Applying ∂_t to Eq. 4. and ∂_i to Eq. 5. , then substituting from the first into the second leads to the equation

$$\nabla^2 p - \bar{\partial}_t^2 \rho = -\partial_i\partial_j(\rho v_i v_j) - \partial_3(\rho'_\omega/\rho)g . \qquad (10)$$

$\underline{5}$ $\quad$ $\underline{6}$ $\quad$ $\underline{7}$ $\quad$ $\underline{8}$

Turning off the sound and applying assumptions $\underline{4}$. and $\underline{5}$. to Eqs. 9. and 10. leads to the equations for the tubulence alone

$$\bar{\partial}_t w'_i + w'_j\partial_j w'_i = -(1/\rho_\omega)\partial_i p'_\omega - (\rho'_\omega/\rho_0)\, g\, \delta_{i3} \qquad (11)$$

$\underline{1}'$ $\quad$ $\underline{2}'$ $\quad$ $\underline{3}'$ $\quad$ $\underline{4}'$

and

$$\nabla^2 p'_\omega - \bar{\partial}_t^2 \rho'_\omega = -\partial_i\partial_j(\rho_\omega w'_i w'_j) - \partial_3(\rho'_\omega/\rho_0)g. \qquad (12)$$

$\underline{5}'$ $\quad$ $\underline{6}'$ $\quad$ $\underline{7}'$ $\quad$ $\underline{8}'$

As an interesting comment, rewriting the first term of Eq. 4. as $\bar{\partial}_t\rho_\alpha + \bar{\partial}_t\rho_\omega$ and substituting for the last term from Eq. 7. gives

$$\bar{\partial}_t \rho_\alpha + v_j\partial_j \rho_\alpha + \rho\, \partial_j v_j = 0 , \qquad (13)$$

and turning off the sound field in this case reduces Eq. 13. to

$$\partial_j w_j = \nabla \cdot w = 0 , \qquad (14)$$

that is, the usual condition of zero divergence for incompressible turbulent fields.

Now simplification can proceed: subtract Eqs. 11. and 12. from Eqs. 9. and 10., and use the linear acoustics condition ($A^2 \approx 0$) and the linear acoustic/turbulence interaction condition ($AT^2 \approx 0$). Introduce the following normalized variables to help in the use of the last three conditions : $m = w/c_0$, $n = c_0/c$, (n represents the acoustic refractive index, and $n = 1 + n'$), $r = \rho_\omega/\rho_0$, $\sigma = s/C_V$, $\underline{t} = c_0 t$, $\underline{u} = u/c_0$, $\underline{\rho}_\alpha = \rho_\alpha/\rho_0$, $\underline{p} = p/(\rho_0 c_0^2)$, $\bar{\partial}_{\underline{t}} = \partial_{\underline{t}} + \bar{m}_i\partial_i$. Taking each term in sequence in Eqs. 9. and 11, and showing discarded items by the appropriate reference to A^2 , AT^2 , *etc.*, :

$\underline{1}. - \underline{1}' : \bar{\partial}_t v_i - \bar{\partial}_t w'_i \rightarrow \bar{\partial}_{\underline{t}}\, \underline{u}_i$,

$\underline{2}. - \underline{2}'$: $v_j\partial_j v_i - w'_j\partial_j w'_i \quad \rightarrow \quad A^2 [\underline{u}_j\partial_j\underline{u}_i] + \underline{u}_j\partial_j m'_i + m'_j\partial_j\underline{u}_i$,

$\underline{3}. - \underline{3}'$: $(1/\rho)\partial_i p - (1/\rho_\omega)\partial_i p'_\omega \rightarrow A^2[(-\underline{\rho}_\alpha\partial_i\underline{p}_\alpha] + AT^2[-\underline{\rho}_\alpha\partial_i\underline{p}'_\omega] + (1-r')\partial_i\underline{p}_\alpha$.

The approximation implies that $\underline{4}. - \underline{4}'. \rightarrow \approx 0$ (gravity does not affect sound) and $\rho^{-1} = (\rho_0 + \rho'_\omega + \rho_\alpha)^{-1} = \rho_0^{-1}(1 + r' + \underline{\rho}_\alpha)^{-1} \approx \rho_0^{-1}(1 - r' - \underline{\rho}_\alpha)$. Also, recall that normalization has divided the entire equation by c_0^2. Thus simplified, Eq. 9. becomes

$$\underline{\partial}_t \underline{u}_i + \underline{u}_j\partial_j m'_i + m'_j\partial_j\underline{u}_i = -(1-r')\partial_i\underline{p}_\alpha . \tag{15}$$

Apply the same approximations to Eqs. 10. and 12. and use the fact that $\rho_\omega = \rho_0(1 + r')$ has a different origin - the assumed gradient - than the other turbulence variables. Then, the rate of change $\underline{\partial}_t r'$ given by Eq. 7., must separate into a part $[\underline{\partial}_t r']_\omega = -m'_i\partial_i r'$ caused by turbulence, and a part $[\underline{\partial}_t r']_\alpha = -\underline{u}_i\partial_i r'$ by the forced sloshing back and forth of the density fluctuations in the passage of the sound wave. Thus the terms become :

$\underline{5}. - \underline{5}'.$: $\nabla^2 p - \nabla^2 p'_\omega \quad \rightarrow \quad \nabla^2\underline{p}_\alpha$,

$\underline{6} - \underline{6}'.$: $\underline{\partial}_t^2 \rho - \underline{\partial}_t^2 \rho'_\omega \quad \rightarrow \quad \underline{\partial}_t^2 \underline{\rho}_\alpha - \underline{\partial}_t (\underline{u}_j\partial_j r')$;

But, rewrite the second term on the RHS as $(\partial_i r')\underline{\partial}_t\underline{u}_i + \underline{u}_i\partial_i\underline{\partial}_t r'$, discard the second term of this expression since from Eq. 7. it is either AT^2 or A^2T, and substitute for $\underline{\partial}_t\underline{u}_i$ in the remaining first term from Eq. 15. with all turbulence fluctuations removed since it is already AT. This changes the term to $-(\partial_i r')\partial_i\underline{p}_\alpha$, and reduces $\underline{6}.-\underline{6}'.$ to :

$\underline{6}. - \underline{6}'.$: $\quad \rightarrow \quad \underline{\partial}_t^2\underline{\rho}_\alpha + (\partial_i r')\partial_i\underline{p}_\alpha$,

$$7. - 7'. : \quad \partial_i\partial_j(\rho v_i v_j) - \partial_i\partial_j(\rho_\omega w'_i w'_j)$$

$$\rightarrow \quad A^3[\partial_i\partial_j(\underline{\rho}_\alpha \underline{u}_i \underline{u}_j)] + A^2[2\partial_i\partial_j(\underline{\rho}_\alpha m'_i \underline{u}_j)]$$

$$+ A^2[\partial_i\partial_j(r'\underline{u}_i\underline{u}_j)] + AT^2[\partial_i\partial_j(\underline{\rho}_\alpha m'_i m'_j)]$$

$$+ AT^2[2\partial_i\partial_j(r' m'_i \underline{u}_j)] + 2\partial_i\partial_j(m'_j \underline{u}_i),$$

or

$$7. - 7'. : \qquad \rightarrow \quad 2\partial_i\partial_j(m'_j \underline{u}_i).$$

The above simplification again used the normalization implied by $\rho = \rho_0 (1 + r' + \underline{\rho}_\alpha)$. Also, the terms with a factor 2 represent pairs of terms for which the symmetry of $\partial_i\partial_j$ and the interchange of the dummy indices $i,j \rightarrow j,i$ allow combination. Introducing the above results leads to

$$\nabla^2 \underline{p}_\alpha - \underline{\partial}_t^2 \underline{\rho}_\alpha = (\partial_i r')(\partial_i \underline{p}_\alpha) - 2\partial_i(m'_j \partial_j \underline{u}_i). \tag{16}$$

Now consider p_α and ρ_α as thermodynamic variables: $p = p(\rho,s)$, so $dp = (\partial p/\partial \rho)_s d\rho + (\partial p/\partial s)_\rho ds = (\partial p/\partial \rho)_s d\rho = c^2 d\rho$, by the adiabatic condition. Then the normalized variables satisfy

$$\underline{\partial}_t \underline{\rho}_\alpha = n^2 \underline{\partial}_t \underline{p}_\alpha , \tag{17}$$

and so

$$\underline{\partial}_t^2 \underline{\rho}_\alpha = n^2 \underline{\partial}_t^2 \underline{p}_\alpha + AT^2[(\underline{\partial}_t n^2)\underline{\partial}_t \underline{p}_\alpha] , \tag{18}$$

or

$$\underline{\partial}_t^2 \underline{\rho}_\alpha = (1 + 2n')\underline{\partial}_t^2 \underline{p}_\alpha . \tag{19}$$

Introducing this result in Eq. 16. and rearranging gives

$$\nabla^2 \underline{p}_\alpha - \underline{\partial}_t^2 \underline{p}_\alpha = 2n' \underline{\partial}_t^2 \underline{p}_\alpha + (\partial_i r')(\partial_i \underline{p}_\alpha) - 2\partial_i\partial_j(m'_i \underline{u}_j) . \tag{20}$$

Note that the RHS has already become AT, so that one can modify it by substituting the equation *itself* in the RHS, that is, by setting $\bar{d}_{\underline{t}}^2\, \underline{p}_\alpha \approx \nabla^2 \underline{p}_\alpha$ in the first term of the RHS. Thus, the equation becomes

$$\nabla^2 \underline{p}_\alpha - \bar{d}_{\underline{t}}^2\, \underline{p}_\alpha = 2n'\, \nabla^2 \underline{p}_\alpha + (\partial_i r')(\partial_i \underline{p}_\alpha) - 2\partial_i\partial_j(m'_i \underline{u}_j). \tag{21}$$

Finally, one can achieve the goal of a single unknown (modifying $\underline{u}_j$) by resorting to the Fourier time transform of the equation and its quantities. Define the time transform as

$$F(k) = (2\pi)^{-1}\int d\underline{t}\; e^{ik\underline{t}}\, f(\underline{t}) \;, \tag{22}$$

and its convolution product as

$$F * G = \int dk'\; F(k-k')G(k'). \tag{23}$$

The corresponding properties become $\underline{p}_\alpha \to P$, $\underline{u}_i \to U_i$, $m'_i \to M'_i$, $n' \to N'$, $r' \to R'$, $\bar{m}_i \to \bar{M}_i\delta(k)$, $\bar{d}_{\underline{t}} \to -ik + \bar{M}_i\partial_i$, and, when one can neglect third order quantities involving $\bar{m}_i$, $\partial_{\underline{t}} \to -ik$, where $\mathbf{i}$ represents $(-1)^{-1/2}$ and $\delta(k)$ is the "Dirac delta". In addition, the stripped-down Euler equation becomes

$$\partial_{\underline{t}}\, \underline{u}_i = -\,\partial_i \underline{p}_\alpha \;\to\; U_i = -ik^{-1}\partial_i P\,. \tag{24}$$

Using all the above changes finally gives the acoustic equation

$$\nabla^2 P + (k + \mathbf{i}\bar{M}_i\partial_i)^2\, P = 2N' * \nabla^2 P + (\partial_i R') * (\partial_i P) + 2\mathbf{i}\partial_i\partial_j[M'_i * (k^{-1}\partial_j P)]. \tag{25}$$

2.2 *Applications.*

a. When only a potential temperature gradient contributed to the fluctuations in R' , Batchelor[10] found that, for the ocean, the second term on the RHS of Eq. 25. remained much smaller than the first term on the RHS. Neglecting it provides the equation for the **Ocean**

$$\nabla^2 P + (k + i\overline{M}_i\partial_i)^2\, P = 2N' * \nabla^2 P + 2i\partial_i\partial_j[M'_i * (k^{-1}\partial_j P)], \tag{26}$$

as found by Batchelor, together with the revision including the mean current, given by Brown and Hall[1] .

b. In the atmosphere the equation for a perfect gas requires that $\partial_i R' = 2\partial_i N'$. Then combining the first and second terms on the RHS of Eq. 25. gives for the **Atmosphere**

$$\nabla^2 P + (k + i\overline{M}_i\partial_i)^2\, P = 2\partial_i(N' * \partial_i P) + 2i\partial_i\partial_j[M'_i * (k^{-1}\partial_j P)], \tag{27}$$

as found by Tatarskii[17] , again with the addition of a mean wind from Brown and Hall[1].

c. A third and apparently unexplored possibility would apply in the case where the potential temperature remained approximately constant (as holds in atmospheric fogs), but random fluctuations in droplet density caused by turbulence with the Kolmogorov-Obukhov spectrum or, for the ocean, random fluctuations in suspended particulates or layers of plankton, caused by internal waves with the Garrett-Munk[28] specrum, suggests dropping the first term on the RHS of Eq. 25. and keeping the second. Then for **Particulates**

$$\nabla^2 P + (k + i\overline{M}_i\partial_i)^2\, P = (\partial_i R') * (\partial_i P) + 2i\partial_i\partial_j[M'_i * (k^{-1}\partial_j P)]. \tag{28}$$

---

[10] G.K.Batchelor, *loc. cit.*

[17] V.I.Tatarskii, *loc. cit.*

[28] C.Garrett and W.Munk, Internal wave spectra in the presence of fine structure, J.Phys.Oceanogs.,1, 196-202 (1971).

d. The difference between the acoustic equation for the atmosphere and that for the ocean leads to a significant difference also in the scattering cross-sections σ_s : for the
Atmosphere

$$\sigma_s = 2\pi k^4 \cos^2\theta \frac{\Phi_T}{4T^2} , \tag{29}$$

but, for the
Ocean

$$\sigma_s = 6\pi k^4 \frac{\Phi_T}{4T^2} , \tag{30}$$

where Φ_T represents the spectrum of the temperature fluctuations. The lack of $\cos^2\theta$ in the ocean scattering cross-section implies very different behavior in the vicinity of 90°. A similar formula for the cross-section due to R' has not yet appeared.

e. Inclusion of σ_s in the "echosonde equation" provides the connection between theory and experiment. Writing W_R for the received electrical power, e_R for the efficiency of acoustic-to-electrical conversion, W_T for the transmitted power, e_T for the efficiency of electrical-to-acoustic conversion, α for the effective intensity attenuation coefficient in nepers per meter, $\Omega = A/R^2$ for the solid angle subtended by the area of the antenna aperture A at the range R, γ for the antenna directivity gain factor, l_p for pulse length, e a unit vector in the direction of the transmitted wave, and $\mathbf{M}$ the mean wind Mach vector, leads to the
Echosonde equation

$$w_R/e_R = e_T W_T \exp(-2\alpha R)(1 + e\cdot\mathbf{M})(l_p/2)(\gamma\ \Omega) \cdot \sigma_s(\theta_s). \tag{31}$$

By varying the angles of experiments to obtain varying power measurements from Eq. 31., the researcher can obtain values for the individual spectra of fluctuations of temperature, density, and velocity , as well as values for mean wind or ocean currents from the Doppler shift derived from $e\cdot\mathbf{M}$.

ACOUSTIC FIELD SENSING WITH OPTICAL FIBERS

HELGE E. ENGAN
Dept. of Electrical Engineering and Computer Science
University of Trondheim
N-7034 Trondheim, Norway

ABSTRACT

Properties of optical fiber components and systems relevant for acoustic sensing are reviewed. Characteristics with special interest for underwater sound detection are emphasized. Propagation in optical fibers is considered. A general optical sensor classification is discussed followed by descriptions of acoustic field sensors and multiplexing techniques.

1. Introduction

Acoustic fields can be sensed in a multitude of ways depending on a number of different physical effects. Among some more recently emerging technologies is sensing by optical fibers. This means that the acoustic field to be measured modulates the light propagating in an optical fiber. This modulation can then be measured by optical detection methods.

The question arises why optical fibers are used. In Table I some advantageous characteristics of optical fiber sensors have been listed. Several of these characteristics will be discussed here.

In the following, basic properties of optical fiber propagation will be reviewed. A general classification of optical fiber sensors will be performed followed by a discussion of specialized acoustic field sensor principles and applications.

2. Optical fiber propagation

Basic behaviour of optical waves propagating in optical fibers will be considered. It is emphasized that only a limited number of topics and configurations will be discussed here. Rigorous introductions to this field can be found in standard textbooks on the subject[1].

A typical optical fiber is sketched in Figure 1. It consists of a center glass part propagating the light and a plastic jacket that is applied to protect the fiber. The glass

Table I Fiber Optic Sensor Advantages

Dielectric sensor:
Passive
Electric insulation from measured field
Immunity to electromagnetic interference
Small size
Low weight
Large bandwidth
High sensitivity
Wide dynamic range
Environmentally rugged:
Vibration
Shock
Temperature
Multiplexing possibilities
Low cost potential

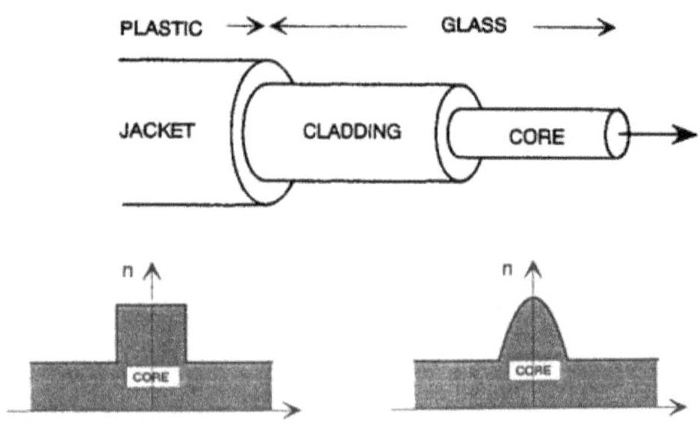

Figure 1. Optical fiber and index of refraction profiles.

part consists of the center core of a high index of refraction and a cladding of a slightly lower index. The cladding has a typical diameter of 100 μm or slightly more. The core diameter varies in the range 10 to somewhat less than 100 μm. The transition between the core and the cladding region can be either abrupt (step index fibers) or occur gradually (graded index fibers). The shape of this index profile will not be discussed here. The material of the center part propagates light slower than the cladding. A trapping or guiding of light in the fiber can therefore occur. This phenomenon will be discussed in some detail.

Consider a step index fiber with a core of radius a and index of refraction n_1 surrounded by a cladding with index of refraction n_2 and a large outer radius compared

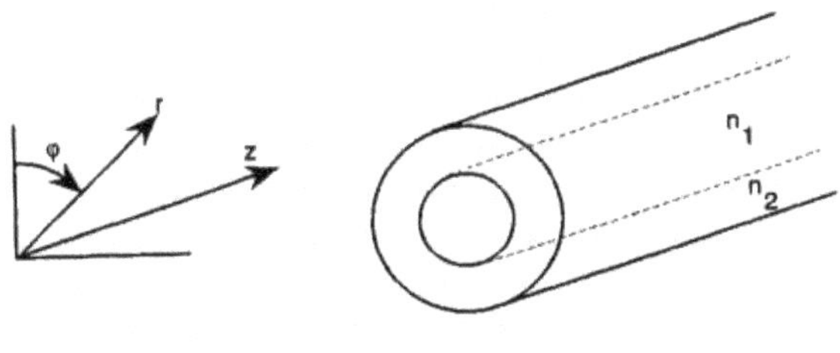

Figure 2. Coordinate system of cylindrical waveguide.

to a, see Figure 2. Using the cylindrical coordinates indicated in this figure, the wave propagation as described by the electric field E is determined by Maxwell's equation as

$$\nabla^2 E = \frac{1}{r}\frac{\partial}{\partial r}\left(r\frac{\partial E}{\partial r}\right) + \frac{1}{r^2}\frac{\partial^2 E}{\partial \varphi^2} + \frac{\partial^2 E}{\partial z^2} = \mu\varepsilon\frac{\partial^2 E}{\partial t^2} \tag{1}$$

where μ and ε are the permeability and permittivity of the medium, respectively.

We try a solution that considers propagation in the z direction of a wave with frequency ω and assumes the solution to be separable in the r and the φ coordinates , i.e.,

$$E = E_r(r)E_\varphi(\varphi)\exp[i(\beta z - \omega t)] \tag{2}$$

If we further assume that E_φ is a periodic function of φ of the form $E_\varphi(\varphi) = \exp(\pm il\varphi)$, where l is an integer, Eq. (1) becomes

$$\frac{\partial^2 E_r}{\partial r^2} + \frac{1}{r}\frac{\partial E_r}{\partial r} + \left(n^2k^2 - \beta^2 - \frac{l^2}{r^2}\right)E_r = 0 \tag{3}$$

Here we recognize Bessel's equation whose solutions are Bessel's functions. We have introduced the propagation constant for free space k where $k^2 = \varepsilon_0\mu_0\omega^2$. It turns out that the propagation velocity of the guided mode will always be between the unguided velocities of the two media, i.e., $n_1k \geq \beta \geq n_2k$. Substituting n_1 and n_2 for n in the two regions, we introduce the "transverse" wave numbers q and p, given by

$$n_1^2k^2-\beta^2=q^2$$
$$\beta^2-n_2^2k^2=p^2 \quad (4)$$

This results in the following equations for the two regions:

$$\frac{\partial^2 E_r}{\partial r^2}+\frac{1}{r}\frac{\partial E_r}{\partial r}+\left(q^2-\frac{l^2}{r^2}\right)E_r=0 \qquad r\leq a \quad (5)$$

$$\frac{\partial^2 E_r}{\partial r^2}+\frac{1}{r}\frac{\partial E_r}{\partial r}-\left(p^2+\frac{l^2}{r^2}\right)E_r=0 \qquad r>a \quad (6)$$

These equations have solutions

$$E_r=A_cJ_l(qr) \qquad r\leq a$$
$$E_r=A_{cl}K_l(pr) \qquad r>a \quad (7)$$

where J_l and K_l are the Bessel function of the first kind and the modified Bessel function of the second kind, respectively.

We have now arrived at the following expressions for the electric field,

$$E=A_cJ_l(qr)\exp(\pm il\varphi)\exp[i(\beta z-\omega t)] \qquad r\leq a \quad (8)$$

$$E=A_{cl}K_l(pr)\exp(\pm il\varphi)\exp[i(\beta z-\omega t)] \qquad r>a \quad (9)$$

The boundary conditions at $r=a$ determine p, q, and β, and also the ratio of the amplitudes A_c and A_{cl}. Often the two indices of refraction differ by a small amount, on the order of 1 percent. In this case the <u>weakly guiding</u> approximation is valid, an approximation implying that the electric field is directed close to the transverse direction. The modes are then linearly polarized and denoted by LP modes.

Solving these boundary conditions for the above equations in the weakly guiding approximation will give solutions of the mode patterns a few examples of which are sketched in Figure 3. Each mode is dispersive and has a propagation velocity in general different from other modes. It should be noted that there are two orthogonal polarizations and two spatially orthogonal distributions in the case of circular symmetry. This leads to degeneracy which can be of importance in many cases. The degeneracy can be broken by introducing some deviation from the circular symmetry of the fiber.

If the core diameter is large (approaching 100 μm), the fiber can usually propagate a large number of modes. In the diameter range around 10 μm the core

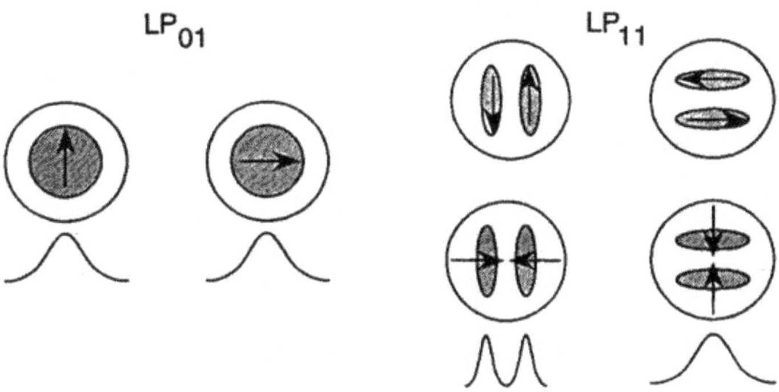

Figure 3. Mode patterns and intensity distributions for some low order LP modes. Arrows indicate electric polarization.

propagates one or a few modes only. These two regimes have fundamentally different aspects from a sensor point of view. In single mode propagation the beam behaviour is much more well defined than in the multimode case. It is substantially more difficult to keep track of the optical phase in multimode propagation than in the single mode case. This is because the difference in velocity of the individual modes contribute to the total electromagnetic field with a phase and amplitude that are highly unpredictable, dependent on the details of both propagation path and excitation conditions at the fiber entrance.

3. Classification of optical fiber sensors

Before proceeding to a specialized discussion of acoustic sensing with optical fibers, it is worthwhile to put this subject into a somewhat wider perspective.

There are many ways to classify fiberoptic sensors. A rough classification scheme is shown in Figure 4. A fiberoptic sensor in the broad sense is based on the principle that light propagating in a fiber will contribute to the process of measuring some environmental parameter. This is the case when the light is modulated by the measurand in some manner on its way from generation to detection. In hybrid sensors the sensing element is some conventional electric transducing device, and a fiberoptic link is used to transmit the signal from the sensor to the processing station. We shall not discuss such devices here and only include sensors where the light propagated in a fiber is modified by the measurand.

The characteristics of the light can be modified not only while it is travelling along the fiber. In many cases the light is propagating for some distance partly or totally

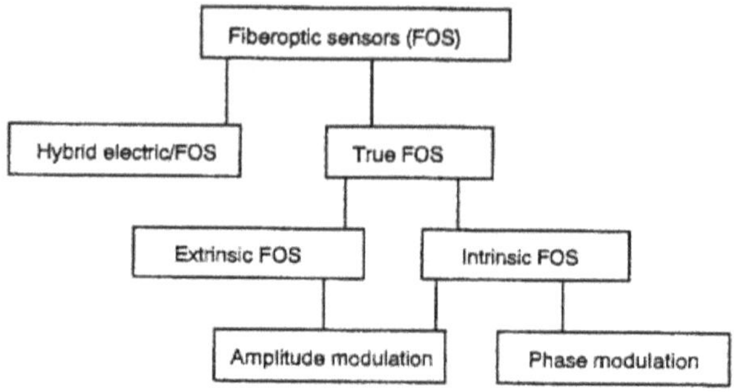

Figure 4. Classification of fiberoptic sensors.

outside the fiber in order to obtain a modulation that can be used for measurements. In this case the sensor is extrinsic as opposed to the intrinsic case where the light does not leave the fiber between generation and detection.

The main characteristics of light propagating in an optical fiber are:

- Amplitude (or intensity)
- Phase
- Frequency (or wavelength)
- Polarization
- Spatial mode

These characteristics can all, alone or in combination, be modified by the environment and thus in principle be used for determining a specific measurand. Changes in polarization states or in the spatial mode distribution are usually transformed into a measurement of amplitude or phase. Direct measurement of frequency modulation caused by a measurand is not much used. We are thus left with the two basic techniques of amplitude and phase modulation as indicated in Figure 4.

Measurement of the light amplitude is usually performed by monitoring the light intensity impinging on a photosensitive element such as a photomultiplier or a semiconductor photodiode. Acoustic sensors based on modulation of the beam amplitude have been reported[2]. Due to its larger potential in sensitivity, however, we shall here focus our attention on phase modulated sensors.

The measurement of phase is predominantly performed by comparing the light propagated along two different paths. This is called an interferometer technique. Many different interferometric schemes have been reported. Here we shall discuss the Mach-Zehnder technique. In Figure 5 a basic Mach-Zehnder interferometric setup is shown. The light generated in a laser is launched into the end of a fiber. The directional coupler divides the optical beam between the two fibers in the signal arm and the reference arm.

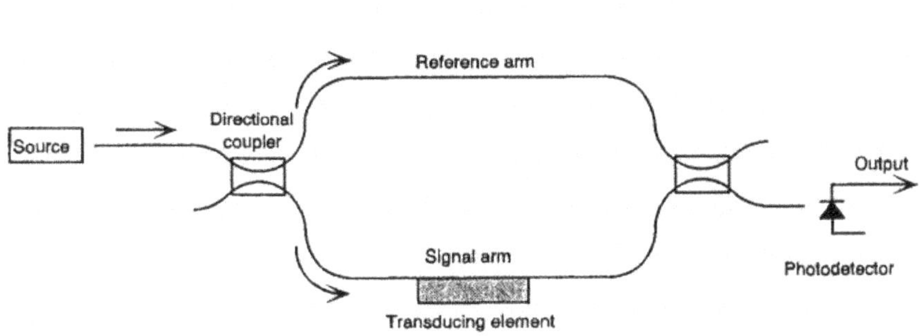

Figure 5. Basic Mach-Zehnder interferometer.

After transmission through these two arms, the light is combined in another directional coupler. At the output the photodetector senses the interference between the beams from the two arms. The phase of the reference arm is held as stable as possible while the phase of the light through the signal arm is modified by the measurand such as for example acoustic pressure.

The output current I_d from the photodetector is proportional to the detected light intensity. A_r and A_s are the amplitudes of the reference and signal arm, respectively. If they are equal in magnitude and with a phase difference φ,

$$I_d \propto (A_r + A_s)^2 \propto (1 + \cos\varphi) \tag{10}$$

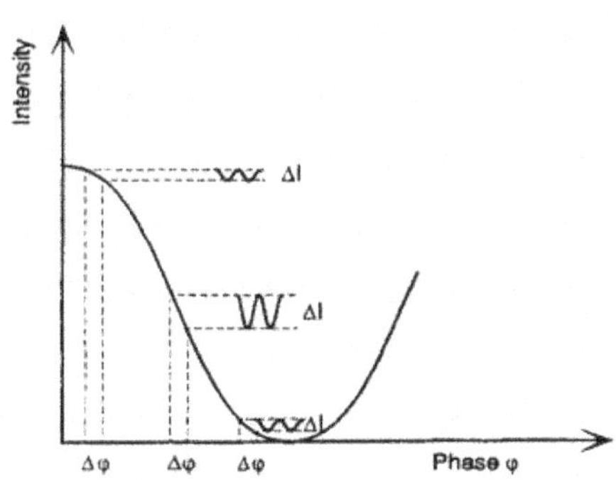

Figure 6. Response of Mach-Zehnder interferometer.

The response according to this equation is shown in Figure 6. This figure demonstrates a major challenge encountered when working with interferometric sensors: The response is strongly nonlinear. Only for very small perturbations the response is approximately linear. Furthermore, the sensitivity to environmental perturbations depends on the bias, i.e. the static phase difference between the two interferometer arms. A drift in this bias will cause a change in sensitivity. The phase for which the sensitivity has its maximum, is labeled the quadrature point.

4. Acoustic field sensors

4.1 Sensing principles

The predominant interest in acoustic field sensing with optical fibers has been in underwater hydrophones. This is a promising area which has still not reached a mature stage. We shall limit our discussion to this application area.

In a Mach-Zehnder interferometer exposed to forces from a surrounding acoustic field the optical length of the signal arm will be modified. The total phase φ of light propagating in a fiber that is uniform along its length L, is given by

$$\varphi = 2\pi \frac{nL}{\lambda} = 2\pi \frac{\nu}{c} nL \tag{11}$$

where n is the effective refractive index of the fiber, ν is the frequency of the light, and λ and c are the wavelength and light velocity of free space, respectively. Small changes in L and n (represented by ΔL and Δn) will cause a phaseshift

$$\Delta\varphi = 2\pi \frac{nL}{\lambda}\left(\frac{\Delta L}{L} + \frac{\Delta n}{n}\right) \tag{12}$$

Changes in the environment, such as the acoustic pressure Δp, will change the length and the refractive index. We shall introduce the transverse strain ε_r and the longitudinal strain $\varepsilon_z = \Delta L/L$. If E and σ are Young's modulus and Poisson's ratio, respectively,[3]

$$\varepsilon_z = \frac{2\sigma\Delta p}{E} \tag{13}$$

$$\varepsilon_r = -\frac{\Delta p(1-\sigma)}{E} \tag{14}$$

From this expression, a low frequency acoustic wave of pressure Δp is found to cause a phaseshift

$$\Delta\varphi=\varphi\left(\varepsilon_z-\frac{n^2}{2}[(P_{11}+P_{12})\varepsilon_r+P_{12}\varepsilon_z]\right) \tag{15}$$

where P_{11} and P_{12} are the elasto-optic Pockel's coefficients of the propagating medium. The first term in the bracket is the direct length change of the fiber. The two other contributions are due to the influence of the strain components on the index of refraction. Using data for fused quartz which is a frequently used material, the following sensitivity value is obtained: $\Delta\varphi/(\Delta p\cdot L) = 4\mathrm{x}10^{-5}$ rad/(Pa·m).

It is apparent from this discussion that the sensitivity of an acoustic sensor increases with length. To compete in sensitivity with conventional hydrophones in underwater applications, one approach is therefore to use a fiber of sufficient length. The actual sensitivity of plastic covered fibers has been measured. Assuming a 1 mW light source, an optical loss of 5 dB/km and quantum limited detection over a 1 Hz bandwidth, the predicted minimum acoustic pressure is as shown in Figure 7[4]. The corresponding

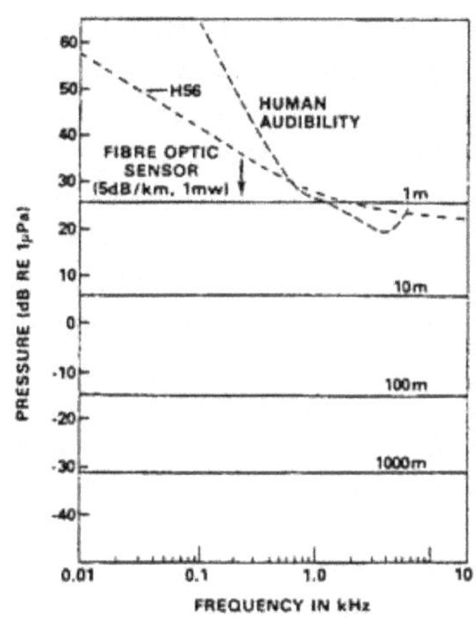

Figure 7. Minimum detectable pressure for four different lengths of plastic coated fiber and for piezoelectric hydrophone H56 calculated for 1 Hz bandwidth and quantum limited operation.

detection threshold in $\Delta\varphi$ for 1 Hz bandwidth is in the range of 0.1 to 1 μrad. Comparison with a low-noise hydrophone shows that high sensitivities are possible even with relatively moderate lengths of fibers. Nevertheless, it has been of great interest to save fiber expenses by applying various additional methods to increase the sensitivity. This will be discussed later.

The coating of the fiber has a strong influence on its effective sensitivity. Qualitatively, it can be seen from Eqs. (13-14) that a softer material such as plastic coatings will modify the strains both by changing the effective values of E and σ. Generally, the optical index (photoelastic) contribution of Eq. (15) is opposite in sign to the length change effect. Also, each contribution varies differently with material parameters and thickness of the layers. A great deal of flexibility therefore exists in designing fibers with specified acoustic responses.

The calculated variation of sensitivity with plastic coating thickness for a typical, commercially available single mode fiber is shown in Figure 8[5]. It is seen that the most

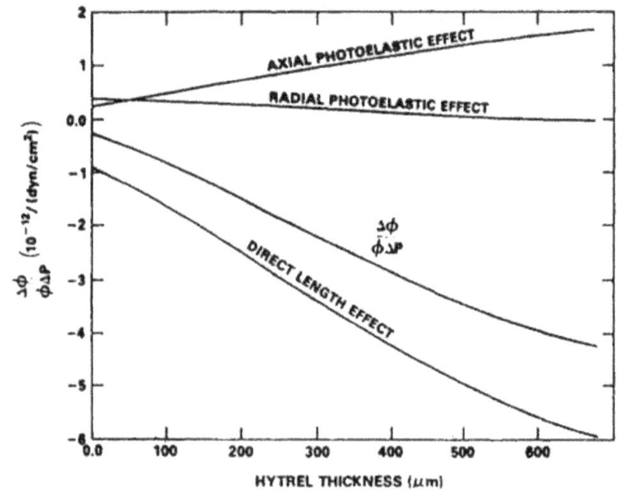

Figure 8. Calculated pressure sensitivity of a typical single mode fiber versus plastic jacket thickness.

significant contribution is due to pressure induced fiber length change. The photoelastic terms give smaller contributions which are of opposite polarity to the first term. It is also seen that the magnitude of the sensitivity increases strongly with thickness primarily due to the direct length effect.

It is desirable to increase the sensitivity of the active part of the optical fiber sensor. In other cases it is important to reduce the sensitivity. For example, in the Mach-Zehnder interferometer it is desirable to make the reference arm as insensitive as possible. Also, in a fiber system the lead fibers should ideally be insensitive to acoustic signals. The sensitivity of a fiber coated with hard materials such as glass and metals has been calculated, and the result is shown in Figure 9[3]. It is seen that for certain coating thicknesses it is theoretically possible to eliminate the acoustic sensitivity. Experimentally, a difference in sensitivity of about 35 dB has been demonstrated between

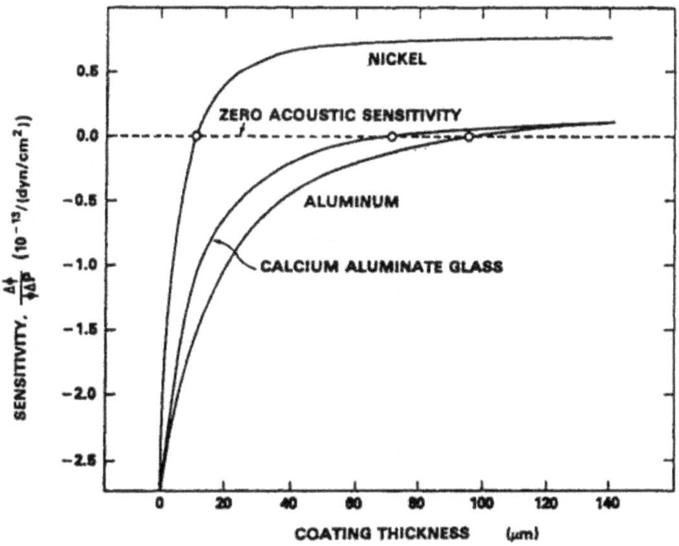

Figure 9. Decreasing sensitivity by applying harder coatings.

plastic-covered and nickel-covered parts of a fiber[6].

We have until now considered sensing of low frequency acoustic fields. This means that the response is identical to the response from hydrostatic pressure. When the frequency is increased, the response is modified due to dimensional effects. An example is shown in Figure 10 which shows the broadband frequency response of a straight piece of a typical fiber immersed in water[4]. This response can be described in terms of three distinct frequency regimes. For low frequencies the fiber response is unconstrained in the axial direction and isotropic in its cross-sectional plane. When the frequency increases and the acoustic wavelength decreases below the length dimension of the fiber, the fiber becomes axially constrained. In this regime, ε_z in Eq. (15) is zero so that sensing is performed only through changes in index of refraction caused by radial strain in the fiber. In the 10 to 100 MHz range the wavelength becomes comparable to the transverse dimensions of the fiber, introducing anisotropic excitation in the cross-sectional plane. This modifies the response through optical birefringence, an effect that also depends on the optical polarization. Changing the optical polarization in this regime can therefore modify the sensitivity significantly.

Other approaches to increase the acoustic field sensitivity of optical fibers have been reported in the literature. One example is shown in Figure 11[7] where a fiber is wound around a mandrel made of a compliant material. The properties of the mandrel can give an increased elongation of the fiber and therefore an improved sensitivity. The

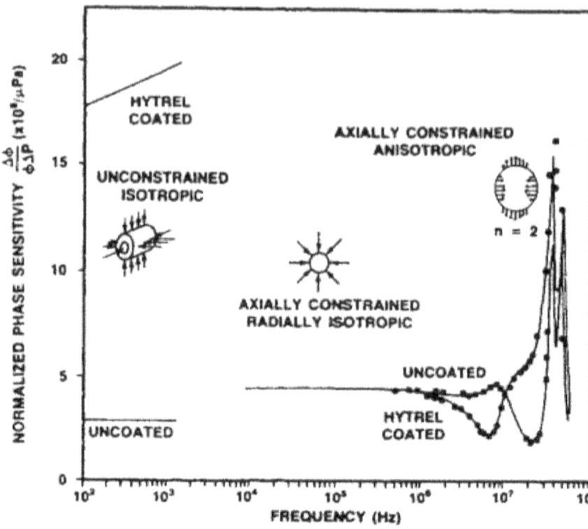

Figure 10. Frequency response of a straight segment of fiber in water.

sensitivity improvement is comparable to what has been obtained with plastic covered fibers. This latter configuration is easy to fabricate and can also be made with a sensitivity that varies along the mandrel. This shading can therefore be used for obtaining desired directional properties of the sensor.

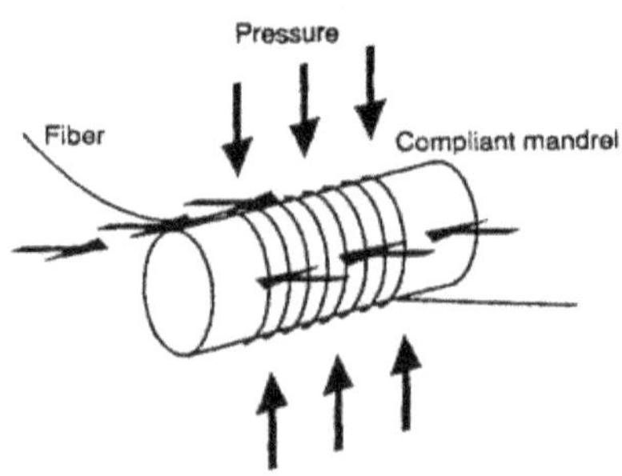

Figure 11. Sensitivity increase by winding fiber on a soft mandrel.

There is a large number of possible geometrical configurations that can be employed to modify various properties of a fiberoptic sensor such as sensitivity and directivity. Some configurations are indicated in Figure 12[8].

If all dimensions of the sensor arrangement are much smaller than an acoustic wavelength, the hydrophone is usually omnidirectional. The sensitivity can be increased linearly by increasing the amount of fiber exposed to the acoustic field. This can be accomplished by winding the fiber into a helix or weaving it into a planar structure. Limitations will be determined by the minimum radius of curvature of the fiber. This minimum radius is usually determined by either mechanical strain limitation or by optical loss in the bend.

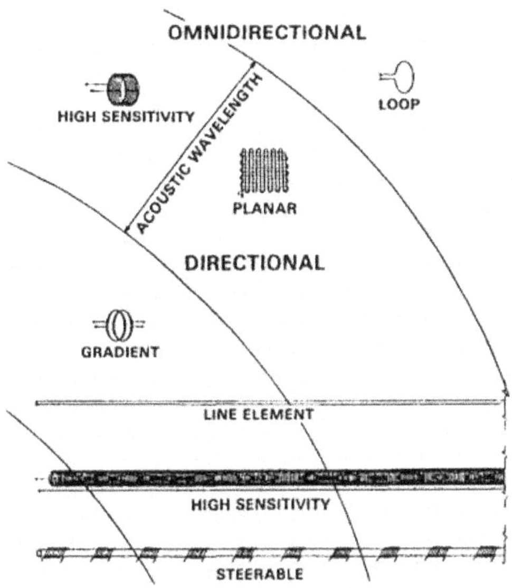

Figure 12. Various sensor configurations.

If the dimensions of the sensor is increased beyond an acoustic wavelength, the sensor becomes directional. A long helix will increase the sensitivity. In this case, the sensitivity is determined by the total length of the fiber, while the directivity is determined by the geometrical length of the helix axis. Several small sensors can be grouped in an array. When the sensors are individually addressed, a steerable sensor array can be obtained.

A directional sensor of small dimensions can be made by placing the reference fiber adjacent to the sensing coil as indicated in Figure 12. This will give a sensitivity proportional to the gradient which automatically gives a directional sensor.

4.2 Detection principles

As mentioned previously, the periodic response of an interferometric sensor is a major challenge. The most sensitive condition (the quadrature point) defines the best working condition for the sensor. However, varying environmental parameters such as temperature and hydrostatic pressure will cause variations in the phase difference that will easily change these conditions. Also, the dynamic range will be limited because of non-linearity when the deviation from the operating point gets too large. We shall describe a couple of principles shown in Figure 13 that overcome these problems.

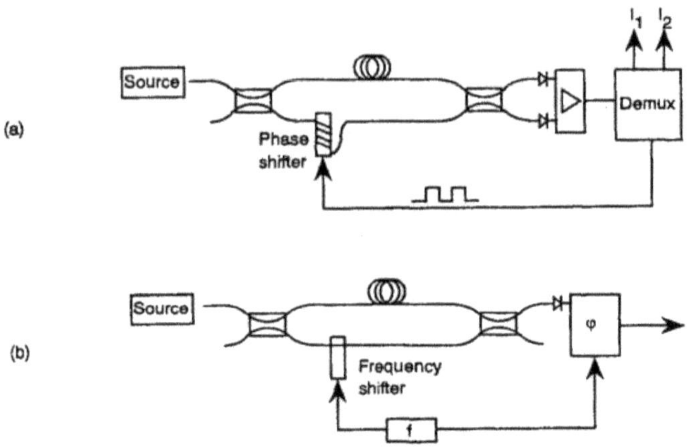

Figure 13. Mach-Zehnder interferometer detection principles. a) Phase switching, b) Heterodyne detection.

The principle of phase switching is illustrated in Figure 13a. A phase modulator is inserted in one of the arms of the Mach-Zehnder interferometer. Such a phase modulator can be easily implemented by winding the fiber tightly on a piezoelectric cylinder. The cylinder is driven by a square wave voltage so that the optical length is periodically changed with steps equal to $\pi/2$. This means that the two outputs I_1 and I_2 from the multiplexer corresponding to the two angles defined by the phase modulator, are proportional to $\sin\varphi$ and $\cos\varphi$, respectively, where $\varphi(t)$ is the phase to be estimated. This phase can be retrieved through a process employing differentiation and cross multiplication, as shown in Figure 14. The signals to the upper and lower inputs of the differential amplifier are, respectively,

$$S_u = I_2\frac{dI_1}{dt} = I^2\cos^2\varphi\frac{d\varphi(t)}{dt} \tag{16}$$

$$S_l = I_1\frac{dI_2}{dt} = -I^2\sin^2\varphi\frac{d\varphi(t)}{dt} \tag{17}$$

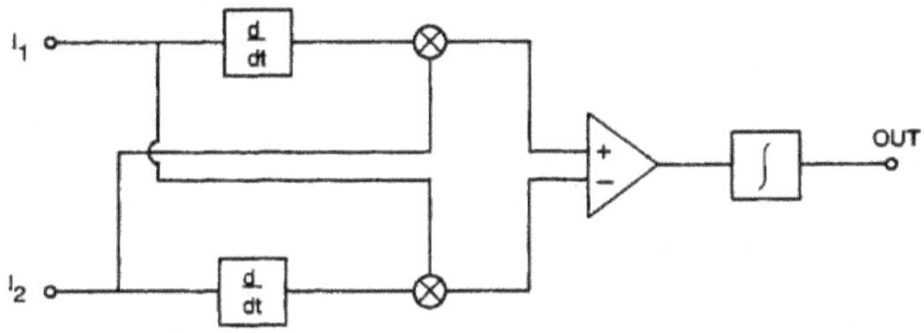

Figure 14. Principle of differentiation and cross multiplication.

The output from the differential amplifier will therefore be

$$S_o = S_u - S_l = I^2 \frac{d\varphi(t)}{dt} \tag{18}$$

Finally, after division with I^2, $\varphi(t)$ can be retrieved by integration as indicated.

A heterodyne detection principle is shown in Figure 13b. The frequency of the light in one interferometer arm is shifted compared to that of the other arm. The detector output contains a frequency equal to the difference of the frequencies of the two light beams entering the directional coupler on the output side. This frequency difference is equal to the frequency shift generated in the reference beam. The phase of the carrier frequency from the detector will change according to the phase modulation in the sensing arm. This phase can be retrieved by a conventional phase meter.

The frequency shifter required by the heterodyne system has been realized mostly by bulk acoustooptic components called Bragg cells. Such bulk optic components require the light to be coupled out of and into the fiber. They are generally unwanted because they contribute to optical instability and loss. More recently, all-fiber frequency shifters have been reported[9]. Such components will contribute to improved performance of acoustic sensors in terms of increased stability, reduced loss and reduced noise level.

5. Sensor multiplexing

It is often of great interest to arrange a number of sensors such as hydrophones in arrays. To manipulate the signals from the sensors individually in a centrally localized processor, it is important to organize the array elements so that they can be individually addressed. For economical reasons it is strongly desired to send the signals along one single fiber that is running along the entire array. For the majority of applications one fiber has sufficient capacity to transmit the required amount of information. To resolve and separate the individual signals, multiplexing is used. A large amount of approaches to this problem have been proposed. We shall describe only one main approach.

The conceptually simplest technique is time division multiplexing. The most straightforward versions of this technique are the ladder structures shown in Figure 15. In Figure 15a a number of Mach-Zehnder interferometers are coupled to the same main (bus-bar) fiber and separated with optical delay coils. The optical input is pulsed with

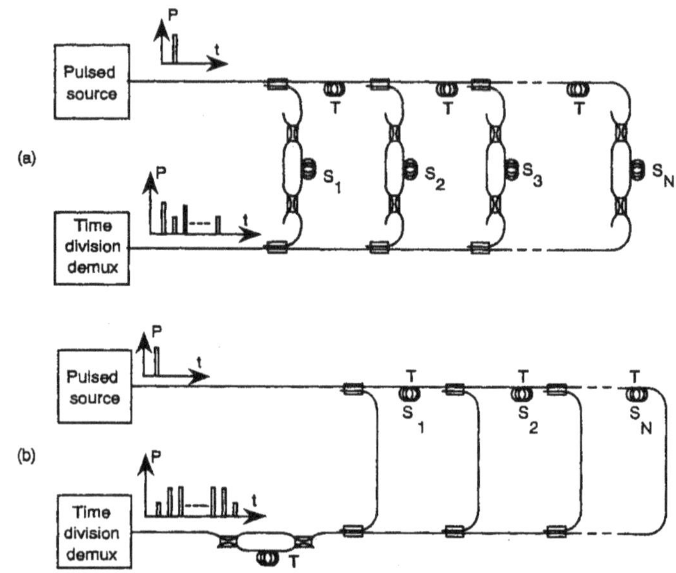

Figure 15. Two examples of time multiplexed sensors. a) Basic configuration. b) Differential path-matched interferometry.

pulse lengths smaller than the optical delay T between two successive sensors. The pulse repetition rate is high compared to the frequencies of interest. The optical detector senses a series of pulses for each input pulse, each pulse representing the output from one interferometric sensor S_i. Each sensor can then be selected in the time domain by a time division demultiplexer.

The configuration shown in Figure 15b applies the method of differential path-matched interferometry[10]. The number of components necessary is reduced compared with the first method. Here, the sensor coil is used for time delay also. The interferometers are not discrete, but are formed by two successive "rungs" of the ladder structure. In a configuration containing N sensors there will be N+1 output pulses for each input pulse. These pulses are applied to a compensator that has a path imbalance equal to the time delay T of each sensor coil. The output from the compensator contains N+2 pulses. Of these, the center N contain relevant information. In addition to having fewer components than the first configuration, it has been shown that this latter configuration has a potentially higher signal-to-noise ratio.

Although we shall not describe other multiplexing approaches here, it is worth noting that it is possible to address individual sensors by coding other parameters of the optical beam such as frequency, wavelength, polarization, and spatial mode[11]. Each of these approaches can be implemented with many different topological schemes.

6. Conclusion

We have discussed only a limited number of aspects of acoustic sensing by optical fibers. These aspects express the mainstream of activities as they have been reported in the literature. We shall now describe a couple of new trends that may influence the further evolution of fiberoptic hydrophones.

There are other interferometers than the Mach-Zehnder configuration that can be used for sensing the acoustic field, although this approach has been predominant. Recently a sensing technique based on a Sagnac interferometer was reported[12]. The

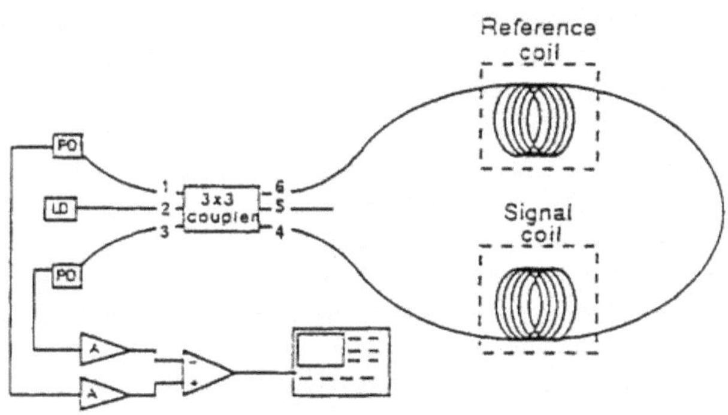

Figure 16. Sagnac interferometer configuration for sensing of underwater sound.

setup is sketched in Figure 16. In a Sagnac interferometer the two interfering beams are propagating in the same fiber but in opposite directions in a loop. The loop is split in a reference coil and a signal coil. The two beams propagating in opposite directions will therefore be in the signal coil at different times. If the sensor is exposed to a time varying field of low frequency, the sensitivity is proportional to frequency. Thus the detection limit is decreasing for increasing frequency. This matches very well with the acoustic noise spectrum in the sea[13]. Investigations of the viability of this approach can therefore be worthwhile.

In the multiplexing schemes the optical power is divided between several individual sensors. There is optical loss associated with this procedure. Optical amplifiers recently reported is a promising approach for compensation of these losses[14]. In such an amplifier an optical pump beam of a different wavelength than the signal beam is propagated along certain sections of fiber that have been doped with Erbium or Neodymium. This causes the signal beam to be amplified. Thus it is possible to amplify the signal beam with no electrical wiring, a feature that will be of great advantage in many cases. This approach seems particularly attractive for applications in hostile environments like the sea. It is believed that by applying such amplifiers, large lengths of fiber with many sensor elements can be used without serious loss in signal to noise ratios.

We have discussed some aspects of acoustic sensing with optical fibers. In particular, the hydrophone concept is the one that has been mainly reported in the literature. The potential of coupling many sensors together in arrays with individually addressed elements has caused great interest. It should be pointed out that operating array systems have been successfully reported[15]. One paper[16] reports tests near Greenland where a fiber hydrophone demonstrated a noise level better than 25 dB below sea state zero. With the advent of fiber amplifiers and possibly new configurations it is most probable that fiberoptic hydrophones will be a very interesting sensor approach in the future.

1. A. W. Snyder and J.D. Love, *Optical Waveguide Theory* (Chapman and Hall, London, 1983).

2. N. Lagakos, P. Macedo, T. Litowitz, R. Mohr, and R. Meister, in *Physics of Fiber Optics II* (American Ceramic Society, 1981) p. 539.

3. R.J. Roark and W.C. Young, *Formulas for Stress and Strain* (McGraw-Hill, New York, 1975).

4. J.A. Bucaro and T.R. Hickman, *Appl. Opt.* **18** (1979) 938.

5. N. Lagakos and J.A. Bucaro, *Appl. Opt.* **20** (1981) 2716.

6. J.A. Bucaro, in *Optical Fiber Sensors*, ed. A.N. Chester, S. Martellucci, and A.M. Verga Scheggi (NATO ASI Series, Martinus Nijhoff, Dordrecht, 1987).

7. J.A. Bucaro, N. Lagakos, J.H. Cole, and T.G. Giallorenzi, in *Physical Acoustics*, ed. W.P. Mason and R.N. Thurston (Academic Press, New York, 1982).

8. J.A. Bucaro, in *Fiber Optics: Advances in Research and Development*, ed. B. Bendow and S.S. Mitra (Plenum Press, New York, 1978).

9. B.Y. Kim, J.N. Blake, H.E. Engan, and H.J. Shaw, *Opt.Lett.* **11** (1986) 389.

10. J. L. Brooks, B. Moslehi, B.Y. Kim, and H.J. Shaw, *J. Lightwave Technol.* **5** (1987) 1014.

11. A.D. Kersey, A. Dandridge, and A.B. Tveten, *Proc. Fiber Optic and Laser Sensors V* **SPIE 838** (1987) 184.

12. K. Kråkenes and K. Bløtekjær, *Opt. Letters* **14** (1989) 1152.

13. R.J. Urick, *Principles of underwater sound* (McGraw-Hill, New York 1983).

14. T. Saitoh and T. Mukai, *IEEE J. Quantum Electron.* **QE-23** (1987) 1010.

15. J.P. Dakin, C.A. Wade, and M.Henning, *Electron. Lett.* **20** (1984) 54.

16. A.M. Yurek, A.B. Tveten, and A. Dandridge, *Proc. 7th Optical Fibre Sensors Conf.* (IREE Australia, Edgecliff, 1990) p. 321.

ACOUSTIC THERMOGRAPHY OF HUMAN BODY: PASSIVE SENSING OF HUMAN BODY BY MEASUREMENTS OF ITS THERMAL ACOUSTIC FIELD

E.E GODIK, YU.V. GULYAEV
V.I. MIRGORODSKY, V.I. PASSECHNIK
Institute of Radioengineering and Electronics
Academy of Sciences of the USSR

ABSTRACT

The main features of acoustothermography are described. Acoustic waves absorption and non uniform distribution of temperature in the human body are properly taken in account, allowing a comparison between sensitivities of acousto- and radio-thermography. Finally some basic experiments and results are reported.

1. Introduction

1974 - V.I. Babiy suggested in analogy with microwave thermography to use the measurements of thermal acoustic noise for determination of temperature in the depth of the Ocean.

1981 - T. Bowen had received a US patent on the system for passive remote measurements of temperature by thermal acoustic radiation.

1985 - Yu.V. Gulyaev, E.E. Godik et al. pointed to the efficiency of use of acoustothermography for human body and other biological objects and gave corresponding theoretical estimates.

1987 - V.I. Mirgorodsky, V.I. Passechnik at al. demonstrated first acoustothermometer for human body.

1987-1991 - More elaborate theory of acoustothermography and several principal experiments were performed in IRE of the USSR Academy of Sciences.

2. Physical Foundations

At thermal equilibrium, in accordance with Plank formula, the total energy of phonons per unit volume is:

$$U_{ac} = \sum_{s,\vec{q}} \frac{\hbar\omega_{s\vec{q}}}{e^{\frac{\hbar\omega_{s\vec{q}}}{kT}} - 1} = \sum_{s} \int_0^\infty u_{ac}(\omega_s)d\omega_s \tag{1}$$

Acoustic wavelength $\lambda_{ac} \ll L$, where L is the size of the detecting transducers used usually.

The number of phonons with $f \leq f_{max}$ ($q \leq q_{max}$) is (Fig. 1):

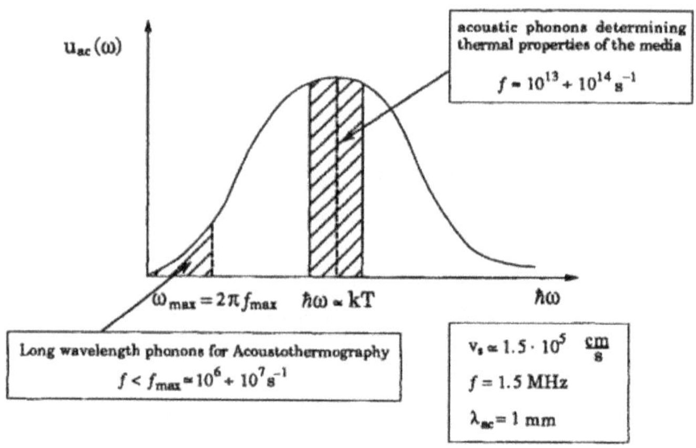

Fig. 1

$$N = \int_0^{q_{max}} \frac{4\pi q^2 dq}{(2\pi)^3} = \frac{q_{max}^3}{6\pi^2} = \frac{4\pi}{3} \cdot \frac{f_{max}^3}{v_s^3}, \tag{2}$$

from which the density of phonon states:

$$\frac{dN}{df} = \frac{4\pi f^2}{v_s^3}. \tag{3}$$

Thus the total phonon energy in unit volume with $f \leq f_{max}$ is giveby:

$$U_{ac}(f_{max}) = \int_0^{f_{max}} \hbar\omega_f \frac{1}{e^{\frac{\hbar\omega_f}{kT}} - 1} \cdot \frac{dN}{df} \cdot df, \tag{4}$$

and for low frequency phonons (thermal acoustic waves), $\frac{\hbar\omega_f}{kT} \ll 1$, we obtain:

$$U_{ac}(f_{max}) = \frac{4\pi}{3} kT \cdot \frac{f_{max}^3}{v_s^3} \left(\frac{\text{erg}}{\text{cm}^3}\right). \tag{5}$$

The part of this energy $U_{ac}(f_{max})$ which belongs to phonons propagating in narrow space angle (Fig. 2):

$$\alpha^2 = \frac{\lambda_{ac}^2}{L^2} \ll 1, \tag{6}$$

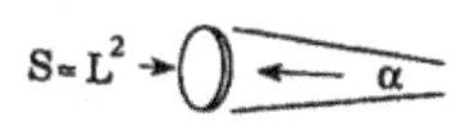

Fig. 2

(wich is the directivity diagram of a transducer) is:

$$U_{ac}(f_{max}) \cdot \alpha^2 = \frac{4\pi}{3} \cdot \frac{kT}{v_s} \cdot \frac{f_{max}}{S}; \quad S = L^2. \tag{7}$$

So, the energy flux (intensity) of acoustic thermal waves propagating to the transducer is:

$$W_{ac}(f_{max}) = \frac{1}{2} U_{ac}(f_{max}) \alpha^2 \cdot v_s = \frac{2\pi}{3} \cdot \frac{kT \cdot f_{max}}{S} \left(\frac{\text{erg}}{\text{s} \cdot \text{cm}^2}\right). \tag{8}$$

Thus, the total thermoacoustic flux power falling down on our transducer with square area S is:

$$W_s \cong kT \cdot f_{max} \left(\frac{\text{erg}}{\text{s}} \text{ or W}\right). \tag{9}$$

For $T = 300$ K, $f_{max} = 10^7$ s^{-1}, $W_s = 4 \cdot 10^{-14}$ W, while a change in temperature $\Delta T = 0.1$ K, with $f_{max} = 10^7$ s^{-1}, yields $\Delta W_s \cong 10^{-17}$ W!

As follows from Eq.(5), phonon energy density with respect to frequency is:

$$\rho_{ac}(f) = \frac{dU_{ac}(f)}{df} = \frac{4\pi kT f^2}{v_s^3} \left(\frac{\text{erg}}{\text{cm}^3} \text{s}\right). \tag{10}$$

So the mean phonon energy in a small frequency interval Δf around f per unit volume is:

$$\overline{U}_{ac}^{\Delta f} = \frac{4\pi kT f^2}{v_s^3} \Delta f \left(\frac{\text{erg}}{\text{cm}^3}\right). \tag{11}$$

This quantity could be expressed through the mean square of acoustic pressure:

$$\overline{U}_{ac}^{\Delta f} = \frac{1}{\rho v_s^2} \cdot \overline{p^2} \tag{12}$$

where ρ is the density of the media and $p\left(\frac{\text{dyn}}{\text{cm}^2} \text{ or Pa}\right)$ is acoustic pressure. So finally for the mean square of thermoacoustic noise pressure we obtain in a frequency interval Δf:

$$\overline{p^2} = \frac{4\pi\rho}{v_s} \cdot f^2 \cdot kT \cdot \Delta f \ (\text{Pa}^2) \text{ or } \left(\frac{\text{dyn}}{\text{cm}^2}\right)^2 . \tag{13}$$

Again, the transducer will feel the acoustic pressure only of those phonons which propagate in the narrow space angle of transducer directivity, $\alpha^2 = \lambda_{ac}^2/S$.

So mean square of thermoacoustic pressure on a transducer is:

$$\overline{p_s^2} = \frac{kT\rho v_s}{S} \cdot \Delta f \left(\frac{\text{dyn}}{\text{cm}^2}\right)^2 \text{ or } (\text{Pa}^2). \tag{14}$$

It is proportional to the frequency interval Δf and does not depend on frequency itself. All these expressions describe the thermal acoustic fluctuations (noise) in infinite uniform media. So they do not take into account directely the acoustic waves absorption and nonuniform temperature distribution in the media.

For thermal equilibrium thermodynamics it is OK.

For acoustothermography one must take all of it into account!

3. Main Formulae of Acoustothermography

We consider a layered medium with layers parallel to xy plane (Fig. 3).

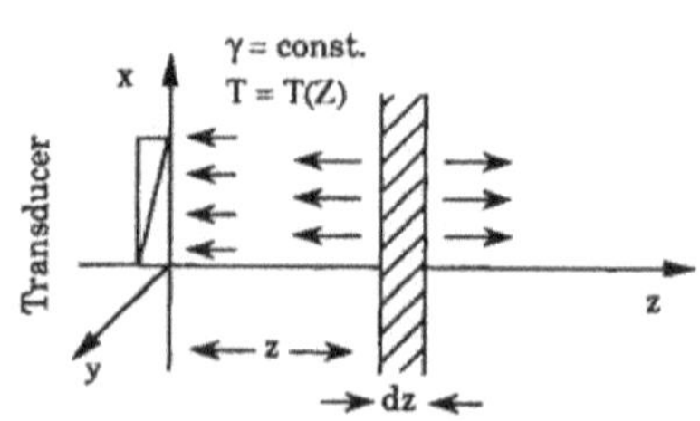

Fig. 3

As it follows from Eq.(8), the thermoacoustic flux intensity generated in unit volume in the direction of the transducer is:

$$W_{ac}(z) = \frac{2\pi k f_{max}}{3S} \cdot T(z) \equiv A \cdot T(z). \tag{15}$$

If γ is the acoustic wave absorption coefficient (generally $\gamma = \gamma(f)$) then

$$\frac{dW_{ac}(z)}{dz} = \gamma \cdot W_{ac}(z), \tag{16}$$

and absorption in the interval dz is:

$$dW_{ac}(z) = \gamma W_{ac}(z)dz. \tag{17}$$

In accordance with Kirchoff law, the radiation of thermoacoustic energy from the interval dz is equal:

$$\gamma \cdot W_{ac}(z)dz = A\gamma \cdot T(z)dz. \tag{18}$$

These thermoacoustic phonons, being injected into other parts of the media, including our transducer, may become nonequilibrium due to nonuniform temperature distribution.

Their intensity at the transducer (z = 0) is: $dW_{ab} = \gamma W_{ac}(z)dz \cdot e^{-\gamma z}$. The total intensity of thermoacoustic phonons of frequency f at the transducer is:

$$W_{ab} = \gamma \int_0^\infty W_{ac}(z)e^{-\gamma z}dz \equiv A\gamma \int_0^\infty T(z)e^{-\gamma z}dz, \tag{19}$$

and it may be called "acoustobrightness".

Introducing the "acoustobrightness temperature", T_{ab}:

$$W_{ab} = A \cdot T_{ab}, \tag{20}$$

we obtain the main formula of acoustothermography:

$$T_{ab} = \gamma \int_0^\infty T(z)e^{-\gamma z}dz, \tag{21}$$

or, taking into account reflection from the transducer, R:

$$T_{ab} = \gamma(1-R) \int_0^\infty T(z)e^{-\gamma z}dz. \tag{22}$$

For completely uniform medium with $T(z) = T_0$ and without reflections

$$T_{ab} = T_0. \tag{23}$$

But in general case

$$T_{ab} \neq T_0, \tag{24}$$

since it is determined not only by thermodynamic temperature, but also by conditions of propagation and absorption of nonequilibrium phonons in nonuniform media, i.e. by the acoustic properties of the medium.

One may say that acoustobrightness temperature T_{ab} is the thermodynamic temperature of such a "black body" which gives the same thermal acoustic flux as we have at our transducer.

Consider two particular geometries which are related to acoustothermometry of human body, Figs. 4 and 5. With reference to the geometry of Fig. 4.

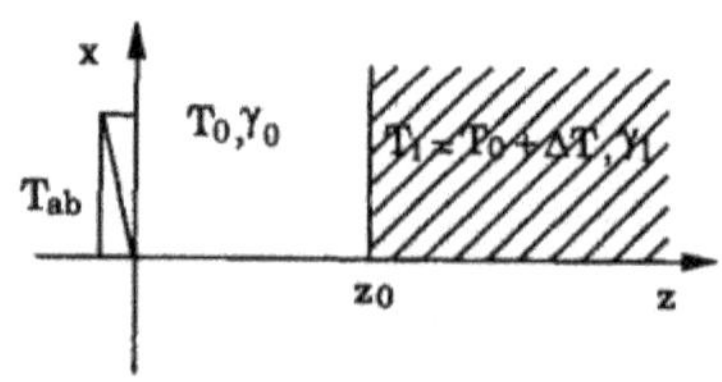

Fig. 4

$$T_{ab} = \gamma \int_0^\infty T(z) e^{-\gamma z} dz, \tag{25}$$

(no reflection), thus

$$T_{ab} = T_0 + \Delta T e^{-\gamma_0 z_0}, \tag{26}$$

and T_{ab} does not depend on γ_1 and if $\gamma_0 \to 0$ (transparent medium):

$$T_{ab} = T_0 + \Delta T \equiv T_1. \tag{27}$$

With the geometry depicted in Fig. 5:

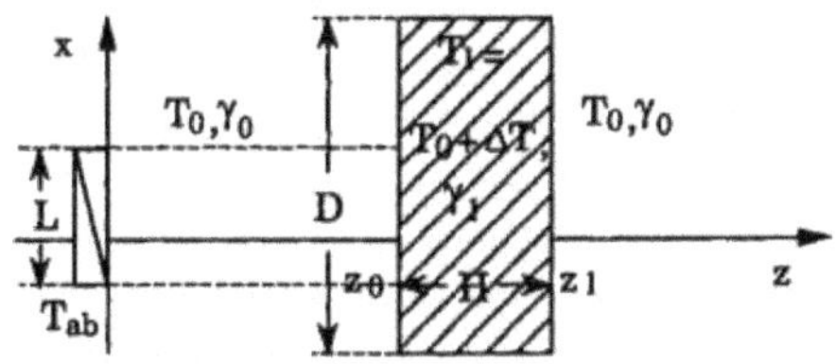

Fig. 5

$$D \gg L \gg \lambda_{ac},$$

$$T_{ab} = \gamma \int_0^{\infty} T(z) e^{-\gamma z} dz,$$

$$T_{ab} = T_0 + \Delta T e^{-\gamma_0 z_0} (1 - e^{-\gamma_1 H}). \tag{28}$$

The factor $(1 - e^{\gamma_1 H})$ could be called "acoustic blackness" of the layer: if $\gamma_1 H \gg 1$, then $1 - e^{-\gamma_1 H} \simeq 1$ - black body. If γ_1 is small it is necessary to have a rather thick layer H, so that it could contribute considerably to acoustobrightness temperature,

$$H \simeq 1/\gamma_1. \tag{29}$$

If this layer imitates human liver then $Z_0 \simeq 1$ cm, $H \simeq 5$ cm.
From measurements in active acoustic thermography:

$$\gamma_0 = 0.14 f_{(MHz)} \mathrm{cm}^{-1}, \quad \gamma_1 = 0.2 f_{(MHz)} \mathrm{cm}^{-1}. \tag{30}$$

Corresponding penetration depths of acoustic waves into the fat on the stomach ($\delta_0 \cong \gamma_0^{-1}$) and into the liver ($\delta_1 \cong \gamma_1^{-1}$) are:

$$\delta_0 = \frac{7.14}{f_{(MHz)}} \mathrm{cm}, \quad \delta_1 = \frac{5}{f_{(MHz)}} \mathrm{cm}. \tag{31}$$

$$T_{ab} \simeq T_0 + 0.6 \cdot \Delta T \tag{32}$$

So, thermal long wavelength acoustic phonons bring considerable information about raise of temperature of liver to the surface of the human body.

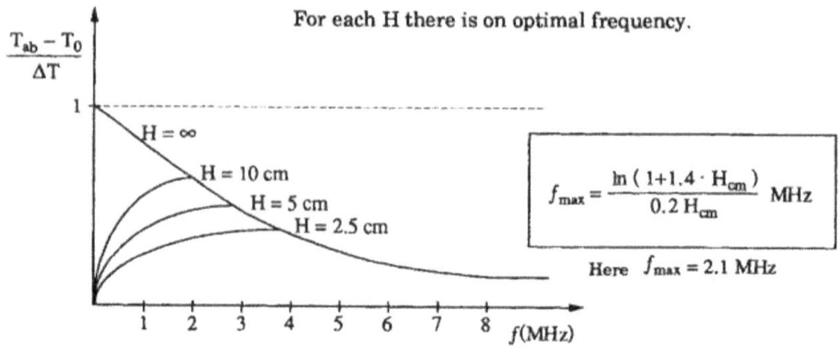

Fig. 6

Let us now estimate the possibilities of acoustothermography (Fig. 7), on the example of simple geometry of Fig. 4.

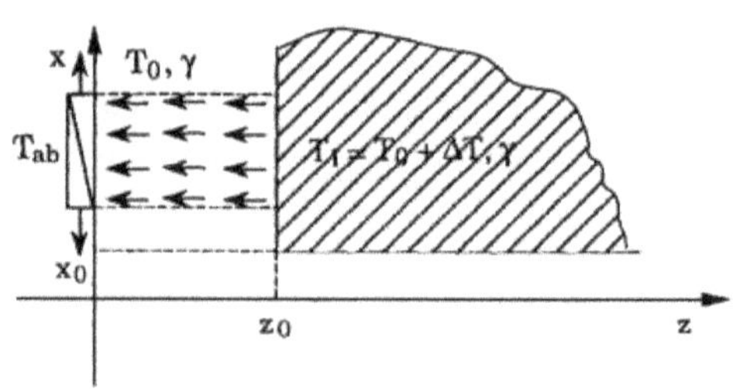

Fig. 7

The main problem of acoustothermography here: to find

$$\Delta T, Z_0 \text{ and } X_0.$$

From Eq.(26):

$$T_{ab}(f) = T_0 + \Delta T \cdot e^{-\gamma(f)Z_0}, \tag{33}$$

if the coordinate of transducer $X > X_0$. ΔT and Z_0 can be determined by making measurements on two different frequencies, f_1 and f_2, and taking into account dependence $\gamma(f)$ (in Eq.(30) $\gamma(f) \sim f$).

Let $T_{ab}(f_1) = T^1_{ab}, T_{ab}(f_2) = T^2_{ab}$ $\gamma(f_1) = \gamma_1, \gamma(f_2) = \gamma_2$, and the corresponding decay depths

$$\delta_1 = \frac{1}{\gamma_1} \text{ and } \delta_2 = \frac{1}{\gamma_2}.$$

Then from Eq.(33) we have:

$$\begin{cases} T^1_{ab} = T_0 + \Delta T \cdot e^{-\gamma_1 z_0} \\ T^2_{ab} = T_0 + \Delta T \cdot e^{-\gamma_2 z_0}. \end{cases} \tag{34}$$

Solving these equations we obtain:

$$\Delta T = \left[\frac{(T^1_{ab} - T_0)^{\delta_1}}{(T^2_{ab} - T_0)^{\delta_2}} \right]^{\frac{1}{\delta_1 - \delta_2}}, \tag{35}$$

$$Z_0 = \frac{\delta_1 \delta_2}{\delta_1 - \delta_2} \cdot ln \frac{T^1_{ab} - T_0}{T^2_{ab} - T_0}. \tag{36}$$

Coordinate X_0 is determined by moving transducer in X-direction and finding the position when the signal is reduced to one half from the maximum value.

The accuracy of the methods of acoustothermography depends on the accuracy of determination of the acoustobrightness temperature, i.e. the mean square of acoustic noise pressure $\overline{p^2}$, on the transducer.

The minimal measurable temperature difference Θ (sensitivity of the method) is determined both by the possibilities of the measurement and by principal restrictions - the presence of electric Nyquist noise on the transducer resistance and different kinds of acoustic losses in the transducer (due to imperfections of construction, reflections, dielectric losses, etc.).

The sensitivity of the measurement method is determined by the nature of the signal: it is "white" noise from which the receiving device "cut off" the signal in the frequency band $\Delta f = \frac{f}{K}$, and averages it during an accumulation time τ.

The minimum measurable temperature difference by this method, Θ, is given by the well known formula:

$$\Theta = T_0 \cdot \frac{\sqrt{2}}{\sqrt{\Delta f \cdot \tau}}. \tag{37}$$

(The same formula is used in radiothermography method).

For $\Delta f = 10^6$ Hz, $\tau = 30$ s, $T_0 = 300$ K

$$\Theta \cong 0.08 \text{ K}. \tag{38}$$

It is 30 times worse in comparison with radiothermography, since there Δf is three orders of magnitude larger. This defect can be partly compensated by taking a matrix made of *mxm* transducers and performing quadratic summing.

Then

$$\Theta_{ac} \simeq \frac{T_0}{m} \cdot \frac{\sqrt{2}}{\sqrt{\Delta f \cdot \tau}}. \tag{39}$$

Maximum probing depth, L_p, is determined by the condition that

$$\Delta T_{ab} = \Delta T \cdot e^{-\frac{L_p}{\delta(f)}} \geq \Theta, \tag{40}$$

from here:

$$L_p = \delta \cdot ln\left(\frac{\Delta T}{\Theta}\right). \tag{41}$$

For the case of human muscles ($\delta \simeq \frac{7}{f_{(MHz)}}$) for $f = 2$ MHz and $\Delta T \simeq 1$ K one has $L_p \simeq 8$ cm.

If T_{ab}^1 and T_{ab}^2 have uncertainity in measurements of the order of Θ it is easy to obtain from Eq.(36) the uncertainity (the error) in determining the depth of the hot object, ΔZ_0:

$$\Delta Z_0 \cong \delta_1 \cdot \frac{\Theta}{\Delta T} \cdot e^{Z_0/\delta_1}. \tag{42}$$

For the case of human muscles and $f = 2$ MHz, $\Delta T = 1$ K, $\Theta = 0,08$ K, one has $\Delta Z_0 \cong 4$ mm.

The uncertainity in determination of X_0 is:

$$\Delta X_0 = D\left(\frac{\Theta}{\Delta T}\right) e^{Z_0/\delta_1}, \tag{43}$$

$D \simeq d$ - size of transducer (~ 1 cm).

For the same case:

$$\Delta X_0 \cong 2 \text{ mm}.$$

In reality, due to factors mentioned above (Nyquist noise, acoustic losses in transducer, etc.), for sensitivity of the AT method instead of $\Theta = T_0 \cdot \frac{\sqrt{2}}{\sqrt{\Delta f \cdot \tau}}$ one must use the more exact expression:

$$\Theta = x \cdot T_n \cdot \frac{\sqrt{2}}{Q}, \tag{44}$$

where $x \simeq 2$ - factor reflecting modulator work, T_n - so called "noise tempe-rature" which is different, generally speaking, from thermodynamic temperature of the transducer, and Q is "accumulation" factor determined by the ratio of frequency bands of AT for high and for low frequencies.

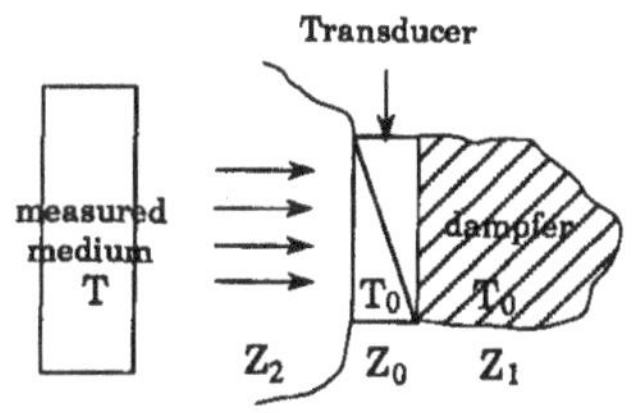

T	- temperature of "hot" object
Z_2	- acoustic impedance of adjacent medium
T_0	- temperature of transducer and dampfer
Z_0 and Z_1	- their acoustic impedances

Fig. 8

Let $Z(f)$ be the electric impedance of the transducer together with its matching circuits.

In the experiment one meatures the mean square of electrical potential, $\overline{u^2(f)}$, on the transducer.

It consists of two parts:

(1) equilibrium one in accordance with Nyquist formula:

$$4kT_0 \cdot ReZ(f) \cdot \Delta f, \tag{45}$$

and

(2) nonequilibrium one, connected with extra mean square of acoustic pressure over (in comparison with) the acoustic noise pressure corresponding transducer temperature T_0. This extra acoustic pressure is due to low frequency phonons coming to transducer from "hot" object in the medium.

If k_p is the transformation coefficient of acoustic pressure into electric potential this term is:

$$|k_p(f)|^2 \cdot \overline{p^2}. \tag{46}$$

For piezoelectric material:

$$|k_p|^2 \simeq K^2 \cdot \frac{4\pi d^2}{\varepsilon \rho v_s^2}, \tag{47}$$

where $K^2 = \frac{4\pi\beta^2}{\varepsilon \rho v_s^2}$ - electromechanical coupling constant, β - piezomodulus, d - the tickness of the transducer, ε - dielectric constant.

Using formulae (14), (45) and (46) one has for $\overline{u^2(f)}$:

$$\overline{u^2(f)} = 4kT_0 ReZ(f)\Delta f + |k_p|^2 \frac{k(T - T_0)Z_2}{S} \Delta f, \tag{48}$$

where $Z_2 = \rho v_s$ in the medium.

Introducing "noise temperature"

$$T_n(f) = T_0 \cdot \frac{4ReZ(f) \cdot S}{Z_2 \cdot |K_p|^2} \equiv \alpha(f) \cdot T_0, \tag{49}$$

one obtains:

$$\overline{u^2(f)} = |k_p|^2 \cdot \frac{k(T - T_0 + T_n(f))Z_2}{S} \Delta f. \tag{50}$$

In fact, $T_n(f)$ is only part of total noise temperature, connected with electric Nyquist noise.

The contribution to noise temperature of different losses in the transducer may be described by impedances Z_2 and Z_1.

For piezoceramic transducer (with impedance Z_0) contacting with human skin (or water):

$$\frac{Z_2}{Z_0} \simeq 0.05. \tag{51}$$

For Z_1 due to other losses one usually has:

$$\frac{Z_1}{Z_0} \simeq 0.08. \tag{52}$$

So, contribution to noise temperature is:

$$T_n = (1 + \frac{Z_1}{Z_2})T_0 \simeq 2.6 \cdot T_0. \tag{53}$$

Thus, to improve sensitivity it is necessary to make better matching and reduce transducer losses. By using special acoustic matching it was in practice possible to obtain threshold temperature sensitivity for $\Delta f \simeq 1$ MHz and $\tau_{acc.} \simeq 30$ s

$$\Theta_{real\ measurement} \simeq 0.12 \text{ K}. \tag{54}$$

In order to improve spatial resolution of acoustothermometry one may use acoustic lenses. This, brings the resolution to about 1 mm.

4. Some Experimental Results

The principal circuit of modulation acoustothermometer is (Fig. 9):

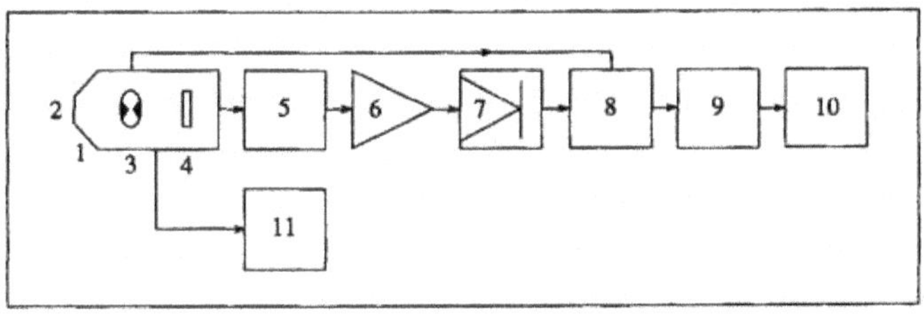

Fig. 9

1. Modulator block filled with water.
2. Acoustically transparent window.
3. Modulator of acoustic flux.
4. Piezoelectric transducer.
5. Matching transformer.
6. High frequency amplifier.
7. Detector.
8. Synchronous detector.
9. Low frequency filter.
10. Recording device.
11. Modulator temperature measuring circuit.

Acoustothermometer works in the following way (Fig. 10):

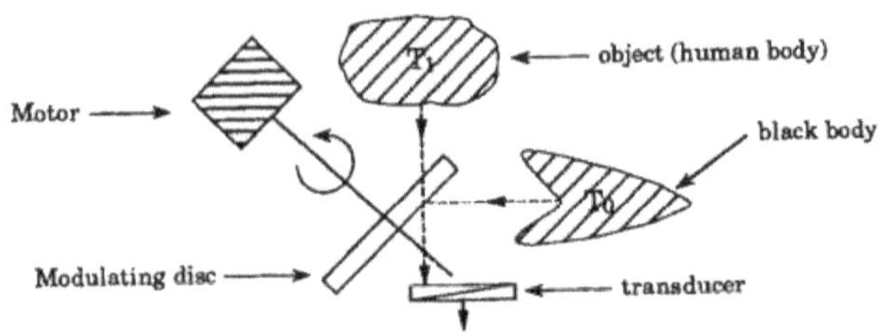

Fig. 10

* The modulator periodically cuts off the acoustic flux coming from the object. * When modulator is closed the acoustic flux from water (or specially made black body) is coming to the transducer. * These signals usually are very close to each other: at temperature difference between object and black body by 1 K the difference of $\overline{p^2}$ is about 0.3 %. * This small difference on the background of large d.c. component is separated by synchronous detector. * The voltage at its output is proportional to the difference of temperatures of the object and the black body (or modulator itself). * So the thermodynamic temperature in the depth of the object under investigation is the sum of "acoustic" temperature and the temperature of modulator (or black body), measured independently. * Low frequency filter performs the accumulation procedure.

The typical parameters of this type acoustothermometer (ΔT) are:

$$\Delta f = 0.8 \div 2 \text{ MHz}, \quad \tau = 30 \div 10 \text{ s}, \Theta = (\frac{1}{6} \text{ of recorder noise"band"}) = 0.12 \text{ K}.$$

4.1. Spatial resolution measurements

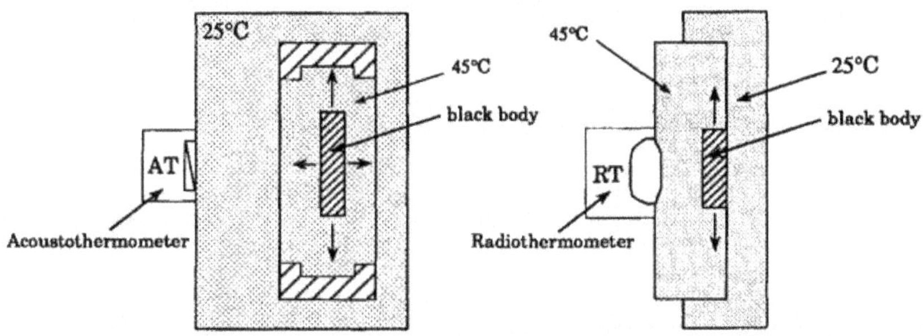

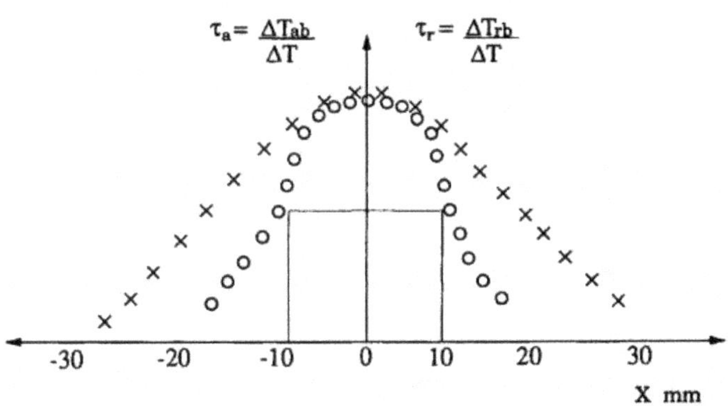

Fig. 11

4.2. First experiment with human body

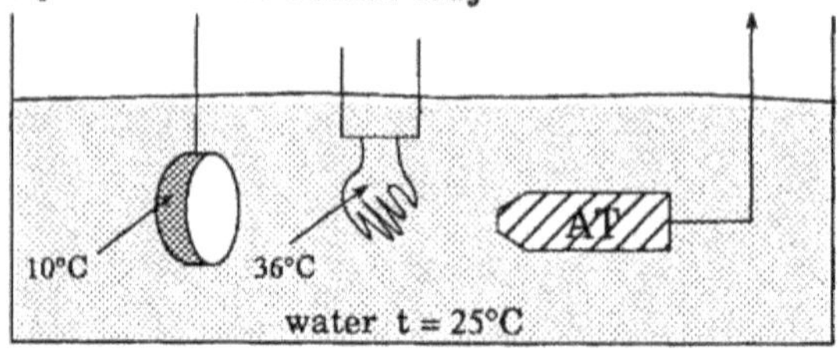

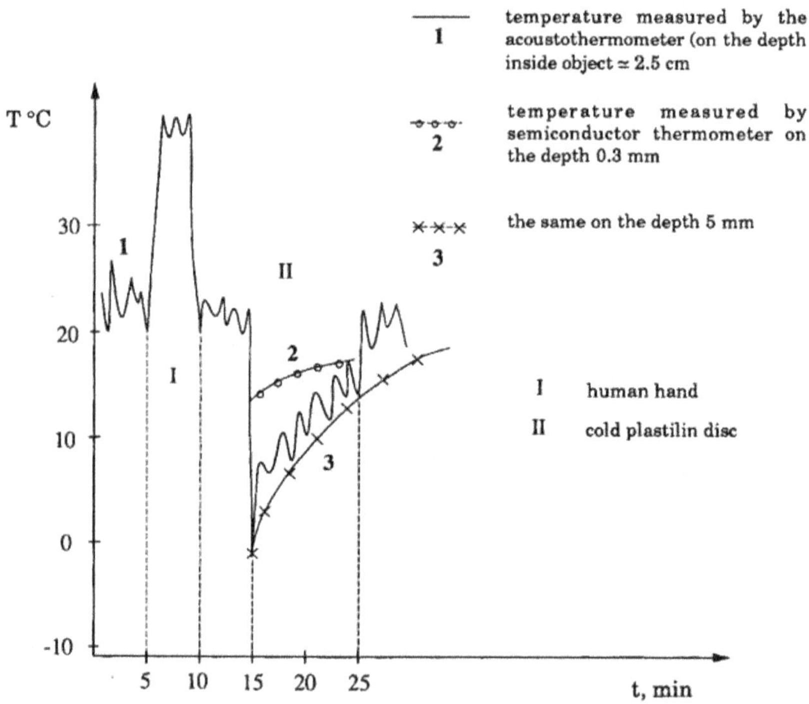

<u>One may see</u>:

1) Temperature inside the human hand is not much changed due to blood supply.
2) AT measurement are close to that of thermometer in depth of the disc.

Fig. 12

4.3. *Diagnostics of thermoregulation system of human leg*

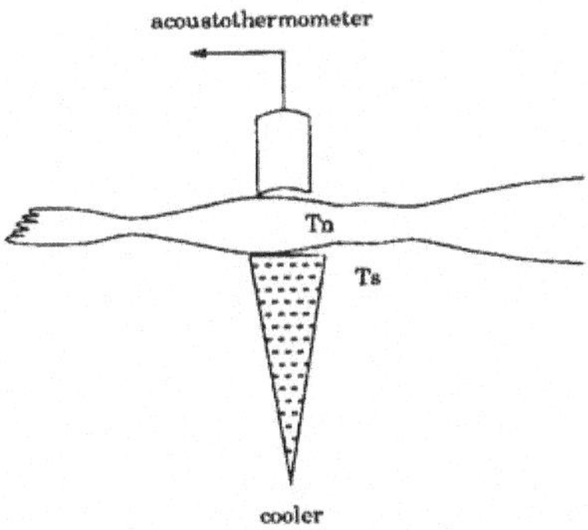

Fig. 13

$$\begin{cases} c\rho\frac{\partial T}{\partial t} - x\nabla^2 T = W(T) \\ W(T) \cong c\rho\nu(T_n - T) \end{cases} \tag{55}$$

c - specific heat, ρ - density of the leg tissue, x - thermoconductivity, ν - the parameter of heat production inside the leg muscle.

The experiment consisted of 4 steps:

I. AT is in thermal equilibrium with surrounding medium ($\sim 23°C$).

II. AT put in contact with the leg ($\sim 36°C$).

III. The cooler ($10°C$) is put into contact from opposite side of the leg.

IV. The cooler is removed.

The temperature measured by AT at the depth of about 2.3 cm:

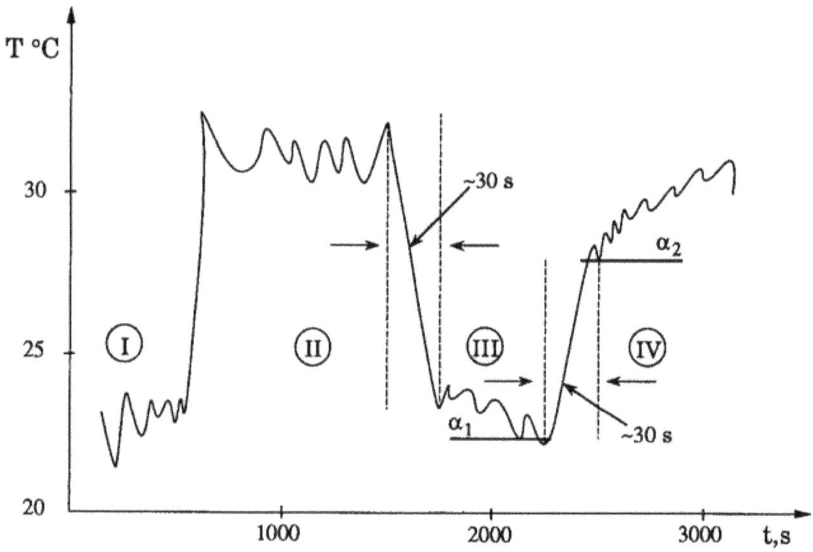

Fig. 14

* Fast steps (~ 30 s) - the direct acoustic (ballistic) influence of cooler on AT:

$$\Delta T = T - T_n = k_t(T_n - T_s)e^{-\alpha d}k_t \cong 0.9, \quad \alpha \simeq 0.08 \div 0.12 \text{ cm}^{-1}.$$

* From experimentally measured slopes α_1 and α_2 and with the help of the solutions of Eq.(55) it was found:
$\chi = (5 \div 7) \cdot 10^{-3}$ cm^2/s - temperature conductivity
$\nu = 10^{-4} \div 5 \cdot 10^{-3}$ s^{-1} - parameter of heat production in the leg.

5. The Problems

1. The development of the theory of passive thermoacoustic tomography.
2. The development of multifrequency experimental technique for passive thermoacoustic tomography.
3. Joint work with medical doctors for interpretation of acoustothermography results.

6. References

1. V. I. Babiy, Sea Hydrophysics Investigations, n.2 (65), p.189, 1974.
2. T. Bowen, US Patent 4, 246,784, Jan 27, 1981.
3. Yu. V. Gulyaev, E. E. Godik, V. V. Dementienko, V. I. Passechnik, A. A. Roubtsov, Soviet Physics - Doklady, v.283, n.6, p.1425, 1985.
4. V. I. Mirgorodsky, V. I. Passechnik, S. V. Peshin, A. A. Roubtsov, E. E. Godik, Yu. V. Gulyaev, Soviet Physics - Doklady, v.297, n.6, p.1370, 1987.
5. V. I. Passechnik, Acoust. J. USSR, v.36, n.4, p.718 and v.36, n.5, p.920, 1990.
6. V. I. Passechnik, Soviet Physics - Radioengineering, n.8, p.77, 1991.
7. V. V. Gerasimov, V. I. Mirgorodsky, S. V. Reshin, V. A. Sablikov, Soviet Physics, Radioengineering, n.9, 1991.

AMPLITUDE AND PHASE ACOUSTIC MICROSCOPY FOR THE INSPECTION OF CERAMIC BEARING BALLS

by

B. T. KHURI-YAKUB, C-H. CHOU, AND P. PARENT

Edward L. Ginzton Laboratory
Stanford University
Stanford, CA 94305-4085

ABSTRACT

Ceramic bearing balls have great potential for replacing steel bearing balls in most applications because of their lower weight, larger strength at high temperatures, and lower lubricant requirements. Bearing balls are subject to high dynamic stresses. Due to the brittleness of ceramic materials, small surface cracks and bulk defects cause catastrophic failure. Therefore, it is of great importance to be able to detect such defects. For this purpose, we developed a scanning acoustic microscope which measures amplitude and phase, and operates at 1-200 MHz. Phase information is more sensitive for the detection of subwavelength depressions. In order to improve the accuracy and stability of the phase measurement, a differential technique is introduced to reduce the effect of noise generated by the vibration of the scanning stage and temperature fluctuations. This paper describes the amplitude and phase measurement system, a theoretical model that relates the amplitude and phase of the measured signal to the physical size (width and depth) of a crack, and it presents some measurements of different types of defects in ceramic bearing balls.

1. INTRODUCTION

Acoustic microscopy is a major nondestructive evaluation tool for imaging, characterization, and detection of defects in structural materials. The concept of imaging with sound was initiated by Sokolov as early as 1949 [1]. The first practical reflection scanning acoustic microscope was demonstrated by Lemons and Quate [2]. In a typical mode of operation, an acoustic wave (either tone-burst or pulsed) is generated by a transducer and focused on the sample through water or some other fluid couplant. The returned signal from the sample is gated, amplified, detected, digitized, and stored in a digital computer. A scanning stage scans either the transducer or the sample, and a point-by-point image of the object is formed. The contrast in the image is due to changes in the elastic properties of the sample or variations in the location of the reflecting interface. To date, most acoustic microscopes measure the amplitude of the reflected signal and display amplitude images of objects.

Since acoustic transducers are linear devices, it is possible to extract the phase information of the sample signal. The added phase information is of essential importance for many applications in image interpretation, signal processing, and material characterization. For most applications, phase variation is more important than amplitude variation for the detection of tiny cracks. Thus, it is possible to work in the 100 MHz frequency range for submicron surface depression detection instead of operating in the GHz range with an amplitude measuring microscope. Therefore, we have developed an amplitude and phase acoustic microscope operating in the 1-200 MHz frequency range; it is used it in various NDE applications, including ceramic bearing ball inspections [3-6].

The major components of an amplitude and phase acoustic microscope are: focused acoustic transducer, amplitude and phase measurement circuitry, scanning apparatus and image display, and a controller computer.

2. FOCUSED ACOUSTIC TRANSDUCER

There are two types of focused acoustic transducers: spherical transducers which are usually used in the frequency range below 20 MHz , and planar transducers with acoustic lenses which are employed at higher frequencies. The spherical transducer focuses the acoustic beam near the center of curvature, according to the Fresnel parameter of the transducer. The Fresnel parameter is defined as:

$$S = \lambda R/a^2 \tag{1}$$

where λ is the wavelength in water (or other couplant), R is the radius of the curvature, and a is the radius (or aperture) of the transducer. If $S \ll 1$, then the focal point is at the center of curvature. As S increases, the focus location moves gradually toward the transducer [7].

A schematic of a planar transducer-lens system is shown in Fig.1. The acoustic wave launched by the transducer propagates through the buffer rod and is refracted into water. The planar transducer can be a longitudinal wave transducer, a shear wave transducer, or a mixed mode transducer which generates both longitudinal and shear waves in the buffer rod. The focal lengths of these different propagation modes are as follows.

For a longitudinal wave transducer:

$$F_L = \frac{R}{1 - V_W/V_L} \tag{2}$$

where R is the radius of the curvature of the lens, and V_W and V_L are the longitudinal wave velocities of water and the buffer rod, respectively.

For a shear wave transducer:

$$F_S = \frac{R}{1 - V_W/V_S} \tag{3}$$

where V_S is the shear wave velocity of the buffer rod.

A mixed mode transducer has three focal points corresponding to longitudinal transmission and reflection in the buffer rod (L-L mode), shear transmission and reflection (S-S mode), and one way longitudinal-one way shear (L-S or S-L mode). The corresponding focal lengths are F_L, F_S and $F_{LS}(=.5(F_L+F_S))$.

The focused transducer-lens system is a critical component of the microscope. It controls the imaging quality, such as the spatial resolution and image contrast, as well as noise level. The transducer operation mode, buffer rod material, focal length, and F-number of the lens, as well as the operation frequency, are all chosen according to the desired application [8,9]. Longitudinal transducers are most commonly used; they give the best image resolution for surface and bulk imaging. Shear wave transducers on buffer rods are most efficient for directional surface wave excitation; therefore, they are best suited for applications related to surface wave properties. Mixed mode transducers provide different modes of acoustic wave propagation at the same time; therefore, they are best suited for relative measurements, such as differential phase measurements, as we will see later.

3. AMPLITUDE AND PHASE MEASUREMENT

The data acquisition scheme of the scanning acoustic microscope is based on amplitude data conversion. In order to measure both the amplitude and phase of the reflected signal, a reference signal is needed, and two or more measurements with phase shifted reference signals are made for each sample point. Increasing the number of measurements improves the accuracy of the phase and amplitude measurements, but increases the image acquisition time. If N is the number of measurements at one location, the phase of the reference signal needs to be shifted by $2\pi/N$ for each measurement. In our apparatus, N is chosen to 10.

Figure 2 shows a block diagram of the amplitude-phase measurement system. In order to generate a reference signal that is phase shifted relative to the sample signal at any frequency in the 1-200 MHz range, we built RF circuits with two output channels to generate 300 MHz signals. The output of channel 1 is mixed with the output of a synthesizer whose frequency can be accurately adjusted in the 301-500 MHz range; the lower sideband of the mixing product is the signal in the desired operation frequency of 1-200 MHz. The signal is switched by a pulser to generate $N(=10)$ tone bursts at each sample location, and is applied to drive the acoustic transducer after proper power amplification. Each burst consists of 10-50 cycles of RF signals. The time interval between two adjacent bursts should be long enough to avoid interference from multiple reflections in the buffer rod or water. For ceramic bearing ball inspection, we operate at a signal frequency of 100 MHz ; the tone burst duration is chosen to be 200-400 ns , and the interval between two adjacent bursts is about 200 μS .

The output of channel 2 is the phase shifted 300 MHz signal. The phase shift is obtained by controlling the time delay of the RF switch in channel 1. The phase of the channel 2 signal shifts by $(i-1)*2\pi/N$ relative to the channel 1 signal when the i-th pulse starts, where i is between 1 and N . The signal of channel 2 is also mixed with the output of the same synthesizer as channel 1. Therefore, a reference signal with the same frequency (1-200 MHz), and with a controlled phase shift relative to the signal driving the acoustic transducer, is generated.

The reflected signal from the sample is selected by a gate circuit and amplified. Then it is mixed with the phase shifted reference signal. Assuming that the reflected signal within each tone burst can be expressed as:

$$S(r,t) = A(r)\cos(\omega t+\phi(r)) \tag{4}$$

and the reference signal corresponding to the i-th burst as:

$$R_i(r,t) = \cos(\omega t+\theta_i) \tag{5}$$

where $\theta_i = (i-1)*2\pi/N$, r is the location of the acoustic beam focus, t is time, ω is the radian frequency, and A and ϕ are the amplitude and phase of the reflected signal.

The measurements at one location produce N pulses with amplitudes A_i , as shown in Fig. 3. These N pulses are passed through a low-pass filter and then digitized and processed to produce the amplitude and phase of the reflected signal at one sample location. In the ideal case, A_i should contain only the $A(r)\cos(\phi(r)-\theta_i)$ term. In practice, mixers and the low-pass filters introduce a DC offset and higher harmonic terms such that:

$$A_i = A(r)\cos(\phi(r)-\theta_i) + DC + \sum_{n=2}^{\infty} E_n \cos(n[\phi(r)-\theta_i]) \tag{6}$$

Using Discrete Fourier Transform (DFT) methods, A(r) and $\phi(r)$ can be calculated, and the DC and all higher harmonics, except the $M*N\pm1$ harmonics, are removed (where M is an arbitrary integer) [10].

4. DIFFERENTIAL PHASE MEASUREMENT

Phase measurement is very sensitive to variations in the distance between the sample and the transducer, as well as the acoustic velocity in the media of wave propagation. For example, the wavelength in water is 15 μm at 100 MHz . Thus, a phase measurement sensitivity of 1° corresponds to a distance change of about 200 Å . Consequently, the phase measurement can be very noisy, due to vibrations of the scanning stage or water temperature fluctuation.

In order to reduce the noise level, we developed a differential phase measurement system employing a mixed mode transducer. The mixed mode transducer generates both longitudinal and shear waves in the buffer rod simultaneously. Because of mode conversion when reflecting from the lens-water interface, the signals travelling inside the buffer rod are separated into three groups, i.e., LL (both ways longitudinal), LS(one way longitudinal, the other shear), and SS waves (both ways shear), as shown in Fig.4. Each group has one signal reflected from the lens-water interface, followed by a weaker signal due to transmission into the water and reflection from the surface of the sample. Since LL, LS, and SS signals have different focal lengths, when one is focused on the sample surface, either of the other two is defocused and can be used as the acoustic reference. In our experiment, we focused the LL wave on the surface of the sample, and used the LS wave as a reference. Because both the LL and LS waves propagate the same distance in the water, the phase noise due to scanning vibration and temperature fluctuation is removed.

For real time differential phase measurement, additional electronic circuits are needed, as shown in Fig.5. The additional circuits separate the LL signal and LS acoustic reference into two channels. The signals are then detected, digitized, and stored in the computer where the amplitude ratio and phase difference are calculated.

5. THE SCANNING APPARATUS

For the inspection of ceramic bearing balls, two different mechanical scanning systems have been used. One is a traditional x-y scanner that scans the transducer over a fixed sample and generates amplitude and phase images. The other is a rotation scanner, which was specially developed for this application. Here, the transducer remains fixed, while the ball is rotated in place, in two directions, using friction with a rubber roller. This system is driven by two stepper motors; the first one rotates the roller which rotates the ball in one direction,while the second one translates the roller, giving the second rotation of the ball in the orthogonal direction. In order to get complete coverage of a given ball, it is important to set the rotation and translation speeds properly, depending on the size of the ball, the data acquisition speed, and the spot size of the acoustic beam.

For differential phase measurement, the rotation scanning system was more suitable than the x-y scanning system because it provides a more stable acoustic reference signal.

6. THE MAIN CONTROLLER AND IMAGE DISPLAYER

An IBM PC-AT computer is used to control the operation of the microscope. The computer sends the commands to the scanner to control the scanning pattern, step size, scanning speed, and the generation of the synchronization pulses which are used to synchronize tone burst generation and data acquisition. It also performs data acquisition, amplitude and phase computation, signal processing, as well as imaging the display on the monitor.

7. PERFORMANCE OF THE MICROSCOPE

For quantitative measurements of small defects, the accuracy of the phase and amplitude measurements is essential. Since phase measurement is more sensitive than amplitude for small surface depressions in ceramic bearing balls, phase accuracy is the key to determining the capability of detection of these submicron size surface depressions.

The accuracy of the amplitude and phase measurements is mainly limited by electronics and scanning vibration noise. To evaluate the noise generated by the electronics, we measure the echo reflected from the lens-water interface because it is independent of the water path. To evaluate the total noise caused by both the electronics and the scanner, we measure the signal reflected from an optical flat. All the measurements, including simple phase and differential phase, were made at a frequency of 120 MHz. Table I summarizes these noise measurements.

TABLE I

	ELECTRONIC NOISE	TOTAL NOISE	
		Simple	Differential
Amplitude	±0.35%	±2%	±2%
Phase	±0.5°	±7.5°	±0.4°

We note that the differentiated phase measurement system reduces the total phase noise from ±7.5° to ±0.4°. This resolution allows us to detect steps of less than 100 Å, and as we demonstrate later, it allows us to detect depressions of the order of 1 μm in width and 1 μm in depth (while the wavelength is 12 μm).

8. DEFECT TYPES AND DETECTION STRATEGIES

There are several different types of defects in ceramic bearing balls. The first type are bulk defects, which can be easily imaged with an acoustic microscope. The second type are deep fissures that exist near the surface and perpendicular to it. These cracks are easily detected using surface waves, which are very sensitive to this kind of perturbation [11]. The third type of defects are surface spalls, which are shallow but relatively large, typically 10 μm or more in size. These are once again very easily detected by measuring the phase of the acoustic signal reflected from the ball at the location of the defect. For an operating frequency of 120 MHz, where the wavelength in water is about 12 μm, we are able to image these defects even if their depth does not exceed 0.5 μm, which corresponds to a phase variation of 28° for a transducer with an F-number of 1 [12], without particular precaution concerning scanner vibrations.

The worst type of defects are shallow cracks of small width with smooth surfaces. These are the most difficult to detect as they are insensitive to surface waves and are small compared to the spot size of the focused acoustic beam. Their size can be of the order of one micron, while the beam size in the 100 MHz range is at least an order of magnitude larger.

We applied a trench theory, originally developed in optics, to predict the relationship of amplitude and phase variations to the physical dimensions of the crack. We will give examples of measurements of surface cracks and compare their response to theoretical predictions.

Finally, in order to enhance the display contrast of the defects in both amplitude and phase images, we employed a signal processing scheme to remove the large phase and amplitude variations caused by the curvature of the balls.

9. SCATTERING THEORY

In order to predict the scanning acoustic microscope's response to narrow surface depressions, we calculate the scattering from trenches using a theory similar to the one dealing with the imaging of trenches with a confocal scanning optical microscope [13].

As shown in Fig. 6, when an acoustic beam illuminates the surface trench (depression) with width L and depth h, the reflected beam can be considered as the sum of a homogeneous wave (i.e., the wave reflected from a perfect plane boundary) and a scattered wave (caused by the depression). That is:

$$V_r = V_{hom} + V_s \tag{7}$$

where V_r, V_{hom}, and V_s are the particle velocities in the z-direction of reflected, homogeneous, and scattered waves, respectively.

The boundary conditions of the continuation of the normal velocity and the acoustic pressure at $z = 0$ are corresponding to the following:

$$V_{tr} = V_{inc} + V_r = V_{inc} + V_{hom} + V_s \tag{8}$$

$$\partial V_{tr} / \partial z = \partial V_{inc} / \partial z + \partial V_r / \partial z = \partial V_{inc} / \partial z + \partial V_{hom} / \partial z + \partial V_s / \partial z \tag{9}$$

where V_{inc} is the particle velocity in the z-direction of the incident wave, and V_{tr} is that of the wave inside the trench.

For the samples with much higher acoustic impedance (such as Si_3N_4) than the working couplant (water), at the boundary, we have:

$$V_{hom} = -V_{inc} \tag{10}$$

$$\partial V_{hom}/\partial z = \partial V_{inc}/\partial z \tag{11}$$

Therefore, Eqs. (8) and (9) can be simplified as:

$$V_{tr} = V_s \tag{12}$$

$$\partial V_{tr} /\partial z = 2\partial V_{inc}/\partial z + \partial V_s/\partial z \tag{13}$$

In order to solve V_s, we decompose V_s at $z = 0$ into spatial frequency components, and then decompose each component into trench modes:

$$V_s = \sum_{n=1}^{N} \int\int A_n(p,q)\sin(n\pi x/L)e^{-jkpx}e^{-jkqy}dpdq \tag{14}$$

where k is the wave number. To satisfy Eq. (12), we can write:

$$V_{tr} = \sum_{n=1}^{N} \iint A_n(p,q)\sin(n\pi x/L)[\sin(\gamma_n(z+h)/(\sin\gamma_n h)]e^{-jkpx}e^{-jkqy}dpdq \tag{15}$$

where:

$$\gamma_n = \sqrt{k^2(1-q^2)-(\frac{n\pi}{L})^2} \tag{16}$$

$\partial V_{inc}/\partial z$ can also be decomposed as:

$$\partial V_{inc}/\partial z = \sum_{n=1}^{N} \iint D_n(p,q)\sin(n\pi x/L)e^{-jkpx}e^{-jkqy}dpdq \tag{17}$$

As in Ref. 13, $A_n(p,q)$ can be solved from the following equation:

$$\sum_{m=1}^{N} H_{mn}(q)A_n(p,q) = D_m(p,q) \tag{18}$$

where $H_{mn}(q) = (1/2)\gamma_n \cot(\gamma_n h)\delta_{mn} - \Gamma_{mn}(q)$.

$$\Gamma_{mn}(q) = \frac{jk^2}{\pi L}\int_{-\infty}^{\infty} \frac{(m\pi/L)(n\pi/L)\sqrt{1-p^2-q^2}}{\left[k^2p^2-(m\pi/L)^2\right]\left[k^2p^2-(n\pi/L)^2\right]}\left[e^{-j(kpL-m\pi)}-1\right]\left[e^{-j(kpL+n\pi)}-1\right]dp \tag{19}$$

The received signal by the transducer of the SAM is:

$$\begin{aligned} v &= v_s + v_h \\ &= 2\int_{s_1} V_s^* \frac{\partial V_{inc}}{\partial z}ds + 2\int_{s_1} V_h^* \frac{\partial V_{inc}}{\partial z}ds \end{aligned} \tag{20}$$

where * denotes a complex conjugate and s_1 is the sample plane $z = 0$. Since the incident wave at $z = 0$ depends on the focal location (x_0, z_0) of the acoustic beam, ϖ is a function of x_0 and z_0.

Figure 7 shows a result of the theoretical calculations as a linescan across a trench, 2 μm in width and 1 μm in depth. The transducer is operated at a frequency of 118 MHz and the lens has an F-number of 0.8. The focus of the lens is placed at the surface of the bearing ball. A maximum phase shift of 2.8 degrees, and a maximum relative amplitude change of 0.2% are predicted by the theoretical calculation for the trench under consideration.

10. IMAGE PROCESSING

The spherical shape of the balls introduces fringes in both the amplitude and phase images. These fringes are due to the change in the distance between the ball and the lens as the sample is scanned. These fringes clutter the image, and can obscure the indication of a defect. In order to overcome this difficulty, we apply the following image processing technique. We take phase and amplitude images of the sample under inspection (i.e., object images). We also take an image of a reference sample without a defect. The reference object could be another part of the same or another bearing ball that is known to have no defects. This type of process is viable for bearing balls which are usually made to tolerances that are a fraction of a micron. Of course, the object and reference images are taken under the same working conditions, including the distance from the lens to the sample. The phase images are carefully unwrapped. Finally, we subtract the reference image from the corresponding object image. Ideally, the spherical features are removed and the defects are enhanced in the processed images.

11. EXPERIMENTAL RESULTS

We have inspected several Si_3N_4 ceramic bearing balls. As an example, Figs. 8-12=1 show the amplitude and phase images of a ball with a diameter of 16 mm. Figures 8 and 9 are the amplitude images of the original object and the processed images, respectively. Figures 10 and 11 are the phase images. The transducer was operated at 118 MHz with the lens of 0.8 F-number. The acoustic beam was focused on the top of the ball. The fields of view are all 1.6 mm x 1.6 mm.

From the figures shown, the processed phase image shows small defects that were not visible without data processing. The processed amplitude image also shows the defects much better than the original image. The remaining fringes are due to the imperfect relocation of the x-y scanner.

The rotation scanner was used to detect the presence of narrow, shallow scratches. The scanner was used, along with the differential phase measurement, in order to enhance the detection sensitivity to 0.4°. Figure 12 shows the amplitude and phase variation of the reflected signal from the ball as it is rolled under the transducer. The linescan is over a small crack whose width was measured with a confocal optical microscope to be 2 μm. The maximum phase variation measured in Fig. 12 is about 2.5°, which indicates that the depth of the crack is around 1 μm according to the theoretical model. The amplitude variation in Fig. 12 is larger than expected, which is probably caused by the roughness of the crack edges.

12. CONCLUSION

We have described the basic operation principle of an amplitude and phase scanning acoustic microscope and its use in various NDE applications. An image processing scheme was developed to enhance defect detection in bearing balls images. We also presented a rotation scanning system capable of rolling bearing balls under a stationary transducer. The system allows the inspection of bearing balls by determining the number of indications where the amplitude and phase of the reflected signal goes above a baseline signal. A theoretical model was developed to determine the size of small defects based on the amplitude and phase of the scattered signals, as measured with a microscope.

13. ACKNOWLEDGMENT

This work was supported by the Department of Energy on Contract No. DE-FG03-84ER45157. The ceramic balls we measured are courtesy of J.Hannoosh and J.Lucek of CERBEC, Inc.

14. REFERENCES

1. S. Y. Sokolov, Dokl. Akad, Nauk. SSSR 64, 333 (1949).
2. R. A. Lemons, and C. F. Quate, "Acoustic Microscope Scanning Version," Appl. Phys. Lett. 24, 163 (Feb. 1974).
3. K. K. Liang, S. D. Bennett, B. T. Khuri-Yakub, and G. S. Kino, "Precise Phase Measurements with the Acoustic Microscope," IEEE Sonics & Ultrasonics SU-32 (2), 266 (March,1985).
4. P. A. Reinholdtsen, C-H.,Chou, and B. T. Khuri-Yakub, "Quantitative Acoustic Microscopy with Amplitude and Phase Imaging", Proc. IEEE Ultrasonics Symposium, Vol. 2, 813-816 (1987).
5. P. Parent, C-H. Chou, and B. T. Khuri-Yakub, "Ball Bearing Inspection with an Acoustic Microscope," Proc. IEEE Ultrasonics Symposium, Vol. 2, 1111-1113 (1988).
6. B. T. Khuri-Yakub, "Acoustic Microscopy," SPIE Scanning Microscopy Technologies and Applications, Vol. 897, 118-128 (1988).
7. G. S. Kino, Acoustic Waves: Devices, Imaging, and Analog Signal Processing, New York: Prentice-Hall (1986) Chapter.3.
8. C-H. Chou, B. T. Khuri-Yakub, and G. S. Kino, "Lens Design for Acoustic Microscopy," IEEE UFFC, Vol. 35, No. 4, 464-469 (July 1988).
9. C-H. Chou, P. Parent, and B. T. Khuri-Yakub, "Acoustic Microscopy with Mixed-Mode Transducers," Review of Progress in Quantitative Nondestructive Evaluation, Vol. 8A, 865-871.
10. P. A. Reinholdtsen, "Image Processing for an Amplitude and Phase Acoustic Microscope," (Ph.D. Dissertation, Department of Applied Physics, Stanford University, 1989).
11. L. R. Clarke, C-H. Chou, B. T. Khuri-Yakub, and D. B. Marshall, "Acoustic Evaluation of Grading Damage in Ceramic Materials," Proc. IEEE Ultrasonics Symposium, Vol.2, 979-982 (1985).

12. K. K. Liang, G. S. Kino, and B. T. Khuri-Yakub, "Material Characterization by the Inversion of V(z)," IEEE Trans. SU-32(2), 213-224 (1985).
13. P. C. D. Hobbs, "Heterodyne Interferometry with a Scanning Optical Microscope," Chapter 6, Ph.D. Dissertation, Stanford University, Stanford, CA. (August 1987).

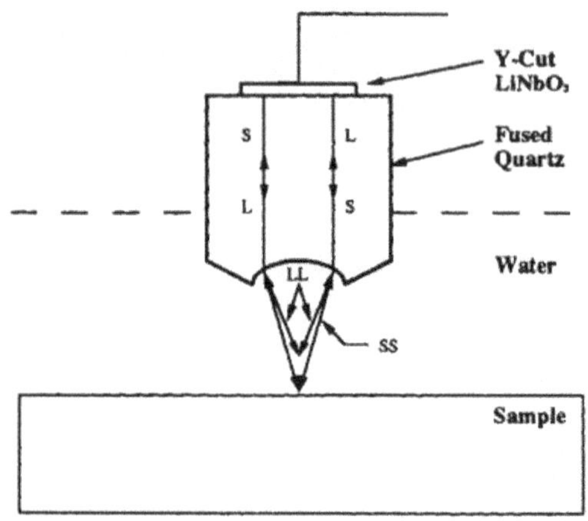

Fig. 1. Schematic of mixed transducer-lens system.

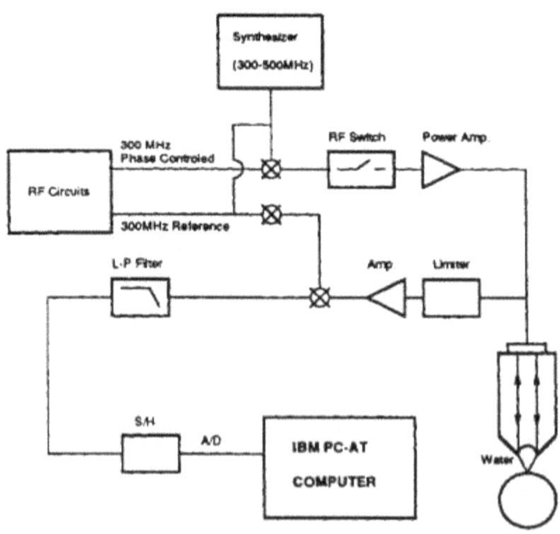

Fig. 2. Block diagram of the amplitude and phase measurement system.

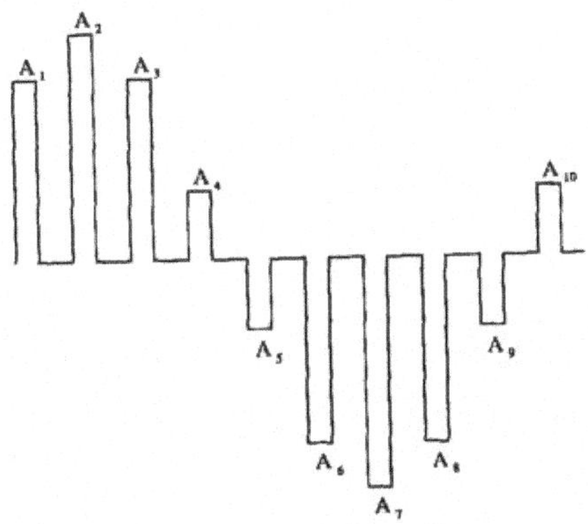

Fig. 3. Schematic of a set of N pulses to digitize at each scan location (N = 10).

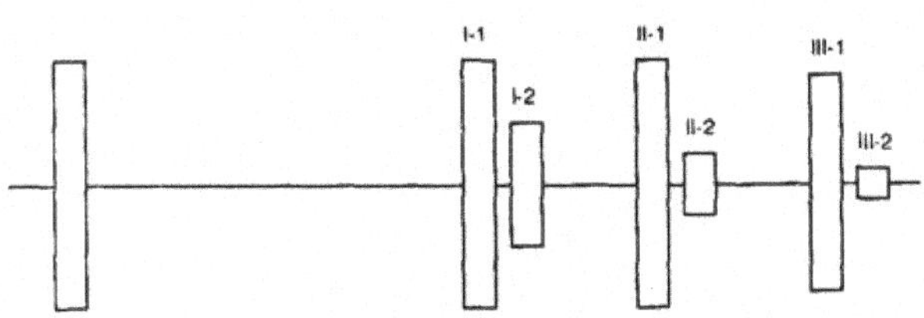

Fig. 4. Schematic of the received signals by the mixed mode transducer. I-1 -- LL wave from the lens-water interface. I-2 -- LL wave from the sample. II-1 -- LS wave from the lens-water interface. II-2 -- LS wave from the sample. I II-1 -- SS wave from the lens-water interface. III-2 -- SS wave from the sample.

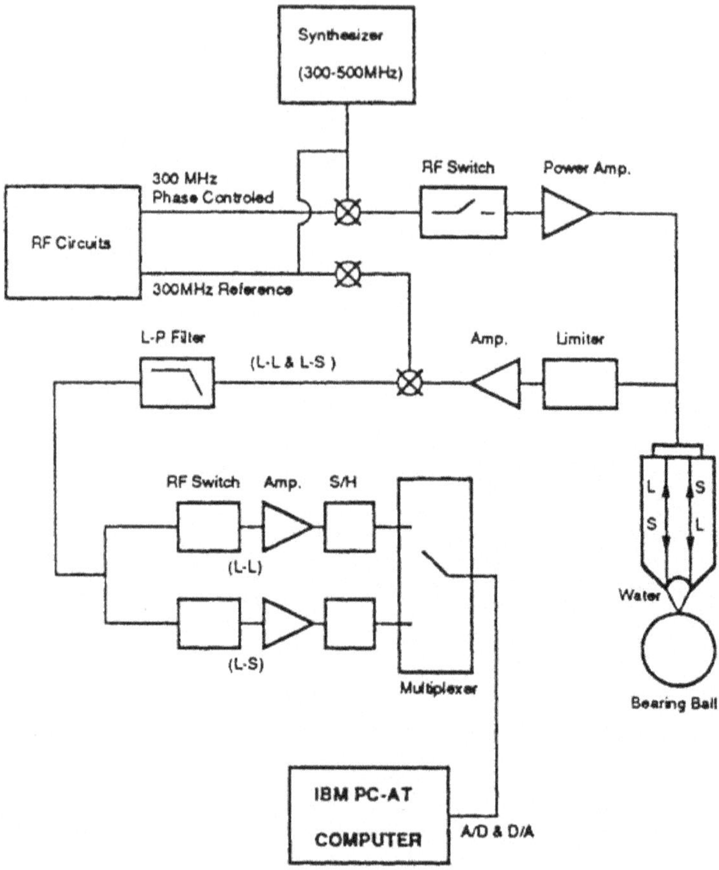

Fig. 5. Block diagram of differential phase measurement system

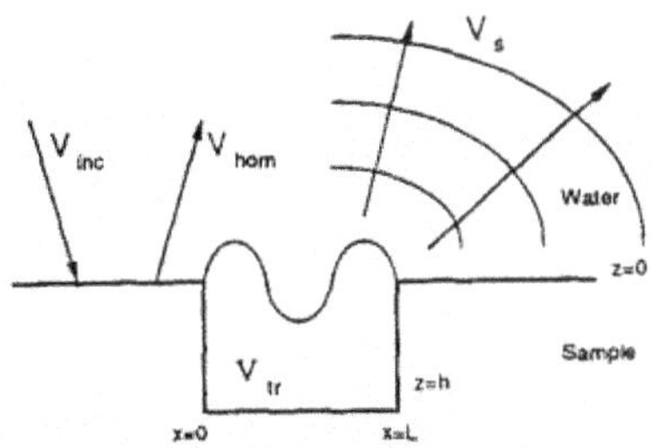

Fig. 6. Trench structure of scattering simulations.

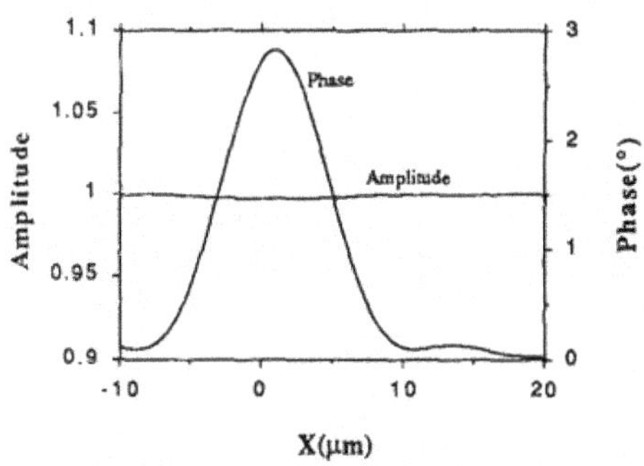

Fig. 7. Theoretical results of linescan of a 2 μm wide, 1 μm deep trench. Operating frequency = 118 MHz, F# of lens = 0.8, focused on top.

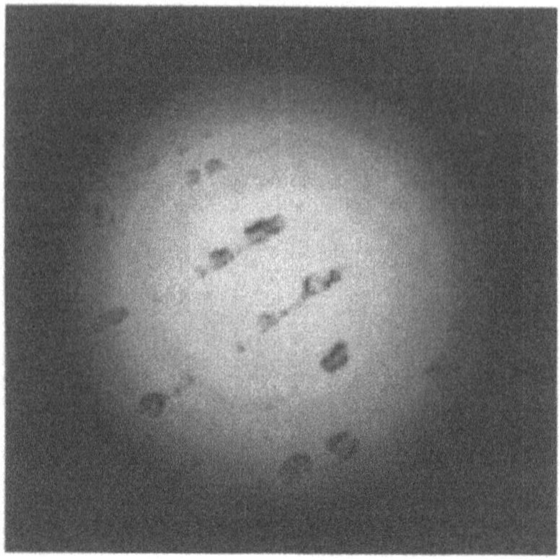

Fig. 8. Amplitude image of a Si_3N_4 bearing ball of diameter 16 mm (original). F-number of lens = 0.8; operating frequency = 118 MHz; field of view = 1.6 mm*1,6 mm.

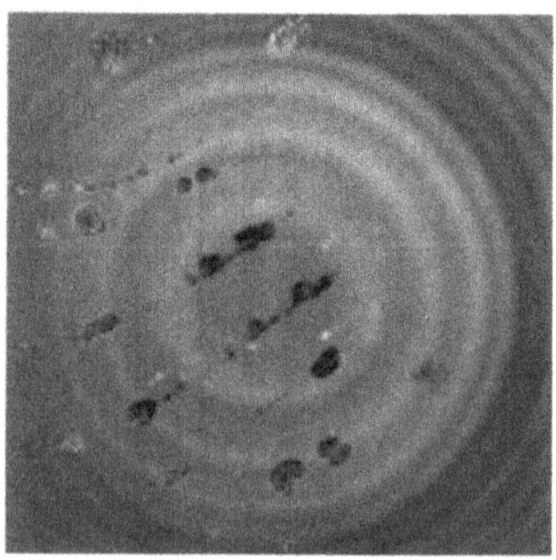

Fig. 9. Processed amplitude image from Fig. 9.

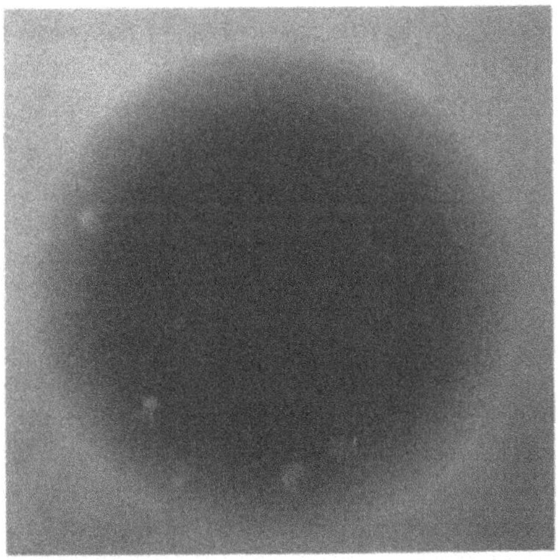

Fig. 10. Phase image of the same object as in Fig. 4 (original).

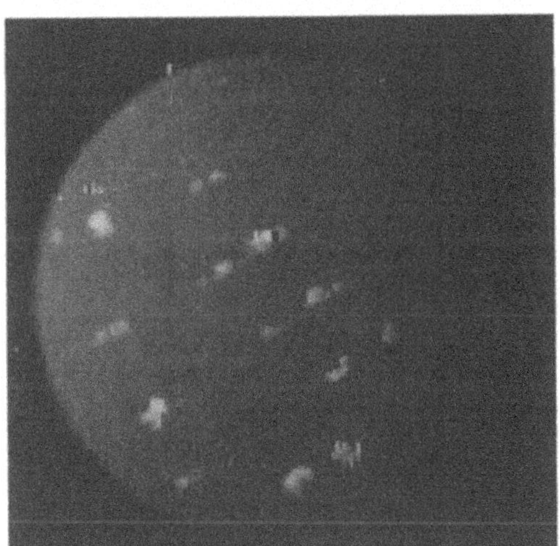

Fig. 11. Processed phase image from Fig. 11.

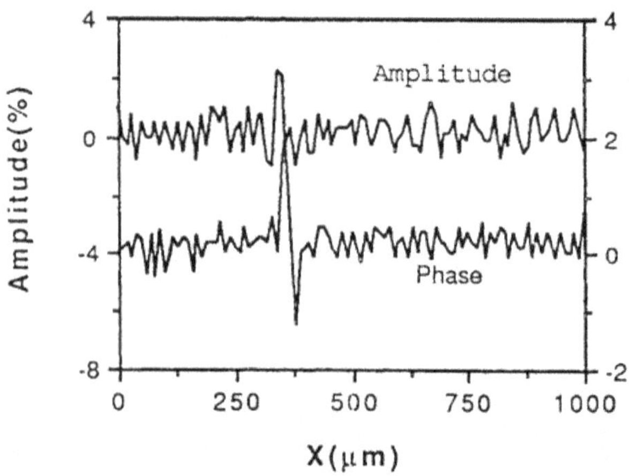

Fig. 12. Experimental results of a linescan across a small surface depression on a Si_3N_4 bearing ball of diameter 16 mm .

DATA PROCESSING TECHNIQUES FROM NONLINEAR DYNAMICS FOR ACOUSTIC SIGNALS

WERNER LAUTERBORN
Institut für Angewandte Physik, Technische Hochschule Darmstadt
Schloßgartenstr. 7, D-6100 Darmstadt, Germany

ABSTRACT

The most important new methods of data processing from nonlinear dynamical systems theory are presented. These are state space analysis with subsequent fractal dimension estimation yielding insight into the number of significant degrees of freedom of a system with irregular output and Lyapunov analysis yielding insight into the dynamical properties of a system. The methods are applied to acoustic cavitation noise giving a low-dimensional noise attractor with one positive Lyapunov exponent.

1. Introduction

The physicist often is confronted with signals that appear irregular. Then noise is said to be present in the system, and it is tried to extract the "meaningful" signal from the noisy one.

The finding that deterministic systems may produce noise intrinsically by their own dynamics has led to new ways of looking at "noisy" systems. Noise from deterministic systems has been termed chaos to distinguish it from other forms of noise where the signal is corrupted from outer influences not coming from the system under investigation itself. Chaos physics thus develops a totally new view of the noise problem. It is a deterministic view. This raises the question of how it can be decided from the signal obtained whether it is of deterministic origin or must be attributed to an involved dynamics not easily describable in simple deterministic terms. Nonlinear dynamical systems theory has developed some new tools, in particular state space analysis with its embedding procedure of a signal (time series) into higher-dimensional spaces ("embedology"). The resulting higher-dimensional objects (point sets) then are characterized by their fractal dimension (dimension spectrum), Lyapunov exponents (Lyapunov spectrum) and other means.

The article is intended to give an insight into these new methods with the example of acoustic cavitation noise where some of the methods have been successfully employed.

2. Relation Between Theory And Experiment

The theory of dynamical systems starts with evolution equations that describe the dynamic development of the state of a system (a model). These may be continuous models

$$\dot{\mathbf{x}} = \mathbf{f}_\mu(\mathbf{x}), \qquad \mathbf{x} \in R^m, \qquad m \geq 1, \tag{1}$$

or discrete ones

$$\mathbf{x}_{n+1} = \mathbf{g}_\mu(\mathbf{x}_n), \quad \mathbf{x}_n \in R^m, \quad m \geq 1, \quad n = 0, 1, \ldots. \tag{2}$$

The state of the system is given by the m dependent variables $\mathbf{x}(t) = [x_1(t), x_2(t), \ldots, x_m(t)]$ or $\mathbf{x}_n = (x_1^{(n)}, x_2^{(n)}, \ldots, x_m^{(n)})$, respectively. The index μ indicates that the system depends on a parameter μ (often it will be several parameters). The dynamic laws (1) or (2) determine how a given state $\mathbf{x}(t)$ or $\mathbf{x}_n$ develops. This evolution can be viewed when the states of the system are displayed as points in a state space R^m. In the continuous case, the temporal (= dynamic) evolution then leads to a curve in this space called trajectory or also orbit (Fig. 1). In the discrete case, a sequence of points is obtained, usually called an orbit.

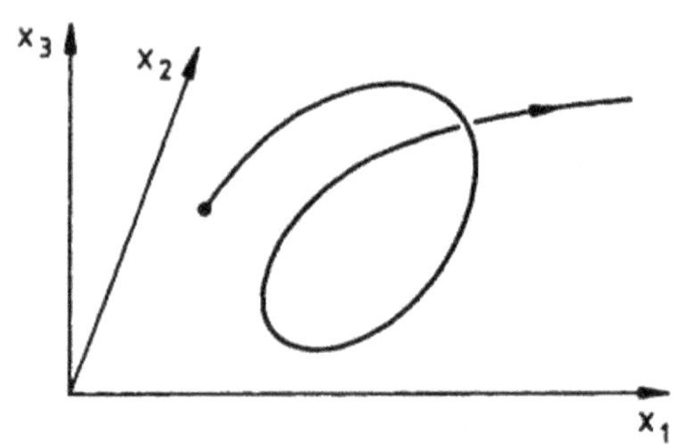

Fig. 1. A trajectory in state space.

An important question is how a set of initial conditions (a volume of the state space R^m) evolves as time proceeds. According to the theorem of Liouville, a volume stays constant in conservative systems whereas it shrinks in dissipative ones. Here, only dissipative systems will be treated. In this case the question almost poses itself, as to how the volume shrinks and how the limit set of points in state space looks, to which a given volume shrinks. This simple question cannot yet be answered in general as obviously an unknown number of different limit sets are possible. The limit sets have been given the name attractor as trajectories out of whole volumina of state space move towards these sets, i. e., seem attracted by them. The set of initial conditions (points in state space) moving upon evolution towards a given attractor is called its basin.

In experiments, the state space is usually not known beforehand and often only one variable of the system can be measured, e. g., a velocity component at a point in the liquid in a Rayleigh-Bénard experiment (a liquid layer heated from below). Thus only a projection of a trajectory of the system with a usually high dimensional state space onto a single coordinate axis is given. One may ask whether, under these circumstances, experiments make sense at all, when only such minor information about a system is obtainable. Fortunately, it can be shown that one variable already contains most of the information about the total system and not just a minor part. Even a simple general solution could be found of how to reconstruct an attractor of a dynamical system from a time series of one (measured) variable only.[1-3] The method has become known as the method of state space analysis (or also phase space analysis). The procedure is described in the next section.

3. State Space Analysis

Theory states: For almost every state variable $p(t)$ and for almost every time interval T, a trajectory in an n-dimensional state space can be constructed from the (measured) values $p(kt_s)$, $k = 1, 2, \ldots, N$ by grouping n values to n-dimensional vectors:

$$\mathbf{p}_k^{(n)} = [p(kt_s), p(kt_s + T), \ldots, p(kt_s + (n-1)T)], \tag{3}$$

where t_s is the sampling interval at which samples of the variable p are taken. It is advisable to choose the delay time T as a multiple of t_s to avoid interpolation. The above construction yields a point set $A^{(n)} = \{\mathbf{p}_k^{(n)}, k = 1, 2, \ldots, N-1\}$ in the embedding space R^n, which represents the attractor. The sampling time t_s and the delay time T must be chosen appropriately according to the problem under investigation. When t_s and also T are too small, then from one sample to the next there is little variation and the points $\mathbf{p}_k^{(n)}$ all lie on a diagonal in the embedding space. On the other hand, when t_s and T are too large, then the point set $A^{(n)}$ and the attractor obtained by connecting consecutive points $\mathbf{p}_k^{(n)}$ get a fuzzy appearance. In both cases, the reconstruction and visualization of the attractor are not very helpful. Figure 2 shows an example of the influence of the delay time T on the reconstruction of a calculated chaotic bubble attractor.[4] A spherical bubble has been set into oscillation by a sound field and the radius has been calculated as a function of time t. For the reconstruction, only the calculated radii of the bubble have been taken and embedded in a three-dimensional state space by using four different delay times T of 1/44, 1/4, 1/2, and 1 in units of the period T_0 of the driving sound field. The bubble had a radius at rest of 100 μm and was driven by a sinusoidal sound wave of amplitude 310 kPa and frequency 22.9 kHz. The reconstruction is best at $T = 1/4\ T_0$. This is the same value as for a simple

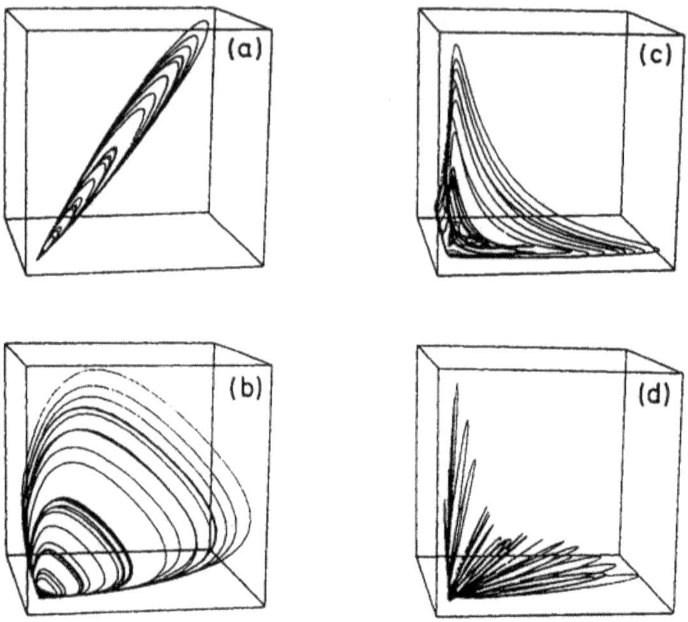

Fig. 2. Four reconstructions of a numerically obtained bubble attractor from radii data only with different delay times T of (a) $1/44\ T_0$, (b) $1/4\ T_0$, (c) $1/2\ T_0$, (d) $1\ T_0$, T_0 being the period of the driving sound wave of frequency 22.9 kHz and amplitude 310 kPa. The bubble has a radius at rest of $R_n = 100\ \mu m$. The sampling time t_s is $1\ \mu s$. For better visualization the transformed coordinates $R^* = \exp(2R/R_n)$ have been used.[4]

harmonic wave of period T_0 where the elongation x and the velocity $v = \dot{x}$ are 90^0 out of phase, i. e. by $1/4\ T_0$.

The chosen dimension n is called embedding dimension. Which dimension n should be taken is not known beforehand when the system is not known sufficiently (as usually in real experiments). Sometimes it happens that a three-dimensional state space is sufficient as embedding space. In the context of acoustics, a three-dimensional embedding space was found sufficient in representing an acoustic cavitation noise attractor.[5-7] In this experiment, a liquid is irradiated with sound of high intensity and the sound output from the liquid is measured. Figure 3 shows the experimental arrangement. The acoustic sound field to obtain cavitation (i. e. rupture of the liquid) is generated by a (hollow) cylindrical transducer of piezoceramic material (PZT4) of 76 mm length, 76 mm inner diameter, and 5 mm wall thickness. The cavitation noise is taken by a hydrophone, digitized and stored for data

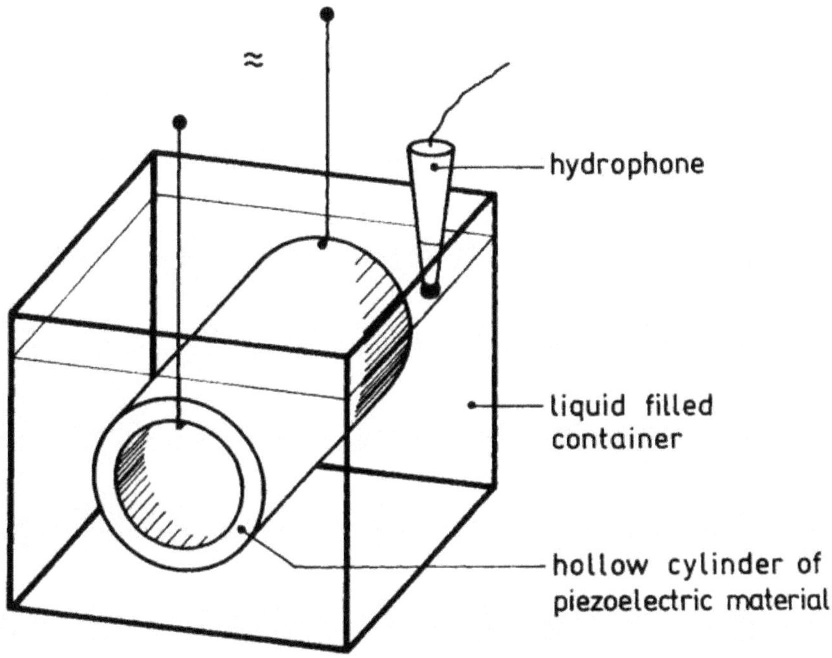

Fig. 3. The experimental arrangement for the generation of acoustic cavitation and the associated noise.

processing. The noise is generated by oscillating bubbles that form upon rupture of the liquid. A typical bubble field observed in the cylinder is given in Fig. 4.

The measurement yields pressure-time samples $p(kt_s)$ that may be used to construct an attractor. An example is given in Fig. 5. The reconstructed attractor in a three-dimensional embedding space is viewed from different directions to show its three-dimensional structure and almost flat overall appearance. The very pronounced structure suggests that the measured acoustic noise is of quite simple deterministic origin. A model with a three-dimensional state space should be sufficient. To deduct the structure of the equations from the attractor, however, is beyond the state of present knowledge. Once an attractor has been reconstructed, its properties can be investigated. Of special interest is its (fractal) dimension. The estimation of the dimension of the object obtained after embedding is discussed in the next section.

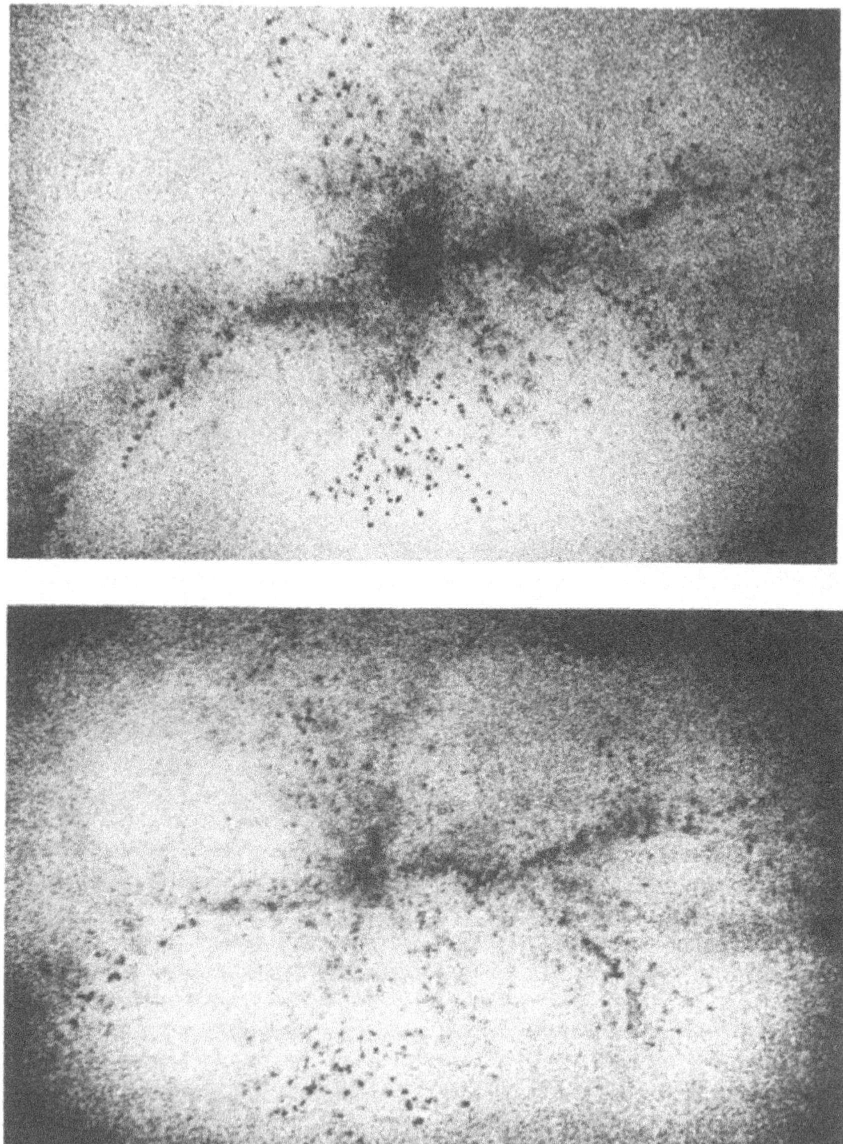

Fig. 4. An acoustically generated bubble field in water (two planes from a high speed hologram) emitting a noise-like signal.

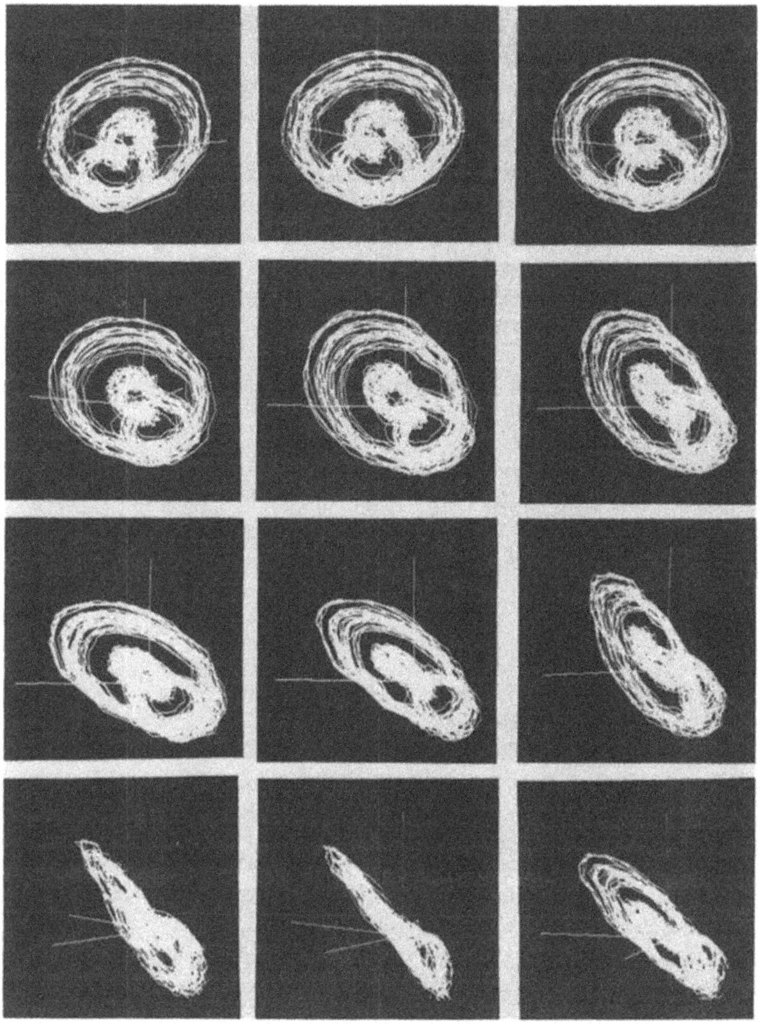

Fig. 5. Example of a reconstructed acoustic cavitation noise attractor from sampled pressure values in a three-dimensional embedding space. The attractor is shown from different directions for visualizing its spatial structure. Sampling time is $t_s = 1\mu s$, and delay time is $T = 5\mu s$.

4. Fractal Dimension Estimation

The notion of the dimension of an object (physical or mathematical) had long occupied physicists and mathematicians until a solution came into sight. Hausdorff[8] was the first to give a satisfactory definition. His definition of the dimension of a set of points naturally leads to fractal dimensions, i. e., dimensions that are not just natural numbers. For a long time, sets with these properties (fractals) were thought to be purely mathematical objects until their obviously abundant occurrence in nature was demonstrated by Mandelbrot.[9] Chaos physics has strongly expanded the importance of fractals and their dimensions. Chaotic attractors are known for their usually peculiar shapes, which point to fractal dimensions. This raises the question of how to determine fractal dimensions, in particular when they are encountered in experiments. Indeed, concepts have been developed whereby fractal dimensions can be de termined both from numerically calculated strange attractors and from measured data.[10-13]

An infinity of different definitions of a dimension has been introduced having their value in describing the inhomogeneity of the attractor. Only a simplified version of the Hausdorff dimension, called capacity d_0, and the most often used correlation dimension d_2 will be discussed in more detail here.

First, the definition of the dimension d_0 of a set of points $A \subset R^m$ is given:

$$d_0 = \lim_{r \to 0} \log \frac{M(r)}{\log(1/r)}, \tag{4}$$

where r is the edge length of an m-dimensional cube and $M(r)$ is the lowest number of m-dimensional cubes of edge length r to cover the given set A (a chaotic attractor, e. g.). When this definition is applied to a point, a line, an area, and a volume, the dimensions $d_0 = 0$, 1, 2, and 3 are obtained as they should be. But the definition is much more powerful. Also, Cantor sets now get a dimension, usually fractal as it turns out. Figure 6 gives the standard example of a Cantor set whereby, starting with the unit interval [0,1], the middle third without its end points is successively taken out of the remaining intervals. In the figure, to the left, the edge length r is given that is conveniently taken to cover the set, and to the right the number M of cubes (intervals) that is needed to cover the set. According to the definition of d_0, one then easily gets, by simply inserting the sequence as given in the figure into Eq. 4,

$$d_0 = \lim_{k \to \infty} \frac{\log 2^k}{\log 3^k} = \frac{\log 2}{\log 3} = 0.6309..., \tag{5}$$

a noninteger number. This is the fractal dimension of the Cantor set.

It turns out that for systems with a high-dimensional state space the direct use of the definition of d_0 (box counting) is not practical in numerical applications. From the next members of dimensions, the information dimension d_1, the correlation dimension d_2, and the higher-order dimensions $d_3, d_4, \ldots$, the correlation dimension

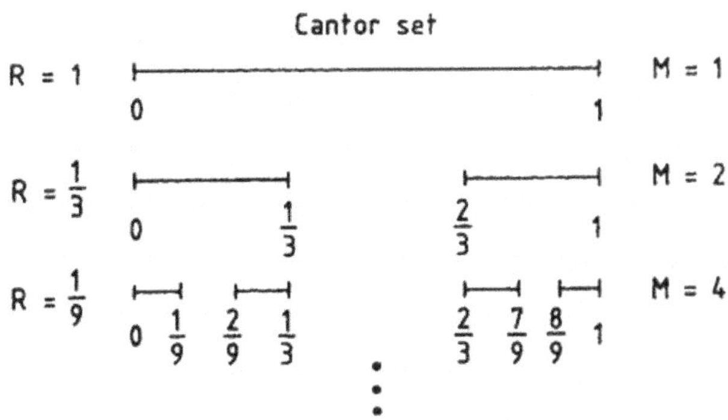

Fig. 6. Cantor set construction by successively taking out the middle third without its end points.

d_2 is the most attractive from the experimentalist's point of view.[10,11] For the sequence of dimensions, the relations $\ldots d_n \geq d_{n-1} \geq \ldots \geq d_2 \geq d_1 \geq d_0$ hold, and often the d_n are nearly all the same. Thus d_2, which is very convenient to determine numerically, is a good estimate of "the" fractal dimension of a strange attractor.

The correlation dimension is defined as

$$d_2 = \lim_{r \to 0} \log \frac{C_m(r)}{\log(r)}, \tag{6}$$

where r again is the edge length of an m-dimensional cube and $C_m(r)$ is the so-called correlation sum

$$C_m(r) = \lim_{N \to \infty} \frac{1}{N^2} \sum H(r - \|\mathbf{p}_k^{(m)} - \mathbf{p}_l^{(m)}\|). \tag{7}$$

In this expression, N is the number of points in R^m of the (strange) attractor available from some calculations or measurements, H is the Heaviside step function $[H(x) = 0$ for $x < 0$, $H(x) = 1$ for $x > 0$ and, used here, $H(0) = 0]$, $\mathbf{p}_k^{(m)}$ are the points of the attractor, and $\|\cdot\|$ is a suitable norm, e. g. the Euclidean norm. The dimension should not depend on the norm; therefore, any norm may be chosen that is most convenient for numerical computation. In our determination of the cavitation noise attractors, we used d_2 together with the maximum norm. Equation 6 immediately shows that d_2 can be determined from the slope of the curve obtained when $C_m(r)$ is plotted versus r, each on a logarithmic scale.

An example for the determination of d_2 is given in Fig. 7 for a numerically calculated strange bubble attractor.[4] In Fig. 7(a), the $\log C(r)$ vs $\log r$ curve is plotted with 100 000 points of the attractor in the Poincaré section plane. Figure 7(b) shows the local slope of the curve in Fig. 7(a) whereby local means a fit of

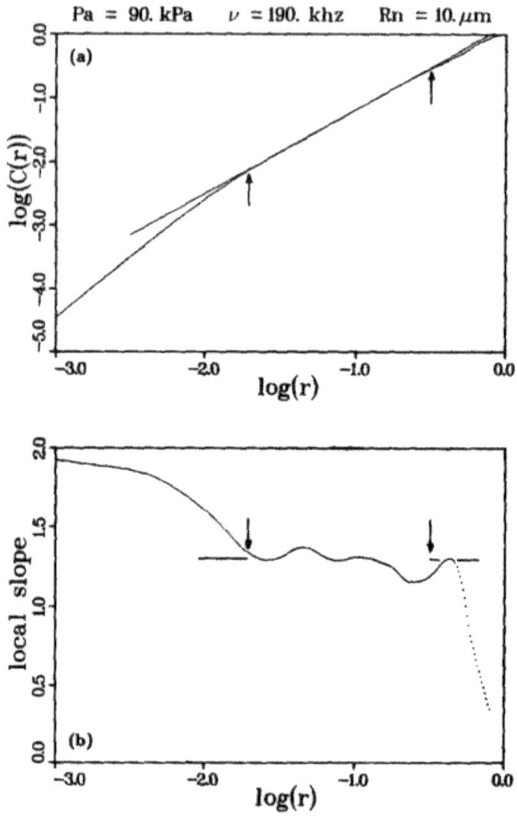

Fig. 7. Determination of the correlation dimension d_2 for a numerically calculated strange bubble attractor.[4] (a) Correlation sum $C(r)$ versus the cube edge length r on a doubly logarithmic scale, (b) local slope (fit to the curve in (a) over the range of a quarter of a decade) versus r. The fit region is from -1.7 to -0.5. In this region the local slope has a plateau giving $d_2 = 1.3 \pm 0.1$.

the slope over a region of a quarter of a decade. It is seen that a plateau in the local slope only occurs for values of $\log r$ between -1.5 and -0.5 giving a fractal dimension of $d_2 = 1.3 \pm 0.1$. For larger r's the gross structure of the attractor naturally leads to a decreasing local slope until at r's above the size of the attractor the slope is zero, because all points of the attractor fit into the cube with edge length r. At low r's the following happens. Because of the finite precision in the calculations, the exact location of the attractor points in the Poincaré section plane is only known to a limited number of digits. Therefore, the attractor looks more and more noisy on smaller and smaller scales. For these reasons the local slope for small r's tends to 2, which is the dimension of the Poincaré section plane. Thus only a region between small and large values of r is available for the determination of the fractal dimension. The fractal dimension of $d_2 = 1.3$ found in this case is valid for the attractor in the Poincaré section plane. To get the fractal dimension of the whole continuous attractor, one dimension has to be added giving $d_2 = 2.3$. Some remarks may be in order to warn the reader who wants to apply this formalism for fractal dimension determination. As can already be detected from Fig. 7(b), the useful region of an approximately constant slope may be small and even almost missing due to a wavy behavior of the slope. The method must therefore be used with great care, in particular with experimental data when used in conjunction with reconstructed attractors. The errors are usually very large.

Thus far we have discussed two types of dimensions, d_0 and d_2, out of the series $d_n, n = 0, 1, 2, \ldots$. The definition of dimension can even be extended to d_q, where q is any real number[14]. To an experimentalist, this generalization may seem rather sophisticated and beyond anything that can be measured. The opposite is true.[15-20] Only the full set $\{d_q, q \in R\}$, or equivalently the (smooth) scaling spectrum $f(\alpha)$ of (local) scaling indices α on the attractor[15] describes the global structure of the attractor satisfactorily and simultaneously in a measurable way. The scaling spectrum $f(\alpha)$ has a quite simple meaning. Take a point of a (strange) set (attractor) and a small sphere around it. Then the number of points of the set inside the sphere will scale with some exponent (index) α when the radius of the sphere goes to zero. Different points of the set may have different indices α. The spectrum $f(\alpha)$ characterizes the strength of a scaling index α or more precisely, $f(\alpha)$ is the global (Hausdorff) dimension of the subset of points of the set (attractor) with scaling index α. The relation $f(\alpha) \leq d_0$ holds, d_0 being the global dimension of the total set. Thus the maximum f_{max} of the spectrum yields the Hausdorff dimension. The spectrum of the Cantor set mentioned above is a line spectrum with just a single line at $\alpha = d_0 = \log 2/\log 3$. The dimension $f(\alpha)$ at $\alpha = d_0$ then must have the same value $f(\alpha = d_0) = d_0$, as every point of the Cantor set scales with the same α. Thus $f(\alpha) = d_0 \int \delta(\alpha - d_0) dx$, where δ is the Dirac δ function. Sets of this kind showing uniform scaling behavior are called self-similar (in a strict sense) and are simple fractal sets.

Strange attractors usually have a more complicated scaling behavior. However, according to the scaling spectrum $f(\alpha)$, they can be thought of as a "superposition" of simple fractal sets given by the subsets of uniform scaling behavior with index α. They are therefore called multifractal sets or, briefly, multifractals. For special cases, the scaling spectrum seems to be universal, i. e., the same for all systems, e. g., at the accumulation point of period doubling.[15,16] The experiments so far confirm the theory,[18–20] and it seems that the experimental determination of $f(\alpha)$[17–20] may be a better way to characterize a strange attractor than a single dimension as, for instance, d_2.

The fractal dimension estimation has been applied to the noise signal from acoustic cavitation noise.[5] As the embedding dimension n is not known beforehand for a real experimental system under investigation, n is successively increased, and the dimension of the point set $A^{(n)}$ is determined. When the system is of deterministic origin with a low-dimensional state space, then at some n the (fractal) dimension of $A^{(n)}$ will stabilize at some definite value smaller than n. The largest fractal dimension of $A^{(n)}$, $n = 1, 2, \ldots$, obtained in this way then determines the relevant number of (nonlinear) degrees of freedom (number of dependent variables) of the dynamic system investigated.

Figure 8 gives four curves of $\log C(r)$ versus $\log r$ for four different embedding

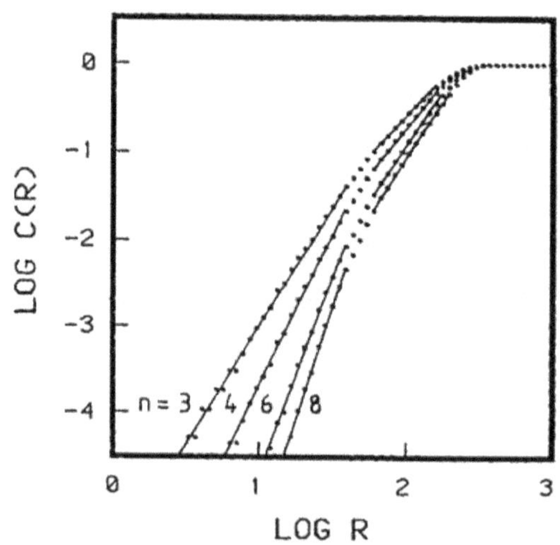

Fig. 8. Log $C(r)$ versus $\log r$ for different embedding dimensions for a cavitation noise attractor.

dimensions of 3, 4, 6, and 8 for measured acoustic cavitation noise. Different slopes are encountered for small and intermediate values of r. The slope at intermediate values of r determines the fractal dimension of the attractor whereas the slope at small r is due to external (true statistical) noise[11] also present in the experiments. These two slopes are plotted versus the embedding dimension in Fig. 9. The dimension of the attractor (dots) levels off at $d_2 \approx 2.5$ for high embedding dimensions whereas the external noise dimension (triangles) does not show convergence to a steady value. The crosses are from an experiment, where noise from a noise generator has been digitized and subject to the same analysis.

The dimension of the cavitation noise can be determined for different values of the sound intensity the liquid (water) is subject to. Then, in a specific case, a function as shown in Fig. 10 is obtained, where the voltage at the transducer is used as a measure of sound intensity. As long as there are periodic oscillations of the bubbles a dimension of one is obtained. At a distinct threshold the dimension of the noise rises to a value of about 2.5, a remarkably low value in view of the many bubbles present in the liquid. The noise has therefore been called acoustic chaos because it is obviously governed by a few variables only.

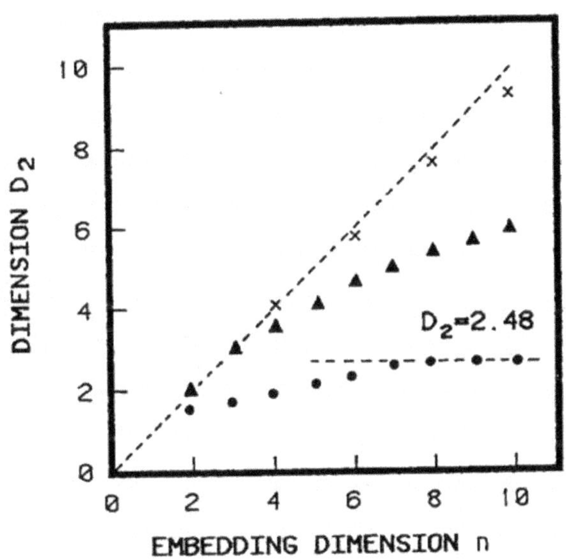

Fig. 9. Correlation dimension versus embedding dimension. Triangles and dots are from cavitation noise data, the crosses from a separate experiment with noise from a noise generator.

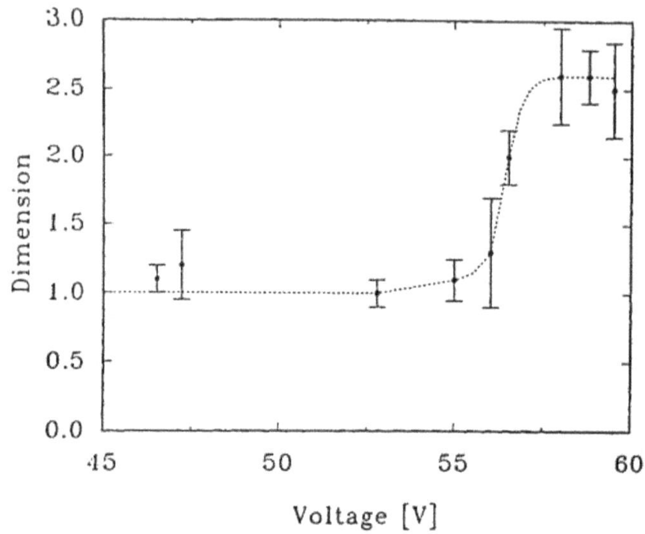

Fig. 10. Fractal dimension estimation versus voltage of the driving.

5. Lyapunov Analysis

Chaotic systems exhibit sensitive dependence on initial conditions. This expression has been introduced to denote the property of a chaotic system, that small differences in the initial conditions, however small, are persistently magnified because of the dynamics of the system, so that in a finite time the system attains totally different states. It is not difficult to envisage this property with systems that are not bounded, as, e. g., unstable linear systems. But physical systems are bounded, and it is not at all obvious how a persistent magnification of small differences is brought about. It seems that a sensitive dependence on initial conditions can only occur through a stretching and folding process of volumes of state space under the action of the dynamics. This process is depicted in Fig. 11. A persistent simple stretching would expand one direction more and more without bounds [Fig. 11(a)]. Neighboring points thus get more and more distant. This expansion in one direction can take place in a bounded volume of state space only when an additional folding process occurs [Fig. 11(b)]. The transformation of a volume $V(t)$ to $V(t')$, $t' > t$ (see Fig. 11), may be viewed as a map, a so-called "horseshoe map". Maps with similar properties are the baker's transformation and Arnold's cat map. When iterated they give the simplest examples of a dynamical system with sensitive dependence on initial conditions in two dimensions. This is the reason why they are intensely studied.

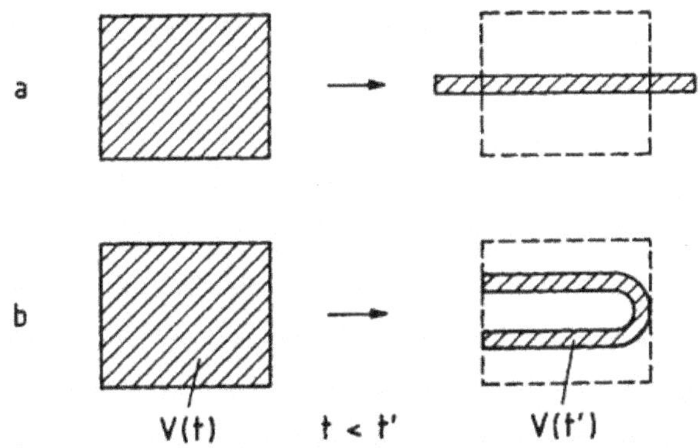

Fig. 11. Stretching and folding of a volume of state space. (a) Stretching only, (b) stretching and folding.

The notion of sensitive dependence on initial conditions is made more precise through the introduction of Lyapunov exponents and Lyapunov spectra. Their definition can nicely be illustrated.[21] Take a small sphere in state space encircling a point of a given trajectory (Fig. 12). The points of the sphere can be viewed as initial points of trajectories. This sphere is shifted in state space and deformed due to the dynamics so that at a later time a deformed sphere is present. To properly make use of this idea, mathematically an infinitesimal sphere is taken and its deformation into an ellipsoid with principal axes $r_i(t)$, $i = 1, 2, \ldots, m$ (m = dimension of the state space), is considered. The Lyapunov exponent λ_i is then defined by

$$\lambda_i = \lim_{t\to\infty} \lim_{r_i(0)\to 0} \frac{1}{t} \log \frac{r_i(t)}{r_i(0)}. \tag{8}$$

The set λ_i, i = 1, ... , m, whereby the λ_i usually are ordered $\lambda_1 \geq \lambda_2 \geq \ldots \geq \lambda_m$, is called the Lyapunov spectrum. It is to be recalled that the $r_i(t)$ should stay infinitesimally small. Then the linearized local dynamics applies that has to be taken along a nonlinear orbit. This is the meaning of $t \to \infty$. A strict mathematical definition resorts to linearized flow maps and may be found in Ref. 22. When $\lambda_1 > 0$, a time-dependent "direction" exists, in which the system expands. The system is then said to be chaotic (when, additionally, it is bounded). All along a trajectory, neighboring trajectories will retreat, on the average. A Lyapunov exponent is a number which, due to the limit $t \to \infty$, is a property of the whole trajectory. To get an idea of how uniformly neighboring trajectories recede from a given trajectory, Lyapunov exponents can be defined for pieces of trajectories to identify the most chaotic parts.

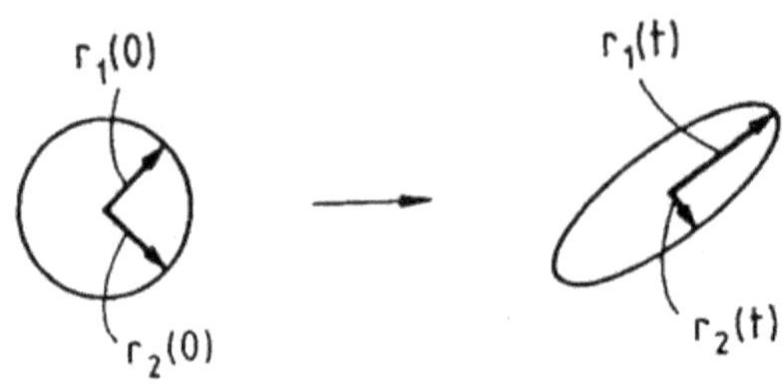

Fig. 12. Notions for the definition of Lyapunov exponents. A small sphere in state space is deformed to an ellipsoid indicating expansion or contraction of neighboring trajectories.

In dissipative systems, the final motion takes place on attractors. Besides the fractal dimensions (or the scaling spectrum), the Lyapunov spectrum may serve to characterize these attractors. With the help of the Lyapunov exponents, the notion of a chaotic attractor can be made more precise, as the following relations and definitions hold for continuous systems with $\lambda_1 \geq \lambda_2 \geq \ldots \geq \lambda_m$:

$\lambda_1 < 0$	fixed point;
$\lambda_1 = 0$	limit cycle ($\lambda_2 < 0$);
$\lambda_1 = \lambda_2 = 0$	quasiperiodic attractor on a twodimensional surface, torus ($\lambda_3 < 0$);
$\lambda_1 > 0$	chaotic attractor ($\lambda_2 = 0$);
$\lambda_1 \geq \lambda_2 > 0$	hyperchaotic attractor.

There exists a connection between the fractal dimension of a strange attractor and its Lyapunov spectrum. A Lyapunov dimension d_L has been defined[23] by

$$d_L = k + \frac{1}{|\lambda_{k+1}|} \sum_{i=1}^{k} \lambda_i \tag{9}$$

with $k \in N$ such that

$$\sum_{i=1}^{k} \lambda_i \geq 0 \quad \text{and} \quad \sum_{i=1}^{k+1} \lambda_i < 0.$$

The Lyapunov dimension d_L is expected to have a value close to the fractal dimension of the attractor. The reason is that a subvolume of the state space where the dynamics takes place expands when confined to k appropriately chosen directions but contracts with $k+1$ dimensions. Thus the fractal dimension of the object must lie between k and $k+1$ and should be approximated by the Lyapunov dimension d_L.

The calculation of the Lyapunov spectrum from a time series of experimental data at present is an involved task and requires great care. Several algorithms have been suggested (e. g. Ref. 21, 22, 24-27) and recently been compared.[28] The Lyapunov spectrum for the acoustic signal from cavitation has been determined via the approximation of the linearized flow in embedding space with an $n \times n$ matrix (n = matrix dimension = embedding dimension).[29] Figure 13 gives an example of a Lyapunov spectrum versus the matrix or embedding dimension obtained for a chaotic acoustic cavitation noise attractor. Just one Lyapunov exponent is positive, one zero, and one negative. The negative exponents are difficult to determine because the data set collapses in that direction and large errors are introduced that way. The one negative exponent has a small plateau for embedding dimensions n = 4 and 5. For higher embedding dimensions spurious exponents necessarily occur, as the matrix dimension follows the embedding dimension and this may become larger than the state space of the system under investigation. They can be singled out, e. g., by time inverting the flow data.[30] The Lyapunov dimension obtained from the values of the Lyapunov exponents yields $d_L \approx 2.6$, in good agreement with the fractal dimension $d_f \approx 2.5$ obtained from the dimension estimation procedure.

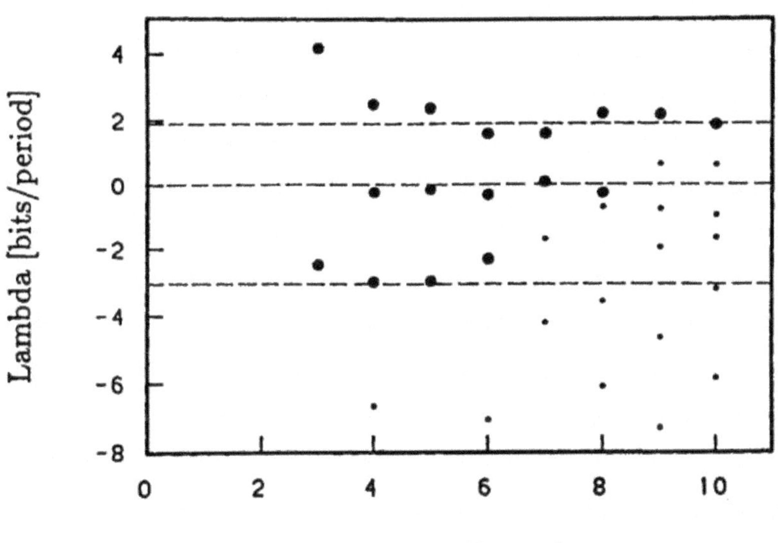

Fig. 13. Lyapunov spectrum versus matrix (=embedding) dimension in the case of a chaotic acoustic cavitation noise attractor.

6. Conclusion

New methods of data analysis from nonlinear dynamical systems theory have been presented in a tutorial manner. For experimental data (time series) the methods rely on the attractor reconstruction method by embedding the samples in sufficiently high-dimensional spaces.[1-3] The point sets obtained can then be analysed to yield their (fractal) dimension(s). The dynamics on the attractor can be characterized by the Lyapunov exponents describing the (averaged) expanding and contracting properties of the dynamics. These methods have been applied to the acoustic output from an acoustically generated cavitation bubble field with the astonishing but so far consistent result that a low-dimensional strange attractor with one positive Lyapunov exponent and a fractal dimension of about 2.5 prevails shortly after the onset of cavitation. Thus the idea that acoustic cavitation is some form of turbulence in a liquid[5,6] has to be abandoned despite its broad band noise emission spectrum. Acoustic cavitation must be a highly cooperative phenomenon governed by only three relevant (nonlinear) degrees of freedom but capable of producing irregular dynamics. It is therefore now called acoustic chaos. Thus the new data processing methods have given new insights in this case and may favorably be employed in other cases, too.

7. Acknowledgements

The work has been sponsered by the Fraunhofer-Gesellschaft, München, and later by the Sonderforschungsbereich 185 - Nonlinear Dynamics of the Deutsche Forschungsgemeinschaft (German Sciene Foundation). I thank the members of the Nonlinear Dynamics Group at the University of Darmstadt, in particular U. Parlitz, J. Holzfuss, T. Kurz, and K.-H. Geist for constant exchange of ideas.

8. References

1. F. Takens, in *Dynamical Systems and Turbulence*, ed. D. A. Rand and L.-S. Young (Springer, Berlin, 1981), pp. 366 - 381.
2. R. Mañé, in Ref. 1, pp. 230 - 242.
3. N. H. Packard, J. P. Crutchfield, J. D. Farmer, and R. S. Shaw, *Phys. Rev. Lett.* **45** (1980) 712.
4. W. Lauterborn and U. Parlitz, *J. Acoust. Soc. Am.* **84** (1988) 1975.
5. W. Lauterborn and J. Holzfuss, *Phys. Lett.* **A115** (1986) 369.
6. W. Lauterborn, in *Frontiers in Physical Acoustics*, ed. D. Sette (North-Holland, Amsterdam, 1986), pp. 123 - 144.
7. W. Lauterborn and J. Holzfuss, *Int. J. Bifurcation and Chaos* **1** (1991) 13.

8. F. Hausdorff, *Math Ann.* **79** (1919) 157.
9. B. B. Mandelbrot, *The Fractal Geometry of Nature* (Freeman, New York, 1983).
10. P. Grassberger and I. Procaccia, *Phys. Rev. Lett.* **50** (1983) 346.
11. A. Ben-Mizrachi, I. Procaccia, and P. Grassberger, *Phys. Rev.* **A29** (1984) 975.
12. J. D. Farmer, E. Ott, and J. A. Yorke, *Physica* **D7** (1983) 153.
13. G. Mayer-Kress (ed.), *Dimensions and Entropies in Chaotic Systems - Quantification of Complex Behavior* (Springer, Berlin, 1986).
14. H. G. E. Hentschel and I. Procaccia, *Physica* **D8** (1983) 435.
15. T. C. Halsey, M. H. Jensen, L. P. Kadanoff, I. Procaccia, and B. I. Shraiman, *Phys. Rev.* **A33** (1986) 1141.
16. D. Bensimon, M. H. Jensen, and L. P. Kadanoff, *Phys. Rev.* **A33** (1986) 3622.
17. M. H. Jensen, L. P. Kadanoff, A. Libchaber, I. Procaccia, and J. Stavans, *Phys. Rev. Lett.* **55** (1985) 2798.
18. J. A. Glazier, M. H. Jensen, A. Libchaber, and J. Stavans, *Phys. Rev.* **A34** (1986) 1621.
19. Z. Su, R. W. Rollins, and E. R. Hunt, *Phys. Rev.* **A36** (1987) 3515.
20. J. A. Glazier, G. Gunaratne, and A. Libchaber, *Phys. Rev.* **A37** (1988) 523.
21. A. Wolf, J. B. Swift, H. L. Swinney, and J. A. Vastano, *Physica* **D16** (1985) 285.
22. J. P. Eckmann and D. Ruelle, *Rev. Mod. Phys.* **57** (1985) 617.
23. J. L. Kaplan and J. A. Yorke, in *Functional Differential Equations and Approximations of Fixed Points*, ed. H.-O. Peitgen and H.-O. Walther (Springer Verlag, Berlin, 1979), pp. 204 - 227.
24. J. P. Eckmann, S. O. Kamphorst, D. Ruelle, and S. Ciliberto, *Phys. Rev.* **A34** (1986) 4971.
25. M. Sano and Y. Sawada, *Phys. Rev. Lett.* **55** (1985) 1082.
26. I. Shimada and T. Nagashima, *Progr. Theor. Phys.* **61** (1979) 1605.
27. G. Benettin, L. Galgani, A. Giorgilli, and J.-M. Strelcyn, *Meccanica* **15** (1980) 9.
28. K. Geist, U. Parlitz, and W. Lauterborn, *Prog. Theor. Phys.* **83** (1990) 875.
29. J. Holzfuss and W. Lauterborn, *Phys. Rev.* **A39** (1998) 2146.
30. U. Parlitz, *Int. J. Bifurcation and Chaos* **2** (1992) in the press.

BASIC PRINCIPLES OF ACOUSTO-OPTICS FOR ULTRASONIC FIELDS INVESTIGATIONS

WALTER G. MAYER
Physics Department, Georgetown University
Washington, DC 20057, U.S.A.

ABSTRACT

An introduction is given into the basic principles of acousto-optic interactions and their usefulness in the non-invasive investigation of ultrasonic fields. Light diffraction as well as Schlieren techniques are discussed and some indicative examples are mentioned.

1. Introduction

Ultrasonic fields can be mapped in a variety of ways, most commonly by the use of small hydrophones. A point-by-point intensity distribution in the field can be obtained; however, frequently one is limited in resolution by the size of the probe and, possibly, the determination of the relative ultrasonic amplitude distribution over the field may be influenced by the electronic arrangement where impedance matching, linearity of the equipment used and the overall transfer functions from mechanical energy of the ultrasonic wave to electrical energy in the receiving apparatus can introduce uncertainties about the real amplitude distribution in the sound field. Thus the use of hydrophones may, in fact, be an invasive method.

Acousto-optic methods, on the other hand, use a light beam which is diffracted by the ultrasonic waves. The diffracted light can then yield information about various characteristics and parameters of the ultrasonic field through which the light had been propagated. The sound field is not changed by the passage of the light; thus one has available a non-invasive method of sound field mapping.

The description of basic principles of acousto-optics to be presented here will serve as a basis for the better understanding of a number of contributions to be presented at this Conference. These contributions deal with very specific applications of the method, thus a general introduction into the principles of light-sound interaction may be helpful.

2. Light Diffraction by Ultrasound

The earliest suggestion that light diffraction could be produced by mechanical waves was offered by Brillouin[1] in 1922. He considered thermal waves or vibration (or high-frequency sound waves) in a transparent medium and a mono-frequency light beam traveling through the medium. The interaction expected by Brillouin was to take place at defined angles of light incidence with respect to the mechanical wave travel direction. He included a graph in his publication which can be redrawn in terms of a Bragg reflection diagram as shown in Fig. 1. Although Brillouin did not specifically mention this, he described the mechanism of photon-phonon interactions. Here k* is the momentum vector of the vibration in the medium, k and k' are the vectors associated with the incident and the reflected (or diffracted) light, respectively. If one assumes visible light and sound waves are to be used then the angle between k and k' will be very small unless k* is big enough to separate k and k' for observation of the incident and diffracted beams. A quick calculation will indicate that the separation (without any special optical arrangements) will be sufficient if visible light and ultrasound in the low Mhz range is used.

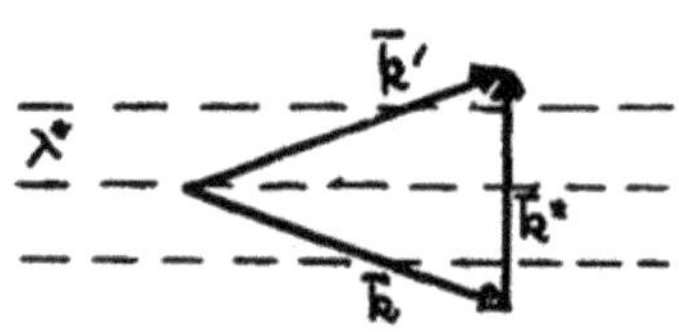

Fig. 1. Vector geometry for light diffraction

Radio-frequency generators to drive transducers in the MHz range were not very common in 1922 so it was not until the 1930s before the first experimental results were obtained by Debye and Sears[2] in Washington and Lucas and Biquard[3] in Paris, practically simultaneously. According to the suggestion of Brillouin, one would have to have an arrangement as shown in Fig. 2a.

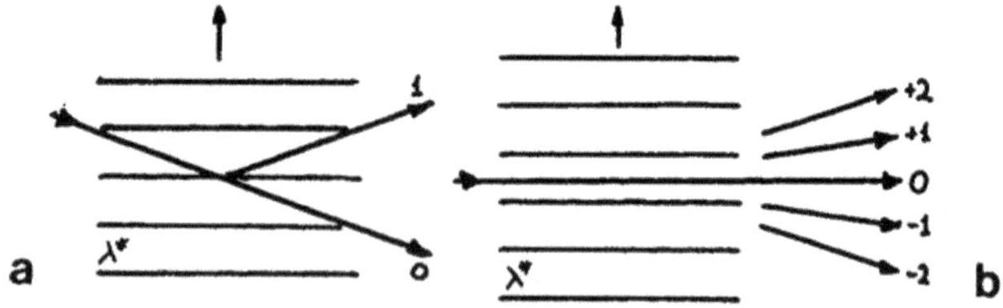

Fig. 2. Light diffraction arrangements. (a) Setup following Brillouin, (b) Setup successful for low MHz ultrasound.

However, neither Lucas and Biquard nor Debye and Sears were able to obtain light diffraction with this arrangement. Instead, when they used a setup as sketched in Fig. 2b, they did succeed in producing a diffraction pattern which showed not one "reflected" beam but evenly spaced orders of varying intensity. The orders were located on either side of the central order. Making use of an optical setup in which a lamp illuminated a slit and the light emerging from the slit being collimated, then allowed to travel through the ultrasound field, and being focused again into a slit image produced the diffraction pattern shown in Fig. 3. The two patterns were obtained by Lucas and Biquard with a frequency of 11.5 MHz. The top portion of the photograph shown was obtained using monochromatic light, the lower portion with light from a mercury arc lamp. Although no general theory existed at that time, three things were noted, i.e., (1) the spacing of the orders increases with increasing frequency, (2) the spacing depends on the light wavelength, and (3) the number of orders becomes bigger and the light distribution within the diffraction pattern changes as the ultrasonic amplitude is changed. However, no functional relation was apparent at that time which would relate sound amplitude to light intensity in the orders.

Fig. 3. Diffraction orders (after Lucas and Biquard[3])

The first theoretical explanation for the interaction of a progressive ultrasonic wave at low frequency (about 1 MHz) with light incident normal to the sound propagation direction, and for low sonic amplitudes, was given by Raman and Nath[4]. This basic theory shows how sound frequency, light wavelength, and sound amplitude as well as direction propagation are related. So this theory was used to interpret light diffraction in order to obtain, noninvasively, information about the character of ultrasonic fields.

However, the original Raman and Nath theory does not hold when the sound frequency becomes too high, the amplitude increases so nonlinear effects become apparent, when the wave is no longer a pure progressive wave or when the light sound propagation directions are not normal to each other.

To remedy some of the limitations an extended theory was developed by Klein, Cook, and Mayer[5] to describe the general aspects of the light-sound interaction process. This theory also includes discussions of Bragg diffraction and light diffraction by nonsinusoidal (finite amplitude) sound beams. These aspects will not be discussed here; instead, only a short derivation and the important results for low-MHz sinusoidal waves will now be considered.

3. Theory of light-sound interaction

Considering the general case of a monofrequency light beam traveling at an arbitrary angle through a transparent medium (liquid or solid) and observing that under certain circumstances there will be a diffraction of the light beam into evenly spaced orders, one can define the coordinates of the problem as shown in Fig. 4, and one can then solve the problem of finding the magnitude and direction of the electric intensity of the incident and the diffracted light. Knowledge of this then enables one to make statements about the relative integrated sound pressure amplitude in the location where the sound field and the light beam interact, as well as about other important experimental parameters that influence the light intensity distribution in the patterns produced.

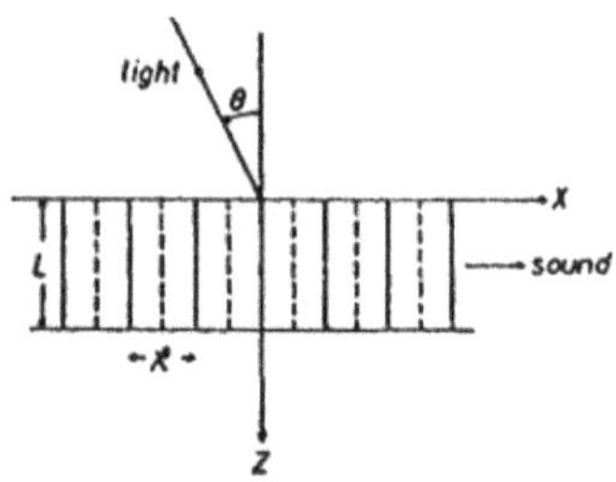

Fig. 4. Interaction diagram.

The initial electric intensity of the incident light beam can be represented by

$$E = E_o \exp[i(\omega t - \mu_o k \cdot r)], \tag{1}$$

where ω and k are the frequency and wave vector, respectively of the light before it enters the medium whose index of refraction is denoted by μ_o. However, in the presence of a sound wave, the index of refraction of the medium changes with the change being periodic with the pressure changes produced by the sound wave.

If the ultrasonic wave travels in the x-direction only, then the local index of refraction will be periodic in k^*x, where k^* is the wave vector of the ultrasonic wave. This implies that the constant E_o in Eq. (1) will have to be replaced by a quantity which is periodic in k^*x. This new quantity, $A(x,z,t)$ can be expanded in a Fourier series

$$A(x,z,t) = \sum \Phi_n(z,t) \exp[-ink^* x] \tag{2}$$

in which the summation includes all n from $-\infty$ to $+\infty$. Substituting Eq. (2) into Eq. (1) yields

$$E = \sum \Phi_n(z,t) \exp[-ink^* x + i(\omega t \mu_o k \cdot r)], \tag{3}$$

where Φ_n denote the amplitudes of the partial light waves created inside the sound beam. Equation (3) shows that every partial wave labeled n has a distinct wave vector whose direction differs from the undiffracted light beam, whose wave vector is $\mu_o k$, by nk^*. This is in keeping with the suggestion by Brillouin as sketched in Fig. 1, except one now has more than a single diffraction order, i.e., there are n positive and n negative orders, all defined by

$$k_n = \mu_o k + nk^* \tag{4}$$

Since the angular separation of the partial waves emerging from the sound field are very small one can express this separation in terms of the wavelength of the two waves as

$$\sin\beta = n\lambda/\lambda^*, \tag{5}$$

where the starred quantities, as before, refer to the sound wave.

Equation (5) shows that the angular spacing of the diffraction orders yield a measured value of the sound wave velocity in the medium if the wavelength of the light and the sonic frequency are known.

However the light intensity in the diffraction orders is not known unless one can solve the wave equation for E as defined previously. This equation can be written, in general terms, as

$$\nabla^2 E = \left(\frac{\mu_o}{c}\right)^2 \frac{\partial^2 E}{\partial t^2} \tag{6}$$

with E defined in Eq. (3). Assuming that the light incidence is normal to the sound wave propagation direction, then the light amplitude is only a function of z, and with a variable change [see Eq. (3)] from E to Φ_n, one has

$$\nabla^2 E \rightarrow \frac{\partial^2 \Phi_n}{\partial z^2} \tag{7}$$

This results in a difference-differential equation which can be written in the form

$$\frac{d\Phi_n}{dz} + \frac{v}{2L}[\Phi_{n-1} - \Phi_{n+1}] = i\frac{nQ}{2L}(n - 2\alpha)\Phi_n \tag{8}$$

Here L denotes the width of the sound field, starting at z = 0 and extending to z = L. The other parameters are defined as

$$v = k\mu L$$

$$Q = (k^*)^2 L/k\mu_o$$

$$\alpha = -(k\mu_o/k^*)\sin\theta$$

where the first parameter, v, is proportional to the sound pressure, the second, Q, proportional to frequency squared, and the third, α, defines the light incidence angle.

One can now calculate the value of the light amplitude in any order labeled n by starting with the initial values for the Φ at z = 0, namely:

$$\Phi_n = 0 \quad \text{for } n \neq 0$$
$$\Phi_n = 1 \quad \text{for } n = 0$$

The results one obtains for the various Φ_n at z = L depends to a large extent on the value of v (the sound pressure, or the Raman-Nath parameter), the value of Q, and on α.

In general, there are two distinct regions: when $Q \ll 1$ and $\Theta = 0$, one finds a symmetric light diffraction pattern, and considering that the square of the absolute value of the light amplitude Φ_n is the measurable light intensity in the nth diffraction order, the result is that by changing the value of the parameter v one changes the light intensity in the nth order, $I_{\pm n}$, according to the square of the nth order Bessel function of order n and argument v. This result,

$$I_{\pm n} = J_n^2(v)$$

is identical to the one obtained by Raman and Nath.

Another interesting case can be considered via Eq. (8). This would be for the case where the frequency is in the mid- to high MHz region, in which case $Q \gg 1$ and $\alpha = n/2$. This now corresponds to the situation suggested by Brillouin [Fig. (1)]. If the sound amplitude is sufficiently high, incremental solutions to z = L in Eq. (8) shows that $I_n \to 1$ while the light intensity in the zeroth order tends to zero. This constitutes Bragg diffraction.

We will not consider Bragg diffraction here. We will limit the discussion to the application in the "Raman-Nath regime" where the frequency of the ultrasound is in the low-MHz region and light incidence is normal to the sound propagation direction. In addition we will assume the value of v to be low so that nonlinearities of the medium can be ignored.

4. Field characterizations with light diffraction patterns

The most commonly used experimental setup for producing light diffraction patterns is shown in Fig. 5. Light from a small laser is focused by a short focal length lens and is then collimated by another lens, L.

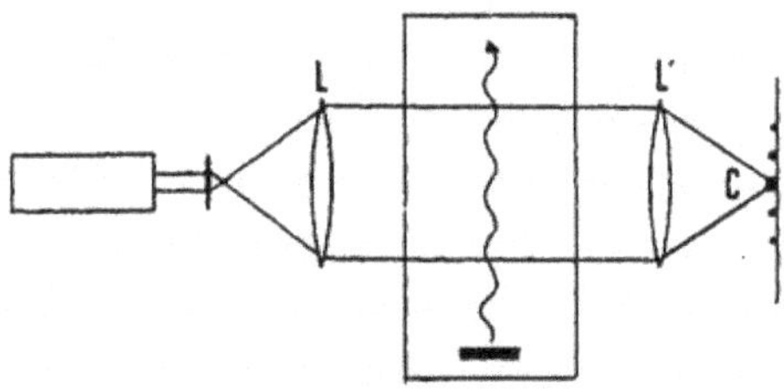

Fig. 5. Typical experimental setup for light diffraction.

The collimated light interacts with the sound field (liquid in a glass tank, e.g.) and lens L' focuses the light in the plane C. If the ultrasonic wave is a progressive wave and the Raman-Nath conditions (see above) are met, the light intensity in the diffraction orders will follow the square of the Bessel function relation given above. The distribution as a function of the parameter v (proportional to sound pressure) is shown in Fig 6. It is evident that the central order will be extinguished when v reaches the value 2.41, the first orders will disappear when v = 3.83, etc. Therefore,

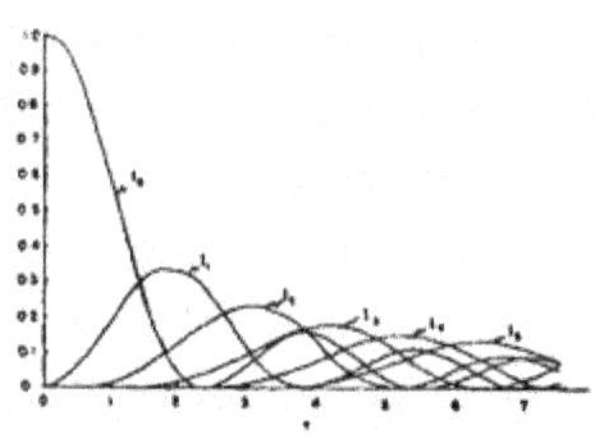

Fig. 6. Light intensity in the orders as a function of v.

the first piece of information which can be obtained from observing the diffraction pattern is the spacing of the orders. According to Eq. (5), the wavelength of the sound can be found, and from a knowledge of the sound frequency, one immediately has a measure of the sound velocity in the medium.

The second piece of information one can find is the absorption in the medium. This can be done by producing a light diffraction pattern by taking a diffraction pattern close to the face of the transducer and recording the light intensity distribution. Moving the transducer face a known distance away from the light beam causes the light to transverse a region where the pressure is lower than it was near the transducer face. The second diffraction pattern will be "weaker", and from the light distribution one can determine the new v-value. Thus one has a direct measure of the pressure decrease per unit length of medium. A variation of this method was first mentioned by Biquard[6].

Light diffraction patterns produced in transparent single crystals can also be used to determine the values of elastic constants. It is evident that the diffraction orders (points) on the screen lie on a straight line which is parallel to the propagation direction of the ultrasonic wave.

This fact was used as early as 1934 by Schaefer and Bergmann[7] who propagated ultrasonic waves in transparent solids. Multiply reflected waves and waves converted to shear waves upon internal reflections in the solid sample created diffraction pattern in a multiplicity of directions. From the spacing of the orders, the velocities of the waves as a function of direction could be found. If the solid is isotropic, the diffraction pattern consists of circles with two definite radii: one for the longitudinal wave and one produced by the shear wave. The radii correspond to the angular spacings given by Eq. (5), one has a convenient method of determining the elastic moduli of the sample.

If, however, the sample is anisotropic, the sound velocity depends on the crystallographic direction. Acousto-optic interaction in a transparent crystal will therefore not produce simple circular patterns, but a much more complicated pattern results, as is shown in Fig. 7. Here again, one determines the sound velocity as a function of crystallographic orientation from the distance between the central order and any point on the

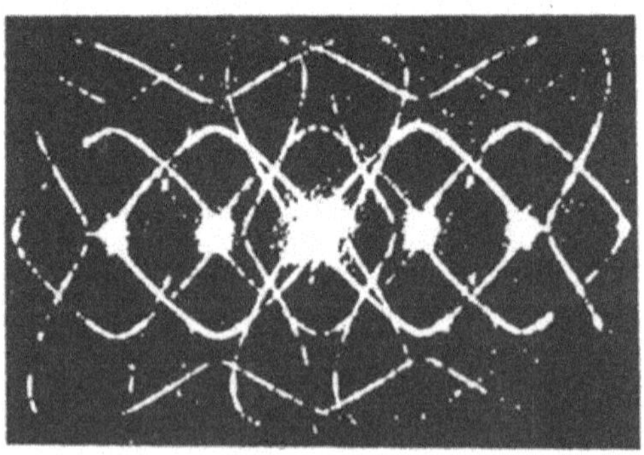

Fig. 7. Diffraction pattern of ultrasonic waves in a RbFl single crystal. (Photograph private communication and gift by Professor Haussühl)

visible pattern. The relation between elastic constants and sound velocities are give by the Christoffel equations, which for a crystal with cubic symmetry are relatively simple. For the [100] direction this equation has the simple form

$$(c_{11} - \rho V^2)\,(c_{44} - \rho V^2)\,(c_{44} - \rho V^2) = 0$$

This indicates that there is one longitudinal and two identical shear wave velocities, V, which are the three roots of this equation. Clearly, the evaluation via acousto-optic methods of complicated symmetries becomes rather involved since the number of dependent elastic constants to be determined increases as the crystal symmetry becomes more complex.

Nevertheless, use of this acousto-optic method frequently makes it possible to obtain all the elastic constants from single photographs of the patterns produced in two samples of different orientations.

There are, of course, many more applications of making field evaluations by means of diffraction pattern measurements, for instance, the determination of the nonlinearity parameters in transparent liquids and solids, the evaluation of ultrasonic pulsed signals (above discussion is concerned only with continuous waves), diffraction by non-bulk waves, like surface waves, etc. Rather than describing these, we will now mention the techniques used to image a portion of an ultrasonic field.

5. Schlieren imaging technique

In the previous section we mentioned that the diffraction pattern evaluation techniques can yield, without interfering with the sound field, information about sound velocity, sound absorption, elastic constants, and we mentioned the possibilities to investigate nonlinear propagation and pulsed propagation.

Actual field mapping was not one of the possibilities directly available through this method. However, a slight change in the optical setup shown in Fig. 5 makes it possible to create an optical image of the sonic field which is transversed by the light beam.

From the discussion in sections 3 and 4 it is evident that the ultrasonic field will affect only those parts of the light beam which travel through the sound field. This means that any light which is diffracted out of the zero order has passed through a portion of the medium which has been insonified. This is indicated schematically in Fig. 8a which shows a cross section of a light beam illuminating an incident, reflected, and trans-

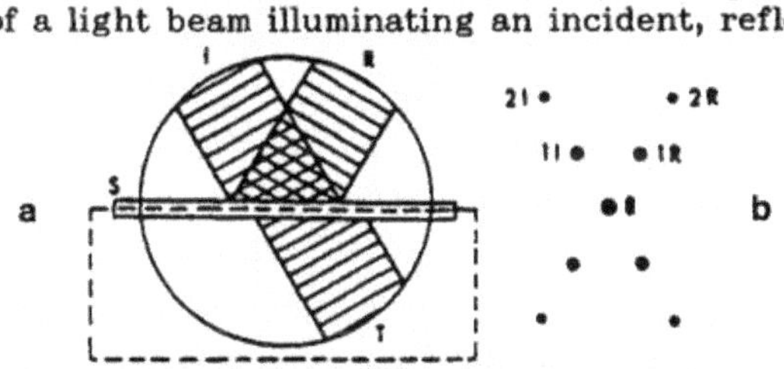

Fig. 8. Schematic diagrams of (a) circular light beam crossing sound beam reflected from a plate S and (b) diffraction pattern orders from incident and transmitted (I, T) and reflected (R) beams.

mitted ultrasonic beam and a solid plate in the center of the sketch. The incident and transmitted beams have the same k-vector direction which is different from the direction of the reflected beam k-vector. Therefore two diffraction patterns are produced; the alignment of the patterns is determined by the sound travel direction as indicated in Fig. 8b.

In order to produce an optical image of these beams one blocks the light in the central order but allows the light in the other orders to continue. As shown in Fig. 9, a lens, L", behind the diffraction plane creates

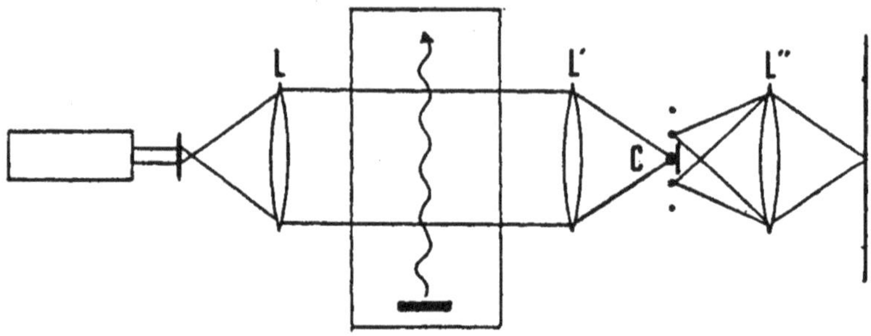

Fig. 9. Diagram of the Schlieren imaging setup

an image of the entire sound field which was illuminated by the expanded light beam. The light intensity at any point in such a Schlieren image is related to the sound pressure at that point via the local value of the parameter v.

The light intensity distribution in such an image is very much influenced by the choice one makes to use one or more non-zero orders to create the image. One does not easily obtain an accurate reading of the exact pressure levels which exist in the field. Generally, however, the brightest portions of the image correspond to the highest intensities in the field and the dark portions clearly are the locations where no sound is present. This system produces an integrated quantitative intensity map of the ultrasonic field in the planes perpendicular to the light propagation direction.

A number of representative images are shown in Fig. 10. The advantage of the Schlieren imaging technique is that the structure of the ultrasonic field can be seen in its entirety.

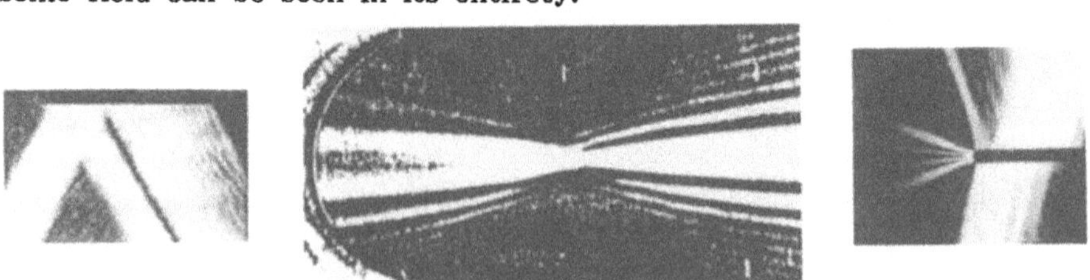

Fig. 10. Examples of Schlieren images of various ultrasonic fields.

6. Conclusion

The two acousto-optic interaction methods, diffraction pattern evaluation and Schlieren imaging, are non-invasive techniques because either mechanical nor electronic devices are used which could interfere with the structure of the sonic field to be evaluated.

The diffraction pattern technique can be used to determine a number of parameters characterizing the sound field, among them being the sound velocity, attenuation coefficient, elastic constants, and others where the spacing of the diffraction orders and the light intensity distribution in the orders of the pattern are the indicators that are evaluated.

The Schlieren imaging technique ordinarily does not yield quantitative results about the important parameters associated with a given sound field but does yield a complete picture of the shape of the field and its geometrical structure.

Thus both methods have the advantage that the sound field is not interfered with while an observation is being made. Particularly, the Schlieren imaging method produces a real-time picture of an entire sound field configuration which would take extraordinary efforts to create if a point-by-point evaluation were to be done with hydrophone probes.

7. References

1. L. Brillouin, Ann. Physique 17 (1922) 88.
2. P. Debye and F. W. Sears, Proc. Nat. Acad. Sci. (Washington) 18 (1932) 409.
3. R. Lucas and P. Biquard, J. Physique Radium 3 (1932) 464.
4. C. V. Raman and N. S. N. Nath, Proc. Indian Acad. Sci. (A) 2 (1935) 406.
5. W. R. Klein, B. D. Cook, W. G. Mayer, ACUSTICA 15 (1965) 67.
6. P. Biquard, C. R. Acad. Sci Paris 197 (1933) 309.
7. C. Schaefer and L. Bergmann, Naturwiss. 22 (1934) 685.

LASER PROBING OF ACOUSTIC FIELDS

GERARD J. QUENTIN
Groupe de Physique des Solides de l'Université Paris 7, Tour 23, 2 place Jussieu, 75251 Paris Cedex 05, France

ABSTRACT

After presenting a quick overview of methods of quantitative measurement of local ultrasonic fields and discussing their advantages and disadvantages, we focus this paper on three topics already studied in our laboratory. They concern the study of "edge waves" in piezoelectric transducers, the quantitative measurement of ultrasonic pressure inside fluids and, finally, Lamb waves propagation in plates.

1. Introduction

Most ultrasonic measurements done in Ultrasonics are relative ones, in the sense that the absolute value of the acoustic pressure (or displacement, or particles velocity) is very seldom measured. This paper is mainly devoted to methods of absolute measurements of the Acoustic Field. It emphasize the possibilities offered by laser probes to reach this goal. In addition we focus our attention upon local and wide band techniques. Such measurements are needed, for example, in order to determine nonlinearity thresholds, mode conversion factors, transduction efficiency.

2. Some methods used for an absolute probing of Acoustic Fields

2.1 Scattering from a small sphere (Papadakis [1], ...)

This method is one of the oldest because of its facility of use. A small spherical target is placed in the acoustic field at the point where the pressure measurement is to be made. The transducer is used in the transmitter-receiver mode. The principle of the method supposes that the small sphere can be considered as a point reflector.

The main advantages of this technique are the following :

- it is very easy to implement,
- no calibration is needed if the sphere is acting really as a point reflector.

Its main disadvantages are :

- the sphere is, in fact, an elastic scatterer with a resonating behavior,
- the received signal is very weak,
- frequency limitations are important (the maximum frequency is limited to 10 MHz)
- the size of the sphere is not negligible (≥ 0.5 mm) and, consequently, the acoustic field is modified.

1.2 Use of a calibrated hydrophone (Harris [2], Bacon [3], ...)

The small sphere may be replace by a tiny little hydrophone. In this case the hydrophone acts as a receiver. Such hydrophones are commercially available with size as small as 0.5 mm, but it is possible to manufacture, in the laboratory, smaller ones with diameters below 0.2 mm. Such transducers made of piezoelectric ceramics or of PVDF can be calibrated, for example, by using the reciprocity principle and a very thick transmitting transducer.

The main advantages of this technique are :

- it is very wide-band ($f \leq 30$ MHz),
- it has a very good omnidirectionality (quasi-uniform response over 2π steradian).

Its limitations are :

- such hydrophones are aging and a recalibration procedure is frequently needed,
- the size of the active surface cannot be considered as negligible and some perturbation of the pressure field follows.

1.3 Bragg diffraction of a laser beam (Wolf and Mayer [4])

This very powerful technique does not imply a perturbation of the acoustic field because the sensor is a laser beam. It is detailed in another chapter of this book.

1.4 Laser heterodyne interferometer measurements of liquid surface displacements (Baboux [5])

In this method the ultrasonic waves are propagating towards the surface of a water tank and the displacements of this surface are measured using a stabilized laser interferometer.

The main advantage of this technique is the fact that strong focusing of the laser beam can be carried out. The consequent spatial resolution is actually one of the best ($\cong 10$ μm). Its limitation is inherent to the probing of the liquid surface. Under the action of the sound beam and, eventually, for other reasons, many wavelets are formed on the surface and the corresponding displacements are also measured giving rise to large spurious low-frequency components inside the received signal.

1.5 Laser measurements of pressure inside fluids through membrane displacements (Reibold [6], Constantopoulos and Royer [7])

Laser interferometric probing is a very good method for measuring surface displacements. In order to get rid of the limitation of the preceding technique, it is possible to insert a very thin membrane. In the case of the experiments performed by Reibold [6] the membrane is part of the water tank and an homodyne interferometric probe is used. In the experiments recently made by Constantopoulos and Royer [7], the membrane is inserted inside the water tank perpendicularly to the ultrasonic beam axis. Various materials and membranes thicknesses have been used and their frequency response tested both in amplitude and phase. For a stainless steel membrane of 20 μm the bandwidth is limited to 3 MHz but for a nylon membrane of 12 μm it can attain 50 MHz; the case of a silicium membrane of the order of 10 μm lies in-between.

Such technique has been applied to beam shaping of ultrasonic fields of normal or focused transducers. It shows in addition the derivative effect upon the field at a focus [7].

2. Vibration modes in piezoelectric transducers

It is well known that the piston-like approximation of plane piezoelectric transducers which is quite good for the far field is rather invalid in the near field because of the radiation of edge waves from the circumference of the transducer [8].

We have tried to analyze such edge waves using an heterodyne interferometric laser probe. The use of such a probe when the laser beam is crossing water is the source of problems that we address below. Consequently, we have used measurements performed with free vibrations of a piezoelectric disc (undamped and in air). For thick discs we have observed that the wave coming from the periphery of the circular disk exhibits no frequency dispersion and that its velocity is, within experimental errors, equal to the Rayleigh velocity at the surface of the piezoelectric material [9]. If, now, we use a thin piezoelectric transducer, the "edge wave" travels at the velocity of the S_0 Lamb wave and exhibits a certain amount of frequency dispersion [10]. Its velocity is a decreasing function of frequency. The wave converges from the circumference of the plate towards its center, then diverges from it, reflects on the circumference, converges again, etc. We have already verified the value of the phase shift of the wave when crossing the center and it agrees with theoretical predictions.

The experiments performed with thick and thin transducers have been complemented with experiments performed in water using a very small hydrophone and the results agree quite well.

3. Direct quantitative measurements of pressure inside fluids [11]

We have recently applied our laser heterodyne interferometric probe to measurements of pressure in fluids (water and air), in order mainly to test the sensitivity of such probe. The set-up is similar to the one used in Bragg diffraction experiments except that only the "signal beam" of the laser probe crosses the sound beam, is reflected from a miror, crosses again the sound beam and enters the interferometer. The phase of the laser beam is shifted when crossing the sound beam by virtue of the piezo-optic effect. In the experiments performed both in water and in air we used wide band immersion transducers in the MHz range. The laser beam width was 80 μm. We assumed, as a first step, that the beam emitted is a uniform parallel beam. Measurements carried out in water show that a quantitative assessment of the ultrasonic pressure is possible for frequencies smaller than 2 MHz. In the case of air, the pressure emitted by such an immersion transducer is very low but is nevertheless measurable (some 10 Pa). We observe in such measurements, at frequencies larger than 1 MHz, a shift in the measured center frequency of the spectrum. This center is displaced towards low frequencies with respect to the actual center acoustic frequency. The reason of this shift is straightforward. At such frequencies, in air, the ultrasonic wavelength becomes comparable to the laser beam width. Nevertheless, knowing the laser beam shape, which is usually a gaussian one, it is possible to deconvolute the measured signal by this light space/time repartition and to obtain a quantitative measurement of ultrasonic pressures in air at frequencies so high as 5 MHz. The experimental sensitivity of our measurements was of the order of 10 Pa in water and 1 Pa in air.

4. Measurement of Lamb wave displacements in solid elastic plates

Both Rayleigh and Lamb waves exhibit an ellipsoidal displacement of material particles. In order to characterize the two components of this displacement by laser probing, it is necessary to be able to measure in-plane and out of plane components. The interferometer that we used in our preceding studies is the BMI one, invented by Royer and Dieulesaint and is very suitable for out of plane measurements of displacements. In order to measure in-plane components we used the interferometer invented by J.-P. Monchalin and commercialized by Ultraoptic. The experiments were carried out at the Institute of Industrial Materials of the CNRC in Boucheville (Quebec, Canada) and the samples where Aluminum plates with a thickness of 0.6 mm. Monchalin and his team had already performed previously similar experiments on Rayleigh waves propagating on a plane metallic interface and verified the theoretical predictions on such waves [12,13]. The case of Lamb waves is more complicated and interesting because these waves are dispersive ones. The excitation of Lamb waves has been made using a YAG laser in the ablation regime. Once recording, at various distances from the point of generation, both the in-plane and out-of-plane components of the displacement, a pseudo-Wigner-Ville transform has been applied to these time-signals. Such a time-frequency transform is very powerful especially to separate various dispersive waves arriving at the same time at the same point. The most important part of our signal corresponded to the A_0 Lamb wave. The experimental dispersion curve is in very close agreement with the theoretical one. We have then measured experimentally the phase shift between the two components which is found to be equal to $\pi/2$ within experimental errors. The ratio of the two components of the displacement agrees also very well with theoretical expectations.

5. Conclusion

Absolute measurements of the components of local ultrasonic displacements are now feasible at fluid/vacuum (or air) and solid/vacuum (or air) interfaces with interferometric laser probes. Ultrasonic pressure measurements are also possible inside fluids. For carrying out displacement measurements at solid/water interfaces, some difficulties occur when interface waves are radiated from the solid into water because they are responsible of a phase-shift of the laser beam difficult to separate from the one due to surface displacements.

Laser absolute local measurements of the ultrasonic field contribute to open the experimental field of quantitative ultrasonics.

REFERENCES

[1] E.P. Papadakis, "Ultrasonic transducer evaluation in five domains: time, space, frequency, surface motion and theory", 1977 Ultrasonics Symposium Proceedings, IEEE Cat. **77** CH 1264-ISU, 104-111.

[2] G.R. Harris *et al.*, "Primary calibration of ultrasonic hydrophones using optical interferometry", IEEE Transactions on Sonics and Ultrasonics, **SU30** (1983), pp. 295-303.

[3] D.R. Bacon, "Primary calibration of ultrasonic hydrophones using optical interferometry", IEEE Trans on Ultrasonics, Ferroelectrics and Frequency Control **35** (1988) 152-161.

[4] J. Wolf and W.G. Mayer, "Optical probing of ultrasonic pulses", Ultrasonics **27** (1989) 150-154.
W.G. Mayer, "Non-invasive acousto-optical sensing", in this book.

[5] H. Djelouah, J. Baboux *et al.*, "Pulsed calibration of miniature ultrasonic receivers using a wideband laser interferometer", Ultrasonics **27** (1989) 80-85.

[6] R. Reibold, "Piezoelectric and optical sensors for ultrasound field investigations", in this book.

[7] C. Constantopoulos and D. Royer, "Etude expérimentale par interféromètre optique du champ acoustique émis par un transducteur piézoélectrique". Internal report, DEA Acoustique Physique, Université Paris 7, to be published.

[8] J.C. Baboux, Thèse de Doctorat, Université de Lyon 1, France (1987).

[9] X. Jia, J. Berger and G. Quentin, "Etude des céramiques piezoelectriques circulaires au régime impulsionnel par sonde interférométrique". J. Phys. (France) **51** C2 (1990, 567-574.

[10] X. Jia, J. Berger and G. Quentin, "Experimental investigations of Lamb waves on pulsed piezoelectric transducers and their ultrasonic radiation into liquids". J. Acoust. Soc. Am. **90** (1991) 1181-1183.

[11] X. Jia, G. Quentin, M. Lassoued and J. Berger, "Quantitative Measurements of Ultrasonic Pressures with an Optical Heterodyne Interferometer", 1991 IEEE Ultrasonics Symposium (IEEE Press, 1991) to be published.

[12] J.D. Aussel and J.P. Monchalin, "Study of Surface Acoustic Wave Dispersion using Laser Ultrasonics", Rev. Progress in Quantitative NDE (D.O. Thomson and D.E. Chimenti, eds., Plenum Press, New York), **8A** (1989) 535-542.

[13] J.P. Monchalin, J.D. Aussel, R. Héon, C.K. Jen, A. Boudreault and R. Bernier, "Measurement of in-plane and out-of-plane ultrasonic displacements by optical heterodyne interferometry", J. of Nondestructive Evaluation **8** (1989).

[14] P. Guy, Thèse de Doctorat, Université Paris 7, France (1992).

[15] P. Guy, G. Quentin and J.P. Monchalin, "Study of Lamb waves by laser-ulrasound techniques", in preparation.

PIEZOELECTRIC AND OPTICAL SENSORS FOR ULTRASOUND FIELD INVESTIGATIONS

R. REIBOLD

Physikalisch-Technische Bundesanstalt
Braunschweig, Germany

ABSTRACT

This paper is intended as an introduction to physical phenomena that are potentially quantitative and their application in sound field investigations. Particular emphasis is placed on piezoelectric and optical sensors which make possible the measurement of ultrasound field details in space and time. A variety of possible realizations will be discussed. Requirements for quantitative measurements will be outlined together with a number of usefull calibration techniques.

1. Introduction

Ultrasound is not directly perceptible to human senses, and it is therefore necessary to expoit some physical effects that will allow its detection and measurement. A considerable number of useful physical phenomena exist that are potentially quantitative, which means that the measured quantity is related to the ultrasonic field quantity by a particular physical constant. Of all the various effects, the piezoelectric effect for the detection of sound pressure or particle displacement is of major importance. Another significant effect, the radiation force, can be used for ultrasound power measurements. For a number of years, optical phenomena such as interferometry, holography, light diffraction and light reflection have become of increasing interest. The respective field quantities detected are the particle displacement, the sound intensity and the acoustic pressure. Other usable phenomena are the calorimetric effect for intensity measurements and the capacitance effect for displacement measurements.

Devices which are capable of converting physical quantities to measurable quantities, usually electrical quantities, are called sensors. Depending on the underlying physical effect and the particular requirements for the various kinds of application, some of the acoustic sensors are simple, easy to use, direct and efficient instruments, but they also involve complicated, and subtle experimental arrangements which require a lot of space.

From the variety of existing acoustic sensors we will restrict ourselves to those which use the piezoelectric effect or optical effects as the basic principle and which make possible the measurement of ultrasonic field details in space and time. Although, the sound-transmitting medium is assumed to be a liquid (water), some of the techniques presented can, however, also be used for sound detection in solids and gases. In the first chapter various realizations of sensors will be discussed in some detail, including their basic conceptions and limitations. The second chapter is devoted to requirements for quantitative measurements and in particular to the need for calibration procedures.

2. Ultrasound Sensors

2.1. Piezoelectric Devices

The piezoelectric effect of quartz was discovered by Jacques and Pierre Curie in 1880. It occurs in certain anisotropic materials in which mechanical distortion such as applied pressure or displacement leads to a change in the electrical charge distribution. Of major interest for broadband operation are those materials with a high electro-mechanical coupling factor, a low mechanical quality factor and a characteristic acoustic impedance close to that of water.

In order to obtain some insight into the dynamical behaviour of a piezoelectric sensor we consider a simplified one-dimensional model (Fig. 1).

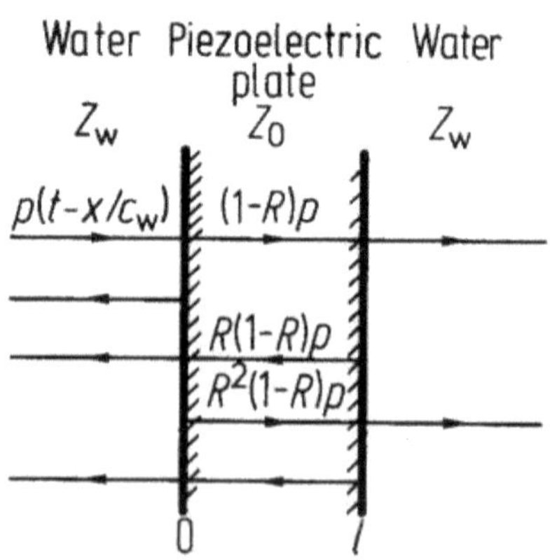

Fig. 1: One-dimensional model of a piezoelectric sensor

It consists of a piezoelectric plate with electrodes on its surfaces and a characteristic acoustic impedance $Z_0 = \rho_0 c_0$,the plate being immersed in water with the characteristic impedance $Z_W = \rho_W c_W$. The incident wave with a pressure of $p(t-x/c_W)$ will be partially reflected by the piezoelectric plate, $R = (Z_W - Z_0) / (Z_W + Z_0)$ being the reflection factor at the interface. The wave entering the plate will then be changed in amplitude by the factor 1-R. Due to successive reflections at the two interfaces the wave travelling in the plate will be diminished in amplitude by the factor R each time. The overall pressure in the plate then reads

$$p(x,t) = (1-R)\,[p(t-\frac{x}{c_0}) + Rp(t-\frac{x}{c_0}-\tau) + R^2 p(t-\frac{x}{c_0}-2\tau) + \ldots] \ , \tag{1}$$

where $\tau = l/c_0$ is the single travel time in the plate. The overall pressure is obviously a complex replica of the incident pressure when multiple reflections in the piezoelectric plate occur. If, however, the piezoelectric plate is matched to the surrounding medium, $Z_0 = Z_W$, or the piezoelectric plate is coupled on its back face to a medium of the same characteristic acoustic impedance, then the overall pressure will be proportional to the incident pressure.

It can be shown that under open-circuit conditions the voltage U at the electrodes of the piezoelectric plate (the reader is referred to the relevant text books) is given by:

$$U(t) \sim \int_0^t (p(0,t') - p(l,t'))\,dt' \ . \tag{2}$$

Here, $p(0,t')$ and $p(l,t')$ signify the pressure at the plate's surfaces. For a very small thickness of the plate the difference in the integrand can be replaced by the differential quotient $\Delta p \approx (\partial p/\partial t)\tau$ and thus the open-circuit voltage will be proportional to the pressure inside the plate. Furthermore, perfect matching assumed, the open-circuit voltage is also proportional to the incident wave.

The requirements discussed for a pressure-sensitive sensor can be essentially achieved by the constructional features shown in Fig. 2. These sensors are in common use and almost all practical realizations are based on these construction principles or modifications of them. In Fig. 2 (top) a piezoelectric disk is mounted on a suitable backing which heavily absorbs the incoming energy before the wave is scattered or reflected back to the disk. Another satisfactory arrangement which was first presented by Reggi et al. [1] is shown in Fig. 2 (bottom). It consists of a stretched polyvinylidene fluoride (PVDF) film ($Z_{PVDF} \approx Z_W$)

on which the sensitive area (≈ 1 mm^2) is at the centre and is formed by the region of overlap of the two electrodes. Permanent piezoelectric behaviour is achieved by polarization under high d.c. electric fields at elevated temperatures.

A number of such devices have been made for use in dosimetry, for non-destructive material testing and for medical application. Lewin [2] and Platte [3] developed so-called needle hydrophones, the basic principle of which is similar to that shown in Fig. 2 (top) with PVDF as the piezoelectric material. Granz et al. [4] and Staudenraus et al. [5] developed special PVDF membrane hydrophones (principle shown in Fig. 2 (bottom)) intended for shock wave investigations.

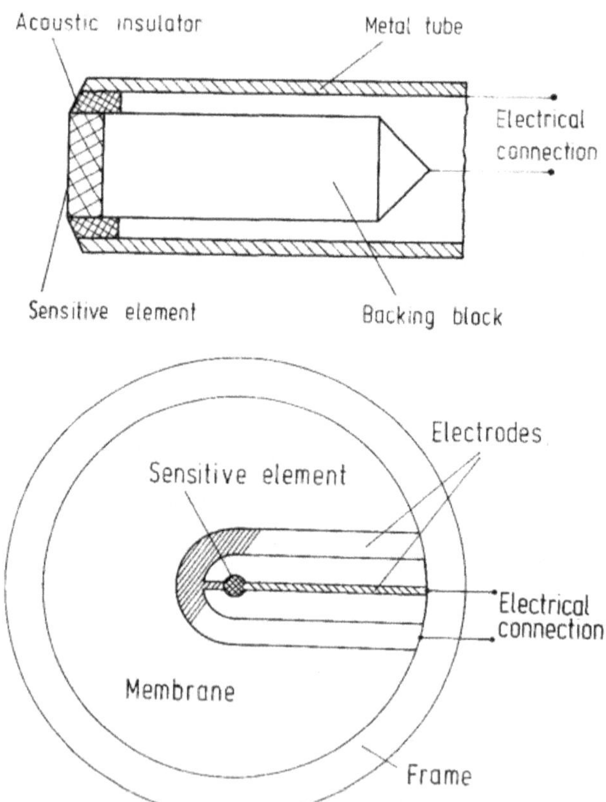

Fig. 2: Principle of design of a probe hydrophone (top) and a membrane hydrophone (bottom)

2.2. *Interferometric Devices*

In the last few years, laser interferometers have been used to an ever increasing extent for the investigation of ultrasound fields. The displacements occurring in the sound-transmitting medium amount to a few nanometers in the case of continuous sound waves and weak pulses, and can reach the micrometer range in the case of shock waves. Depending on the extent of the displacement, the interferometric measuring method must be adapted to the particular measurement problem. The laser interferometer developed at the PTB combines in one optical set-up the stabilized version for displacements of up to $\pm$ 50 nm and the quadrature version for greater displacements [6,7].

The interferometric test set-up is shown in Fig. 3. The design corresponds to that of a modified Michelson interferometer in the case of which the interaction of the light with the laser is suppressed by polarizing optical components.

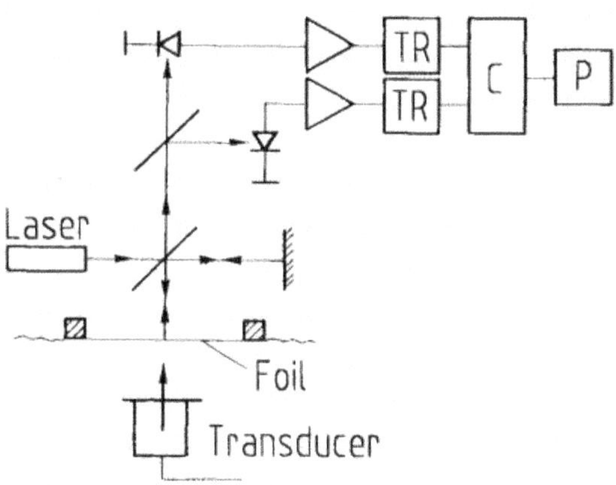

Fig. 3: Diagrammatic view of the interferometric test set-up. The particle displacement of the emitted ultrasonic pulse is measured on a foil at the water surface

In the quadrature interferometer, the signals of both photodiodes are used. For the light intensities $I_1(t)$ and $I_2(t)$, the following particle displacement results:

$$\xi(t) = \frac{\lambda}{4\pi}\left[\arctan\left(\frac{I_2}{I_1}\right) + m\pi\right] \qquad m = 0, \pm 1, \pm 2, \ldots \tag{3}$$

where λ is the light wavelength. The shape of particle displacement is numerically determined from the digitized interferometer signals.

In the case of the stabilized interferometer, the reference mirror is included in a control loop, which stabilizes the interferometer to the zero crossing of the sinusoidal change of the light intensity and at the same time controls low-frequency light path changes caused by environmental effects. In the range of ± 50 nm, an output signal is obtained which is directly proportional to the particle displacement of the object

$$\xi(t) = \frac{1}{2\sqrt{I_0 I_R}} \frac{\lambda}{4\pi} I(t) \quad . \tag{4}$$

In this formula, I_0 is the light intensity in the object branch and I_R that in the reference branch.

The ultrasonic signals are measured on a polypropylene foil, 6 μm thick, its upper side coated with aluminium, tightly stretched over the water surface. The laser light is focussed on the foil in order to attain as high a spatial resolution as possible; the focal point (point of measurement) has a diameter of about 20 μm.

The efficiency of the interferometric procedure is demonstrated by the investigation of a multi-transducer linear array with dynamic focussing and sector scanning. In general, it can be said that the requirements for the spatial resolution of the measuring system are particularly high in the range of the focal point of focussing transducers. Fig. 4 shows the typical pulse of a focussing linear array.

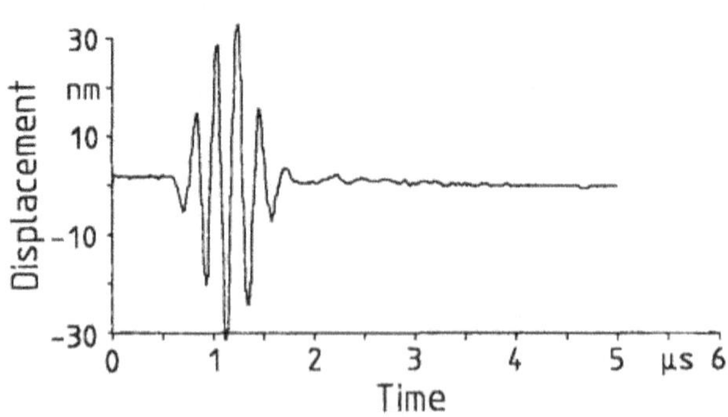

Fig. 4: Typical pulse of a focussing linear array measured in the focus of the sound field

In Fig. 5, the peak value of the particle displacement is represented as a function of the position in the focal plane. In the upper part of the figure, the direction of radiation is 0°,

i. e. normal to the transducer surface. The half-width amounts to about 1 mm. In the lower part of the figure, the transducer radiates at 40° to the normal. Here, the half-width is somewhat greater.

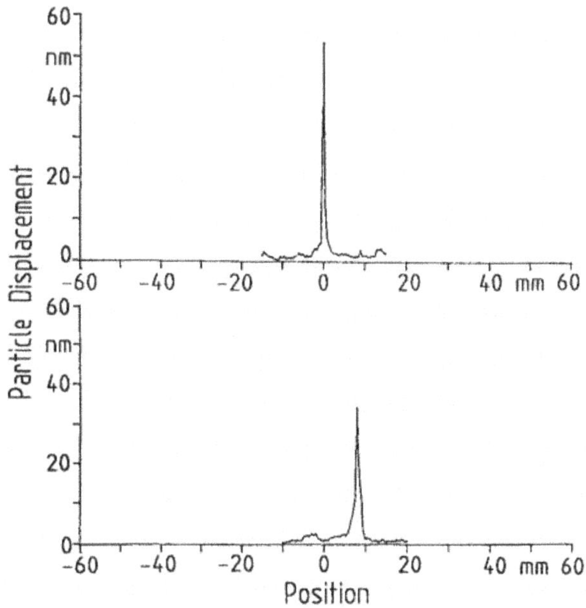

Fig. 5: Peak value of the particle displacement as a function of the position in the focal plane of a focussing linear array (focal length 10 mm).
Upper section: radiation direction 0°
Lower section: radiation direction 40°

A modified interferometer version [8] using a monomode glass fibre as a sensor element is shown in Fig. 6. The polished, gold-coated fibre tip is dipped in water and acts as a mechanical-to-optical converter. Fig. 7 presents some typical time waveforms measured in the focus of the sound field of broadband transducers.

2.3. Fibre Optic Sensors Based on Reflectivity

For the measurement of ultrasonic shock waves Staudenraus and Eisenmenger [9] developed an acousto-optic fibre sensor (Fig. 8). Unlike the interferometric fibre sensor mentioned in the previous section, here the acoustic quantity measured is not the particle displacement but the pressure in the liquid.

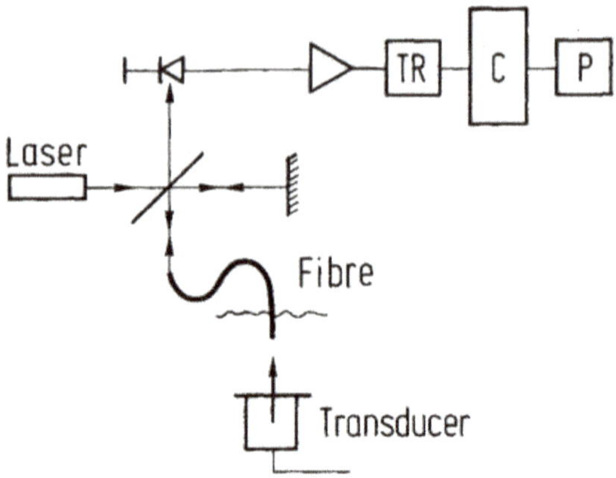

Fig. 6:
Interferometric arrangement using a monomode glass fibre as a sensor element

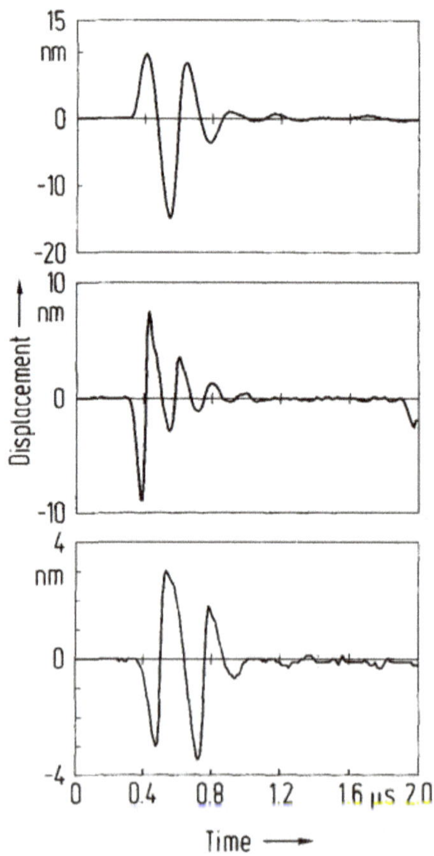

Fig. 7:
Typical time waveforms measured in the focus of the sound field of broadband transducers

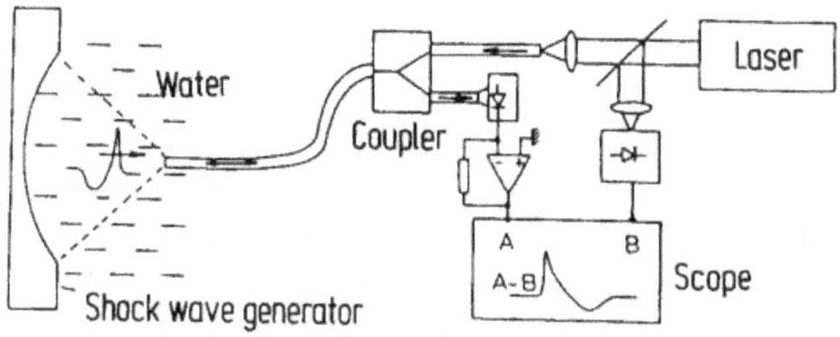

Fig. 8: Acousto-optic fibre sensor after Staudenraus and Eisenmenger [9]

If a shock wave strikes the fibre's tip, the reflection coefficient R in the liquid changes synchronously with the pressure and with this the light thrown back onto the photodetector also changes. Neglecting the change of the refractive index in the fibre, the relationship between the change in the reflection coefficient ΔR and the pressure change p is given by

$$\Delta R = \left(\frac{\partial R}{\partial n}\frac{\partial n}{\partial p}\right)_{p=p_0} p \quad , \tag{5}$$

where n is the refractive index in the liquid.

2.4. *Sensing by Light Diffraction*

The ultrasound sensors hitherto discussed allow with some limitations "point-by-point" measurements to be performed. As opposed to this, light diffraction techniques are characterized by projection values obtained from the sound field from which the complex local pressure must be numerically determined by tomographic back projection. Nevertheless, due to the high spatial resolution obtainable and the absence of disturbances which affect the sound propagation process, light diffraction tomography has proved to be a useful measuring technique [10].

The optical set-up corresponds to that of a common schlieren arrangement (Fig. 9). Lens L images the progressive harmonic wave into the image plane with a certain magnification. The discrete diffraction spectrum of the sound field appears in the focal plane of lens L; all diffraction orders except for the zeroth and first orders are cut out from this spectrum with the aid of a schlieren stop (slit aperture).

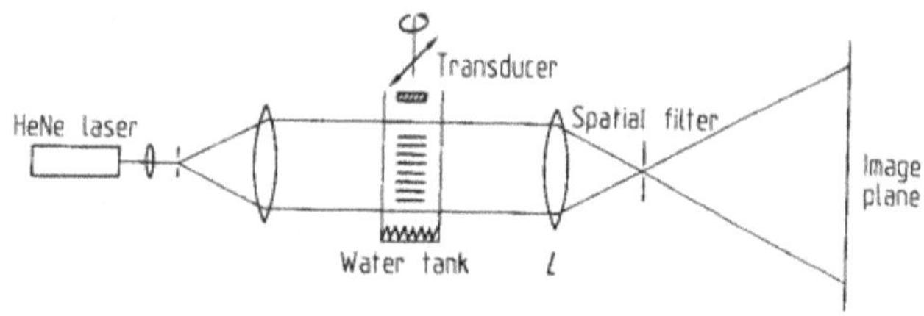

Fig. 9: Schematic representation of the light diffraction arrangement

For the light intensity normalized to the light intensity without the sound field, according to this assumption the following is obtained for a measurement point in the image plane:

$$I(t) = J_0^2(\hat{v}) + J_1^2(\hat{v}) + 2J_0(\hat{v})\,J_1(\hat{v})\cos(\omega^* t + \bar{\varphi}) \quad , \tag{6}$$

where J_0 and J_1 are Bessel functions and ω^* is the angular sound frequency. The Raman Nath parameter v is proportional to the mean sound pressure amplitude and $\bar{\varphi}$ is the mean acoustic phase. Equation (6) contains a direct term $J_0^2 + J_1^2$ and an alternating term produced by interference of the zeroth order and the first order, Doppler shifted by the sound frequency.

During the measurement, the sound field is scanned in a cross-sectional plane in equidistant steps. After each scan, the transducer is turned typically by $\Theta = 3{,}6°$, and the measuring process repeated. From the projection values $v\,(x, \Theta)$ and $\bar{\varphi}\,(x, \Theta)$ obtained in this way, the local quantities $p\,(x, y)$ and $\varphi\,(x, y)$, which depend upon the cross-sectional coordinates x and y, are calculated. Fig. 10 shows typical sound pressure profiles of a circular focussing transducer.

The method of light diffraction tomography is not limited to the investigation of progressive, harmonic sound waves. In a modified form, it can also be applied to stationary [11] and pulsed [12] ultrasound fields.

3. Calibration Techniques

The assumption made in the introduction, i. e. that the measured quantity is related to the field quantity by a particular physical constant, needs some discussion and generalization.

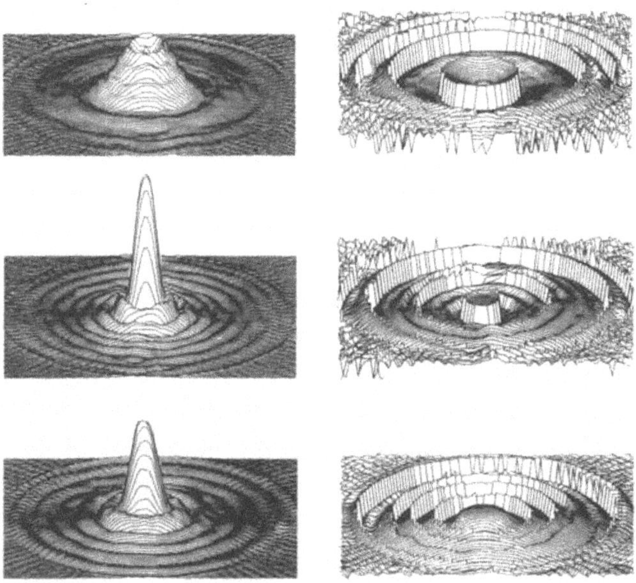

Fig. 10: Filtered back projection of the sound-pressure amplitude (left) and phase (right) by means of discrete convolution. The measured values are sampled at distances of z = 55 mm (top), in the focal plane at z = 110 mm (middle) and at z = 150 mm from the transducer surface. Circular focussing transducer: diameter 20 mm, frequency 4 MHz, area 40 mm x 40 mm

A number of effects exist which considerably violate simple relationships between the quantities of interest. In the case of piezoelectric sensors, especially those made from ceramic and PVDF materials, the state of polarization may be different. The measured signals may be distorted due to diffraction effects at the sensor's tip. All sensors evidently suffer from their limited spatial resolution, which causes frequency and directivity dependence of the sensitivity. Imperfect matching may result in resonance phenomena. Last but not least, the electric cable length and the input impedance and bandwidth of the electronic measuring device may influence the measurement results.

The limitations outlined above make clear that a description of the transfer characteristic of a sensor requires a more general point of view. Within certain limits the sensors can be regarded as linear system. The essence of the respective theory is that the measured time waveform $u(t)$ is given by the convolution of the acoustic quantity with an arbitrary time function $p(t)$ at a point in the field with the impulse response function $h(t)$, the latter being entirely defined by the characteristics of the sensor:

$$u(t) = h(t) * p(t) \quad . \tag{7}$$

A Fourier transformation of this equation yields the simple product

$$U(\omega) = H(\omega)\ P(\omega) \quad . \tag{8}$$

$\omega = 2\pi f$ is the angular frequency. Eq. (8) suggests that a distortion-free measurement of the acoustic quantity can only be obtained if the absolute value of the transfer function $H(\omega)$ is constant and if its phase changes linearly with frequency. A linear phase change is equivalent to a frequency independent time delay. The process of determining the transfer function is referred to as calibration.

3.1. Calibration by Substitution

In the substitution technique the sensor is calibrated by comparison with a secondary standard which has itself been calibrated using an absolute technique. This is achieved by successively placing the two devices at the same point in a series of acoustic fields, chosen to cover the frequency range of interest.

For routine calibrations of ultrasonic hydrophones (piezoelectric sensors), the National Physical Laboratory (UK) employs a multi-frequency substitution method [13]. Due to non-linear distortion, the sound field used contains many frequency components, usually at integer frequencies between 1 MHz and 20 MHz, and can thus be used to calibrate the hydrophone over a wide frequency range simultaneously.

Another substitution technique termed time-delay spectrometry allows the continuous recording of the frequency response of the transmission factor of a hydrophone. This method will be discussed in more detail in the following section.

3.2. Reciprocity Calibration and Time-delay Spectrometry

These techniques clearly require a reciprocal sound source, e.g. a piezoelectric transducer, to be used. If a transmitting transducer is driven with a current I_1, the free-field sound pressure at an on-axis far field point is given by $p = SI_1$, where S is termed the transmission factor. Similarly, if a receiving transducer is placed at that field point and axially aligned, the open-circuit voltage $U_1 = Mp$ is obtained across the transducer terminals. M is the receiver's sensitivity factor. The ratio $J_p = M/S$ is the reciprocity parameter.

In the arrangement shown in Fig. 11 the transducer operates as a projector and receiver. The free-field sound pressure p at the transducer surface is then given by

$$p = \sqrt{\frac{U_1 I_1}{J_p}} \tag{9}$$

If the reflector is turned so that the reference point in the field coincides with the sensor (the hydrophone in Fig. 11), then its sensitivity can be determined.

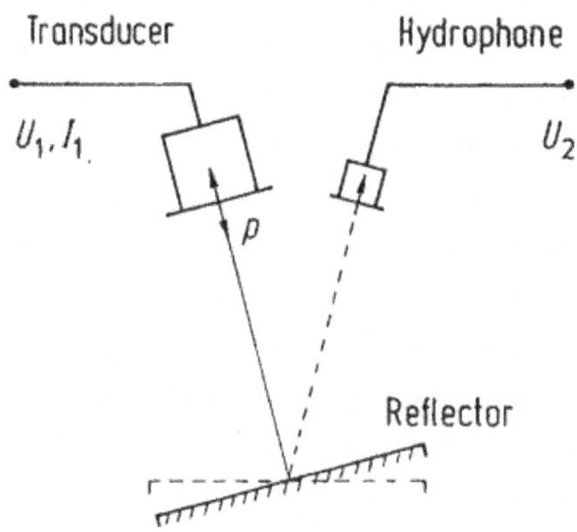

Fig. 11: Schematic diagram of the set-up for reciprocity calibration

The great advantage of the reciprocity method is that the determination of the acoustic quantities can be reduced to the relatively simple measurement of electrical quantities. It must be mentioned that practical application makes a number of corrections necessary, but this will not be explained in detail in this paper.

Reciprocity calibration is used as a primary technique for the absolute calibration of piezoelectric hydrophones in the Physikalisch-Technische Bundesanstalt (Germany) [14]. Moreover, it is the subject of international specification standards and agreed recommendations [15].

Whereas the reciprocity method is restricted to the calibration at discrete frequencies, a combination with Time-Delay Spectrometry (TDS) [16] allows the continuous recording of the frequency response of the transmission factor. The basic component of this apparatus is a spectrum analyzer, the tracking filter of which is detuned in relation to the generator frequency f_{t1} by the travel time of the sound wave t_0 and the sweep rate df/dt of the sweep generator (Fig. 12). This procedure allows continuous operation in the course of which only signals with the adjusted travel time are received; sound waves with longer travel times, such as echos, are filtered out.

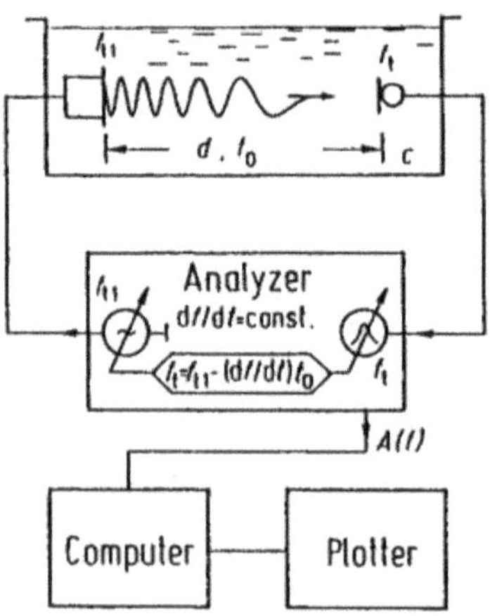

Fig. 12: Principle of Time-Delay Spectrometry

Fig. 13 shows the free-field sensitivity level of a hydrophone for discrete frequencies according to the two-transducer reciprocity method and with continuous frequency response according to the TDS reciprocity method.

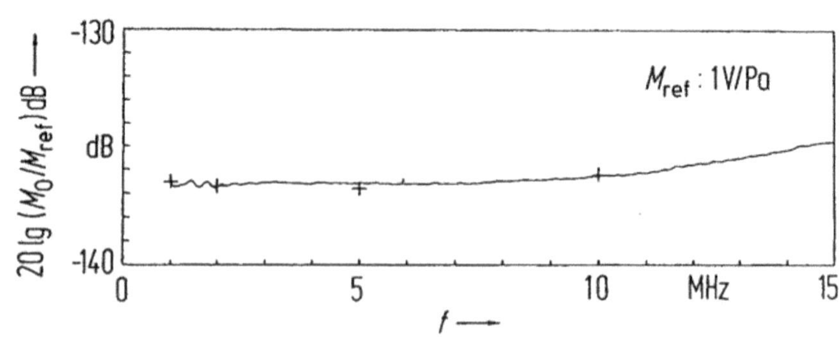

Fig. 13: Free-field sensitivity level of a hydrophone as a function of the frequency.
(——) TDS reciprocity method [14],
(++++) two-transducer reciprocity method [16]

3.3. Calibration Using Laser Interferometry

Laser interferometry might be regarded as the most perfect and convincing absolute calibration technique because of its direct traceability to primary standards of length. But this is not altogether true. The particle displacement caused by ultrasonic waves is usually smaller than the optical wavelength, which requires fractional fringes to be counted.

In the case of the quadrature interferometer (Fig. 3) the precision of the measurement system is usually limited by three systematic errors: phase shift error, non-zero offset of the detectors and unequal gain of the two channels. These errors can be corrected by means of a numerical procedure given by Heydemann [17] in 1981.

The situation is simpler for the stabilized interferometer. Here, knowledge of the intensities I_0 and I_R is not required in practical application. The proportionality factor in eq. (4) and thus the slope in the operating point of the system can be determined by low-frequency excitation of the reference mirror with an amplitude greater than that of the optical wavelength. Accurate measurement of the peak intensity and the period of oscillation is, however, necessary in any case.

Considering the fibre-optic interferometer sensor (Fig. 6) it becomes clear that besides the insufficiencies of the interferometric data aquisition mentioned, the measurement results will also be affected by diffraction effects at the fibre's tip and by optical phase changes due to the wave travelling along the fibre. Fig. 14 shows this behaviour in terms of the normalized time waveforms by means of an ultrasonic pulse measured using the arrangement of Fig. 3 and that of Fig. 6.

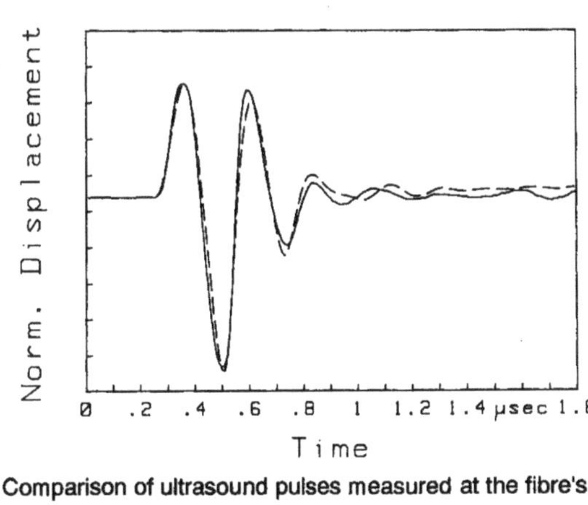

Fig. 14: Comparison of ultrasound pulses measured at the fibre's tip (solid line) and at the pellicle in the water surface (broken line) in terms of the normalized time waveforms

The National Physical Laboratory (UK) carries out primary calibration of hydrophones using laser interferometry in combination with a substitution technique [18]. The particle displacement caused by an ultrasonic wave is detected by a thin plastic pellicle. The hydrophone is then replaced by the pellicle and calibrated by measuring the hydrophone's output voltage. From international intercomparisons it turned out that this calibration technique is very reliable.

3.4. Wideband Standard Acoustic Source

A completely different but efficient and easy-to-handle calibration technique was first described by Eisenmenger [19] in 1962 and then improved by Baboux et al. [19] in 1989 using a wideband standard source. The main part of this device consists of a circular disk of lithium niobate 40 mm in diameter and 18 mm thick. This disk is charged to high d.c. voltage of about 1 kV and then discharged using a series of five avalanche transistors. In this way a very short current impulse and sound impulse of a few nanoseconds is obtained.

As is well known, the sound radiated by a piston-like source can be described by a plane direct wave and the edge diffracted wave. According to the dimensions of the disk chosen it is possible to isolate the direct wave for field points close to the source. This direct wave is then used for the purpose of calibration.

4. Summary

The aim of this paper has been to present some of the general features of ultrasound sensors, with emphasis on piezoelectric and optical devices. The sensors discussed are specifically intended for the measurement of time waveforms and acoustic field distributions. A brief general introduction to the requirements for calibration techniques has been given, and widely-used methods have been discussed in some detail. The list of references is comprehensive but by no means complete. It will, however, offer a useful basis to readers interested in the field.

5. References

1. A. S. De Reggi, S. C. Roth, J. M. Kenney, S. Edelmann and G. R. Harris, *Piezoelectric polymer probe for ultrasonic applications*, J. Acoust. Soc. Am. **69** (1981) 853.
2. P. A. Lewin, *Miniature piezoelectric polymer ultrasonic hydrophone probes*, Ultrasonics **19** (1981) 213.
3. M. Platte, *A polyvinylidene fluoride needle hydrophone for ultrasonic applications*, Ultrasonics **23** (1985) 113.
4. B. Granz, *PVDF hydrophone for the measurement of shock waves*, IEEE Trans. EI **24** (1989) 499.
5. J. Staudenraus, *Erzeugung und Ausbreitung freifeldfokussierter Hochenergiedruckimpulse in Wasser*, Fortschritt-Berichte VDI, Reihe 21 Nr. 89 (VDI, Düsseldorf, 1991).
6. R. Reibold and W. Molkenstruck, *Laser interferometric measurement and computerized evaluation of ultrasonic displacements*, Acustica **49** (1981) 205.
7. R. Reibold and W. Molkenstruck, *Laserinterferometrie zur Untersuchung breitbandiger Ultraschallsignale*, PTB-Ber. PTB-MA-64, p. 93.
8. W. Menssen, W. Molkenstruck and R. Reibold, *Fibre optic sensor system*, to be published in Ultrasonics International '91 (Butterworth Heinemann, Oxford, 1991).
9. J. Staudenraus and W. Eisenmenger, *Optisches Sondenhydrophon*, in Fortschritte der Akustik, DAGA 88 (DPG GmbH, Bad Honnef, 1988) p. 467.
10. R. Reibold and W. Molkenstruck, *Light diffraction tomography applied to the investigation of ultrasonic fields. Part I: Continuous waves*, Acustica **56** (1984) 180.
11. R. Reibold, *Light diffraction tomography applied to the investigation of ultrasonic fields. Part II: Standing waves,* Acustica **63** (1987) 283.
12. R. Reibold, *Investigation of ultrasonic pulses using light diffraction*, in Proceeding of the 5th FASE Congress, ed. J.L.B. Coelho (Lisboa, 1987) p. 19.
13. R.A. Smith and D.R. Bacon, *A multiple-frequency hydrophone calibration technique*, J. Acoust. Soc. Am. **87** (1990) 2231.
14. K. Brendel and G. Ludwig, *Calibration of ultrasonic probe transducers*, Acustica **36** (1976) 203.
15. IEC - *Characteristics and calibration of hydrophones for operation in the frequency range 0.5 MHz to 15 MHz*, IEC 866: 1987
16. G. Ludwig and K. Brendel, *Calibration of hydrophones based on reciprocity and time-delay spectrometry*, IEEE Trans. UFFC **35** (1988) 168.
17. P.L.M. Heydemann, *Determination and correction of quadrature fringe measurement errors in interferometers*, Appl. Optics **20** (1981) 3382.

18. D.R. Bacon, *Primary calibration of ultrasonic hydrophones using optical interferometry*, IEEE Trans. UFFC **35** (1988) 152.
19. W. Eisenmenger, *Eine Kontrollschallquelle für breitbandige Mikrophone zur Messung von Druckimpulsen in Flüssigkeiten*, Acustica **12** (1962) 165.
20. J.C. Baboux, M. Perdrix, F. Lakestani, R. Denis and G.P. Battagin, *Use of a single crystal thick disk for a wide-band standard acoustic source*, in Ultrasonics International'89 (Butterworth, Guildford, 1989) p. 448.

ACOUSTOOPTICAL PROBING OF ANISOTROPIC MEDIA

A.S. ŚLIWIŃSKI
Institute of Experimental Physics
University of Gdańsk
80-952 Gdańsk, Poland

ABSTRACT

The ultrasonic light diffraction (ULD) phenomena in trasparent anisotropic media has been reviewed and discussed in the light existing theories in the limitation of uniaxial crystals. Some results of nematic liquid crystals examination obtained by acousto-optical methods were used to examplify possibilities of anisotropic properties determination from ULD patterns. In liquid crystals under external forces the anisotropy is a subject of dynamical variations and these processes can be studied by acousto-optical probing.

1. Introduction

As acoustooptical (AO) probing of material properties one understands detecting of an optical signal after its interaction with an ultrasonic wave (UW) which was being penetra ted a medium examined. Due to the specific mechanism of light and ultrasonic interaction the transmission of information about a medium properties via ultrasonics (U) to the light (L) detector is possible with a great accuracy and fidelity. The mechanism of L and U interaction is described elsewhere in monographies on acoustooptics [1-9], as well as in many papers and reports, partly published in related proceedings of specialized conferences and Schools [10-14] organized on that topic during the last decade.

The fundamental phenomena appearing in the interaction process between L and U beams are scattering and diffraction of light by the dynamical (often periodical) structure of the medium influenced by the elastic (ultrasonic) wave. In resulting scattered and diffracted light one can observe (and detect depending on the arrangement used) various phenomena like schlieren images, diffraction patterns, deflection and modulation of light in diffraction orders etc. [8,15]. All these optical phenomena carry information about an ultrasonic field build up in the volume of interaction (cross-sections of both, ultrasonic and light beams) depending on the sound propagation properties of the medium. In case of an anisotropic medium in which the sound velocity is a function of orientation, the structure of the ultrasonic field and, consequently the response of the interacting light beam is strongly depending on the anisotropy [2-7, 16-20]. Also, information about irregularities (if present) in homogeneity of the medium may be incorporated in the optical response.

In this paper we limit our considerations to some application of the ultrasonic light diffraction (ULD) phenomena for anisotropic as well as inhomogeneous media examination because often both of these properties are simultaneously present in material samples under study. We shall review shortly about the ULD for anisotropic case examplified by uniaxial crystals discussing possibilities of determination of anisotropic properties from diffraction patterns. Some examples of correlation between light diffraction spectrum and the diffracting structure presented by the ultrasonic field influenced by the anisotropy of a medium in the case of nematic liquid crystals will be given.

2. Diffraction of light by ultrasonic waves in uniaxial crystals

2.1. Theoretical

There exist good developed theories of the ULD phenomenon for single crystals [2-7, 16-20] in which adequate expressions are determined for light intensity distributions in diffraction patterns against AO variables (like the Raman-Nath or Cook parameters characterizing the change of refractive index being proportional to UW pressure and depending on the wavelengths of U and L, respectively) and the orientation of the crystal in respect to the propagation directions of U and L. Also, the state of polarisation of the UW (bulk or shear) and of the light beams plays the essential role.

The classical phenomenon in optically anisotropic crystals is the birefringent effect; so, in AO interaction in such crystals the light beam (depending on the incident angle) is splited into two ones: the ordinary beam (OB) and the extraordinary beam (EB) of refractive indices n_o and n_e, respectively. The geometrical situation of AO interaction in a birefringent uniaxial medium is presented in the Fig.1. The linearly polarized light beam incidenting the uniaxial crystal sample is birefringed and both OB

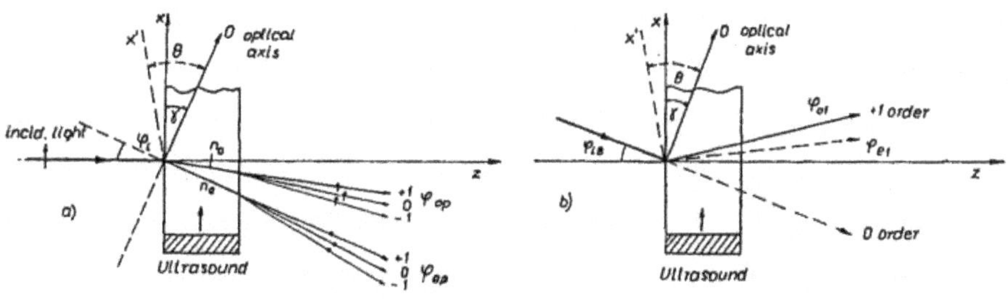

Fig.1. Light diffraction by ultrasonic waves and birefrengent effect in uniaxial medium: a) Raman-Nath case, b) Bragg case

and EB due to the diffraction by the ultrasonic wave create the diffraction patterns. The Fig.1a corresponds to the Raman-Nath and the Fig.1b to the Bragg kinds of diffraction, respectively. The polarization state of the OB is the same as the incidenting beam (|) and of the EB it is perpendicular (·). In the general case during diffraction rotation of the polarization plane may take place. Then the following expression represent the diffraction result in the phase (unit optical path) difference [4,5]:

$$\eta_{\alpha\beta p} = K \, tg \, \varphi_i + \left(p + \frac{1}{2}\right) \frac{K^2}{k_\alpha \cos\varphi_i} - k_\beta \frac{n_\beta^2 - n_\alpha^2}{2 n_\alpha n_\beta \cos\varphi_i} \qquad (1)$$

where $\alpha, \beta = \pm 1$ represent states of polarization of the (α) light beams (β), p-order of diffraction spectrum, n_α, n_β light refractive indices, φ_i - angle of incident light, k_α k_β and K -light and ultrasonic wave nummbers, respectively.

For $n_\alpha = n_\beta$ the formula (1) streams to the expression given by the Raman-Nath theory of ULD [5]. When the 2-nd term dominates, one could expect more influence of the Bragg diffraction instead of Raman-Nath one which appears very weak for greater (a, Q > 1) values of $a = \frac{2\pi \delta n l}{\lambda}$ (Raman-Nath parameter) or $Q = \frac{\Lambda^2 l}{n\lambda}$ (Cook parameter). δn is the amplitude of variation of the light refractive index n caused by UW acoustic pressure, l the width of the UW layer, λ wave-length of light in vaccum, Λ wavelength of UW.

The geometry of the interacting process may be schematically represented using the known diagramms of light beam wave surfaces (Fig. 2) represented by a sphere for OB and a rotary ellipse for EB. The value for OB refractive index $n_o = N_o$ (in uniaxiel crystal).

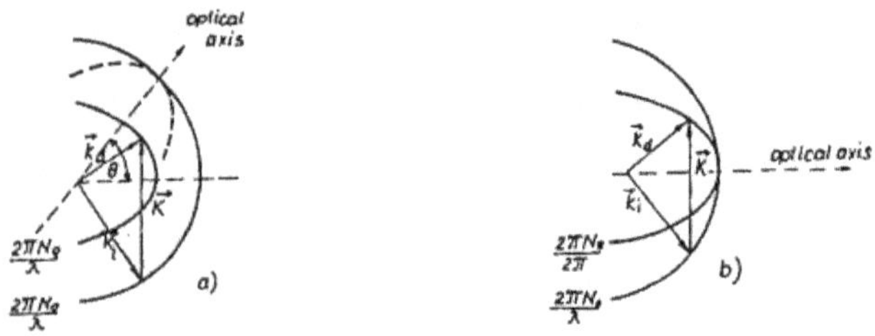

Fig. 2. Light and ultrasonic waves vectors representing Bragg interaction in anisotropic uniaxial medium related to the light wave surfaces of ordinary (N_o) and extraordinary (N_e) beams; a) the interaction plane forms an angle $\theta = 0$ and b) it is parallel $\theta = 0$ to the optical axis.

The value for EB refractive index n_e changes from $n_e = N_o$ for $\theta = 0$ (along the optical axis) to $n_e = N_e$ for $\theta = 90^o$ (perpendicularly to the optical axis). The Fig. 2 represents the Bragg diffraction case in the interacting plane (both the L and U beams lie in this plane): a) situated under angle θ (OX' in Fig. 1) and b) parallel ($\theta = 0$) (OX in Fig. 1) to the optical axis, respectively. k_i and k_d are the light wave vectors of incident and diffracted light beams, K is the wave vector of the UW. The diagramms correspond to the energy and momentum consevation laws in the process of foton-fonon interaction according to the formulae

$$\omega_d = \omega_i \pm \Omega \quad \text{and} \quad k_d = k_i \pm K, \text{ respectively,} \qquad (2)$$

where ω_d is the foton angular frequency after interaction, ω_i angular frequency of the incidenting foton, $\Omega = 2\pi f$ angular frequency of UW. The sign (+) or (-) corresponds to the Stokes or anty-Stokes transition of light frequency (energy) during the foton-fonon interaction.

Relations between the angles are following [4,5]:

$$\sin\varphi_i = -\frac{\lambda f}{2n_i c}\left[1+\frac{c^2}{\lambda^2 f^2}\left(n_i^2-n_d^2\right)\right]$$
$$\sin\varphi_d = \frac{\lambda f}{2n_d c}\left[1-\frac{c^2}{\lambda^2 f^2}\left(n_i^2-n_d^2\right)\right] \qquad (3)$$

where f and c are frequency and velocity of sound, n_i, n_d refractive indices of incident and diffracted light, respectively.

For the isotropic medium i.e. for $n_i = n_d = n$ the expressions (3) give

$$\sin\varphi_i = -\sin\varphi_d = -\frac{\lambda f}{2nc} = \mp\frac{K}{k_i}\ ; \qquad (4)$$

the last term results from (2).

It may be deduced from (4) that the Bragg diffraction in the isotropic medium may take place in the range of frequencies from 0 to $f_{max} = \frac{2nc}{\lambda}$. For the anisotropic medium $n_d = n_o$ for the ordinary and $n_d = n_e$ for extraordinary beam, respectively.

Depending on the angle of incidence and the orientation of the incident light beam in respect to the optical axis, the Bragg relations in anisotropic media may take place only in limited frequency range, which is different for different angles θ between the optical axis and the plane of foton-fonon interaction.

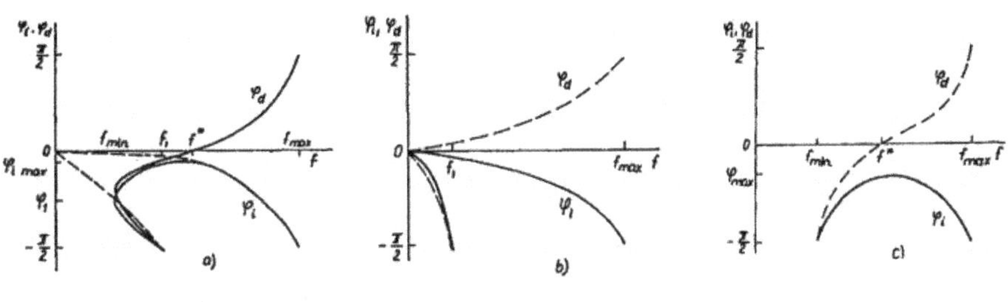

Fig. 3. Dependence of incident and diffraction angles of light against frequency of ultrasonics in Bragg diffraction: a) $\theta \neq 0$, b) $\theta = 0$, c) $\theta = 90^o$ (after [5]).

Fig. 3 represents the dependences of the incident φ_i and diffraction φ_d angles (for the positive crystal case $N_o > N_e$) for the Bragg ULD in the plane under the angle θ (a) in respect, to the optical axis and for $\theta = o$ (b) and $\theta = 90^o$ (c). Fig. 4 represents the optical wave surfaces (a) and dependences of the incident and diffraction angles against the frequency when the UW

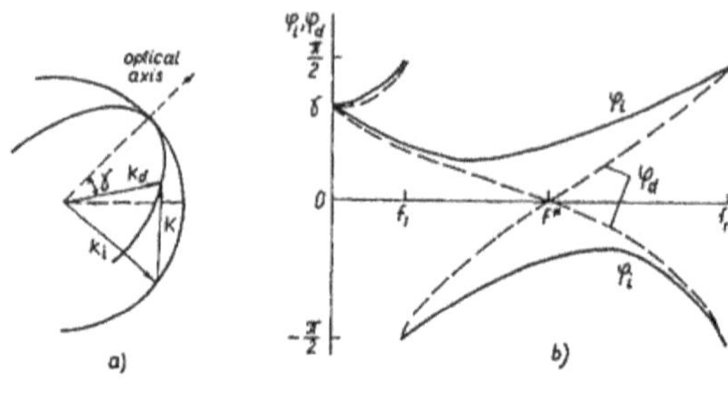

Fig. 1. Diffraction in the plane parallel to the optical axis ($\theta = 0$) in the case the front of ultrasonic wave have an angle γ (after [5]).

forms the angle γ (comp. Fig. 1) with the optical axis.

The range of frequences between f_{min} and f_{max} as well as the characteristic frequencies f_1 and f^* depend on the difference $(N_o - N_e)$ (for minimum) and the sum $(N_o + N_e)$ (for maximum) as well as on the sound velocity c and the light wavelength λ. Also, they depend on the angle θ ; simultaneously the curves $\varphi_i(f)$ and $\varphi_d(f)$ deform themselves depending on θ. The detail discussion about all the dependences may be found in the literature [2-7].

The diffracted light amplitudes distribution in diffraction orders (Fig. 1) may be expressed in the general form as [21, 22]

$$E(x,z,t)=E_0\sum_{\alpha=1,2}\sum_{p=-\infty}^{+\infty}(-i)^p e_\alpha C_{\alpha p}(x)\exp\{i[\omega_p t-\eta_{\alpha p}(k_{\alpha px}x+k_{\alpha pz}z\pm Kz)]\} \tag{5}$$

where E_0 is the amplitude of the incident light, $e_{\alpha p}$ - components of the unit vectors defining the polarisation plane of diffracted light waves, $C_{\alpha p}(x)$- relative amplitudes of diffractional orders, $\omega_p = \omega \pm p\Omega \equiv \omega_d$ (in (2)); the wave numbers should obey recursion relationship which follows from the boundary conditions:

$$\eta_{\alpha p}k_{\alpha pz}-\eta_{\beta,p-1}k_{\beta px}-K=0 \tag{6}$$

and the components k_{px} of the light wave vectors in diffraction orders are related by the expression

$$k^2_{\beta px} = \frac{\omega^2}{c^2}-k^2_{\alpha pz}\ . \tag{7}$$

Amplitudes of diffracted beams satisfy in the general case the system of recurrant differential equations

$$2\frac{dC_{\alpha p}}{dz}=\sum_{\beta=\pm 1}\left[q_{\alpha\beta,p-1}C_{p-1}\exp(i\eta_{\alpha\beta,p-1}z)+q_{\alpha\beta,p+1}C_{p+1}\exp(-i\eta_{\alpha\beta p}z)\right] \tag{8}$$

where $$q_{\alpha\beta p} = \frac{4\pi^2}{\lambda^2 k_{\beta p z}} \cdot \frac{n_{\alpha p}^2 \cdot n_{\beta p}^2}{n_{\alpha p} + n_{\beta p}} \delta n_{\alpha\beta} \tag{9}$$

and $\delta n_{\alpha\beta}$ is the tensor of variations in refractive index induced by the UW acoustic pressure.

It is seen that $q_{\alpha\beta p} \cdot l$ plays the analogical role as the Raman-Nath parameter (mentioned above) used for the ULD in isotropic media.

The light intensity distribution in diffraction orders can be calculated by using the known procedure

$$I(q_{\alpha\beta p} \cdot l) = E \cdot E^*, \tag{10}$$

where * denotes the complex conjugate value.

The general solution for (10) is complicated. However, in the case of the large interaction width l those diffractional orders are only imported for which [see (1)]

$$\eta_{\alpha\beta p} \cdot l \leqslant \pi \tag{11}$$

what corresponds to the 0-th (I_o) and the 1-st (I_1) orders of the Bragg kind of diffraction. The exact condition for the Bragg relations (2) is the requirement of $\eta_{\alpha\beta p}$ = o what also corresponds to the condition (6).

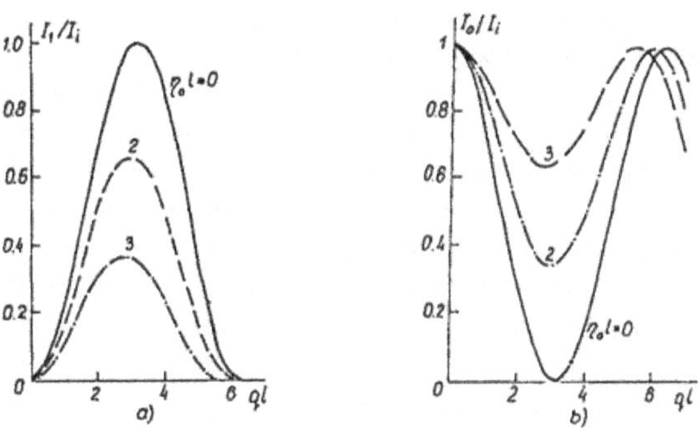

Fig. 5. Dependence of the relative light intensities in the 1-st (a) and 0-th (b) diffraction orders for Bragg diffraction for different ηl against ql (after [5]).

The ratios I_o/I_i and I_1/I_i (where I_i is the intensity of incident light) as finctuons of ql for the Bragg diffraction are presented in the Fig. 5 (after [5]) for few values of ηl. The drawings are for arbitrary medium (isotropic) however they give the idea, how the dependence on acoustic power works.

In anisotropic media these intensities depend on all the pecularities of polarization state variations, incident angles and crystal symmetry discussed above.

In some cases in anisotropic media the multi Bragg scattering may take place. For example the Fig. 6 represents the double

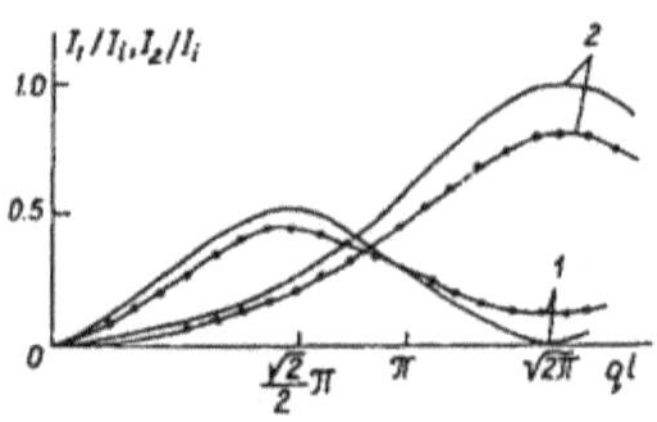

Fig. 6. Dependence of relative light intensities in the 1-st and 2-nd orders for a double Bragg diffraction (after [5]); —— theory, ... experiment.

Bragg scattering against ql for the first (I_1/I_i) and the second (I_2/I_i) orders of diffraction; the theoretical results (solid lines) are compared with the experimental ones (points).

2.2. *Applications for material properties examination examplified by liquid crystals.*

ULD methods are successfully used for determination of UW velocities and related to them elastic constants and fotoelastic constants depending on the orientation of anisotropic material samples like crystals of different crystalographic classes. The methods for proporiate measurements are known [2, 4, 16, 21] we shall not discuss them here. However it is worth mentioning about the Schaefer-Bergmann optical method for elastic constant measurements known since 1935 and described in [21] based on ULD in solids.

In anisotropic media also the attenuation of the UW depends on the orientation. ULD methods are also used for these measurements [2, 23-26]. On the contrary of the solid crystals where the attenuation is small, in the liquid crystals it occurs rather large due to viscosity, which is anisotropic, too.

For the uniaxial crystals to which we limit our considerations a simple measure of anisotropy is the difference of properties along the crystal axis ($\|$) and perpondicularly ($\perp$) to it. For example the anisotropy of sound velocity c and attenuation α is expressed by

$$\Delta c = c_{\parallel} - c_{\perp}, \quad \Delta\alpha = \alpha_{\parallel} - \alpha_{\perp} \quad \text{, respectively,} \tag{12}$$

In similar way the anisotropy of refractive index n may be define as

$$\Delta n = n_{\parallel} - n_{\perp} \quad . \tag{13}$$

As an example of A.O. examination of uniaxial anisotropic material we shall consider nematic liquid crystal which molecules are oriented in one direction by external magnetic field. In the Fig. 7 the experimental geometry for perpendicular interaction between linearly polarized light beam and the ultrasonic one in the nematic liquid crystal is presented. The homeotropic texture sample is located in the range of the light beam passing along the optical axis z. In the situation of the absence of the UW (Fig. 7a) one does not observe any effect of berefringence because the angle between the optical axis and the direction of the light

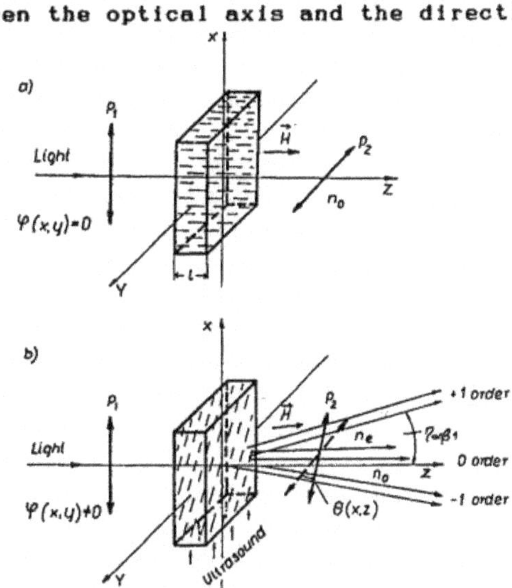

Fig. 7. Light and ultrasonic interaction geometry in an anisotropic layer (nematic liquid crystal of the thickness l - homeotropic texture in magnetic field); a)without ultrasonics, I = 0 with crossed polarizers P_1 and P_2; b)ultrasonics present inducing birefringence and diffraction of light, I $\neq$ 0. θ (x,z) - angle of rotation of polarization plane corresponding to the avarage tilting of long acxes of molecules by ultrasonic.

beam φ (x,z) = 0. In the crossed polarizers the intensity of light I = 0 at the output. On the introduction of the ultrasonic wave into the sample the orientations of the long axes of the liquid crystal molecules change under the influence of the acoustic field and birefringence occur (Fig. 7b), then I $\neq$ 0 at the output. The angle between the axis of the light beam and the direction of the molecule axes $\varphi(x, z) \neq 0$; it is dependent on the parameters of the acoustic field. The refractive index of the ordinary beam n_o does not depend on φ ; only the one of the extraordinary beam n_e will be changed and for that variation of n_e one has:

$$n_e(z, f_{ak}) = \frac{n_o n_e}{\sqrt{n_o^2 \cos[\varphi(z, f_{ak})] + n_e^2 \sin[\varphi(z, f_{ak})]}} \tag{14}$$

where: f_{ak} - characterizes the influence of an ultrasonic field (intensity, frequency, field distribution etc.). The phase shift between the ordinary and extraordinary beams will be:

$$\delta(z) = \frac{2\pi l(f_{ak}) \langle \Delta n(z, f_{ak}) \rangle}{\lambda} \tag{15}$$

where: l - thickeness of the traversed way,
λ - wavelength of light,
$\Delta n = n_o - n_e$, $\langle \ldots \rangle$ - sign of averaging process along the thickeness of the layer.

As seen from the above considerations, the polarization state of the light beam undergoes changes with a simultaneous change of phase relations between the ordinary and extraordinary beam in a liquid-crystalline sample disturbed by an ultrasonic wave - as a result we obtain diffractional images whose shapes depend on the state of orientation degree of the molecule axes evoked the UW. In this case the diffraction is analogous to the simultaneous diffraction of two (Fig. 7b) coherent light beams (with mutually perpendicular polarization shifted in the phase and with different intensities) on the phase diffraction grating with a constant equal (see Fig. 2) to the wave-length Λ_1 of the UW introduced into the liquid crystal. Every beam of diffracted light in diffraction order is splited into two beams (ordinary and extraordinary) due to birefringence effect. Simultaneously rotation of the polarization plane of light takes place. Direction of diffracted beams and variation of polarization state in a given order p in general case for the linearly polarized light beam incidenting at the angle φ_i is determined by the relation (1). For the perpendicular incidence (Fig. 7b) one gets for the 0 and ±1 orders

$$\eta_{\alpha\beta 0} = \frac{1}{2}\frac{K^2}{k_\alpha} - k_\beta \frac{n_\beta^2 - n_\alpha^2}{2n_\alpha n_\beta} \tag{16}$$

and

$$\eta_{\alpha\beta,\pm 1} = \left(\frac{1}{2} \pm 1\right)\frac{K^2}{k_\alpha} - k_\beta \frac{n_\beta^2 - n_\alpha^2}{2n_\alpha n_\beta} \tag{17}$$

α, β = 1, 2 characterize states of polarization.

One can see that the light intensity distribution in diffraction pattern is asymmetric. The first terms determine direction of the beams depending on the acoustic wave vector $\vec{K}$ and the second terms which do not depend on K are describing the birefringence and a phase shift. For the layer of the thickness l one must consider the product $\eta_{\alpha\beta p} \cdot l$ (see (11)).

That picture is different, however, when the ultrasonic beam has great intensity i.e. when its radiation pressure causes a streaming of the liquid. After achieving a threshold of intensity the laminar streaming becomes a turbulent one and characteristic vorteces occur. According to Miyano and Shen [24] in the case of the turbulent reorientation rotation of liquid crystal molecules caused by the acoustic field formed vorteces which represent an additional spatially periodic deformation creating an additional

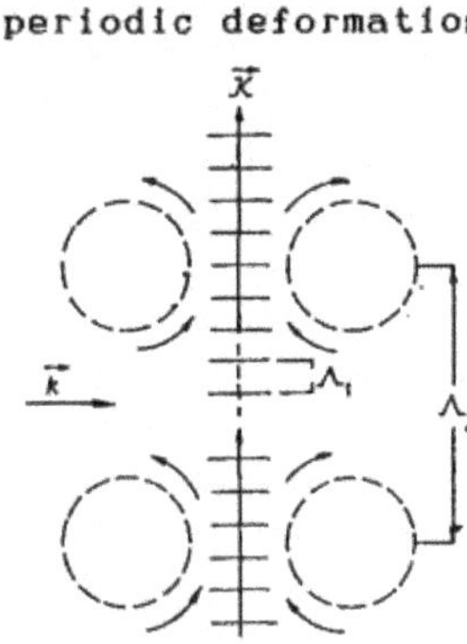

Fig. 8. Vortices structure with distance Λ_2 between them and ultrasonics of the wavelength Λ_1; $\vec{k}$ wave vector of light.

diffractional grating with the constant Λ_2 equal to the distance between the spatial structures (vorteces). The situation depends on the visco-elastic properties of the liquid crystal molecules, their orientation and the parameters of the acoustic field (f_{ak}). Schematically such a model is presented in Fig.8 [23]. The overlapping of these two diffraction gratings should manifest itself by the appearance of additional orders in the diffraction pattern (fine structure). As an example, Fig.9 presents the expected diffraction conditions for an isotropic medium and Fig.10 - analogous diffraction conditions for a birefringent refraction medium (as liquid crystal).

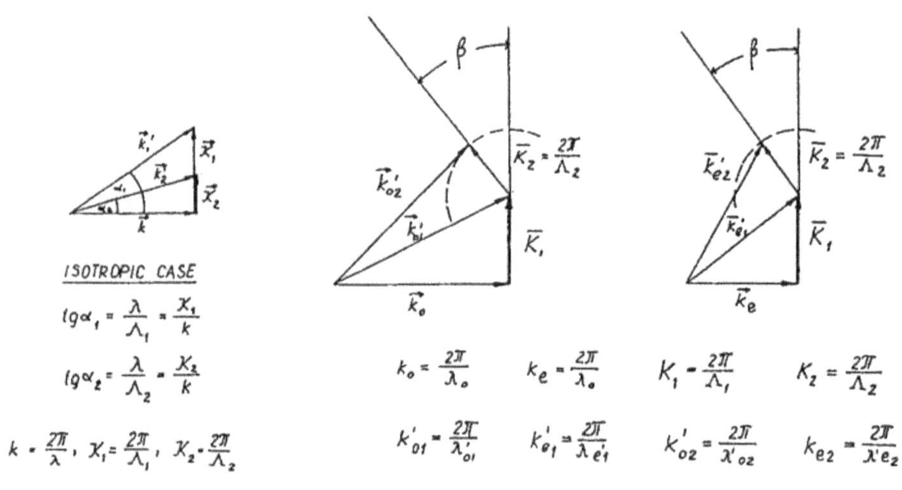

Fig. 9. Fig. 10.
Bragg diagram and diffraction angles in double periodic structure in isotropic (Fig.9) and in anisotropic case(Fig.10).

The summation principle of wave vectors corresponds the equation (2) and is following

$$\vec{k} \pm p\vec{K} = p\vec{k}' \tag{18}$$

where: $\vec{k}$ - wave vector of the interacting light before diffraction (for an ordinary (k_o) or extraordinary (k_e) beam.
$\vec{K}$ - wave vector of acoustic wave
$\vec{k}'$ - wave vector of diffracted light of pth order
p - spectrum order

The expression (18) relates the light diffraction angles and the diffraction gratings constants relevant for the ordinary and extraordinary rays what allows to find those constants formed by spatial structures i.e. ultrasonic periods and distance of vorteces

3. ULD patterns in nematic liquid crystals.

3.1. Experimental setup and procedure.

The measurement system is presented in Fig. 11. The system can be used both for an investigation of molecule arrangement inside the sample by classical orthoscopic optical methods [8,23,26] and investigations of diffraction images formed by the spatial st-ructures of the investigated sample. All experimental

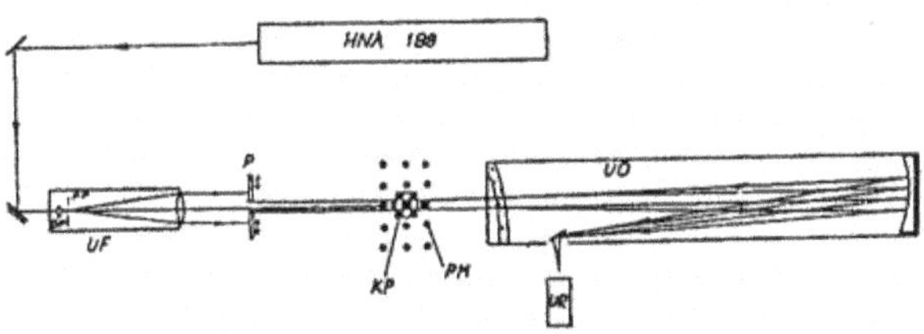

Fig. 11. Experimental setup for acousto-optical investigation of liquid crystal samples; HNA - argon laser, UF - optical beam former, P - polarizer, KP - liquid crystal cell coupled with the ultrasonics transducer and situated in magnetic field PM; UO - mirror refractometer, UR - photooptical detector or photo-camera.

details are described in [25].

In the system the natural anisotropy of the medium can be modified using the external ordering magnetic field, and other external factors changing density, stress, streaming of molecule etc. by exposing the medium to an ultrasonic wave (see Fig. 7b). The measure of the modification may be variation of the angles α and β (Fig. 12) of the molecule orientation axis relating to the initial directions caused by magnetic field only. The angle α corresponds to the molecule rotation respect in the plane yz and β to the rotation to the x axis (direction of ultrasonic beam).

The diffraction images were registered separately for the

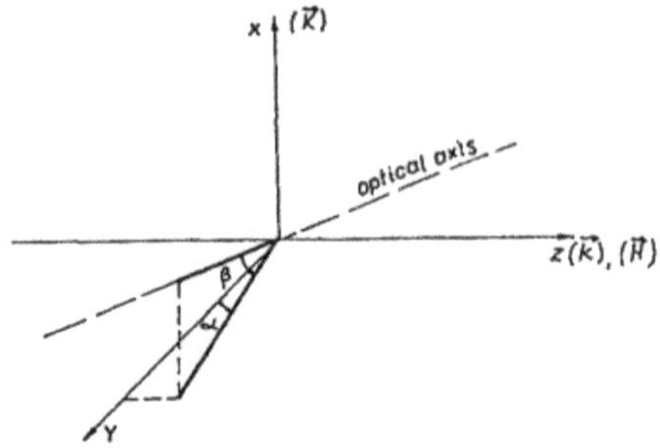

Fig. 12. Positions of a liquid crystal molecule before (along optical axis) and after (under angles α and β) variation of orientation caused by ultrasonics.

polarized light in the polarization plane of the "ordinary" and "extraordinary" ray. The observation of the diffraction images was carried out for liquid-crystalline samples: N-4 hexyloxybenzyliden-p-toluidin, merck 4, merck 5 and W5 mixture for the ultrasonic wave with ultrasonic intensity from 0.0 to 1.0×10^4 W/m^2.

In order to find correlation between the observed diffraction effects and changes of the molecule arrangement of a nematic (which corresponds to the changes of the direction of the sample optical axis), orthoscopic investigations of the sample were made in parallel to diffraction investigations (see in [25]).

3.2. Some examples of experimental results.

The results presented here are only examples of many pictures observed and photographed during examinations.

Fig. 13 presents examplary diffraction patterns obtained for nematic liquid crystal N-4 hexyloxybenzyliden-p-toluidin for the lineary polarized light paralleły to the ordinary beam at ultrasonic frequency f = 6.70 MHz (upper row) and 9.70 MHZ (lower row-). The pictures in the first column (a,a') represent the initial state without ultrasonics. The pictures in the second column (b,b') correspond to the state below the treshold; mainly laminar streaming, molecules reoriented of an angle φ occur; beside that some influence of instabilities is visible when diffraction of light takes place on a single grating (only by ultrasonic wave of

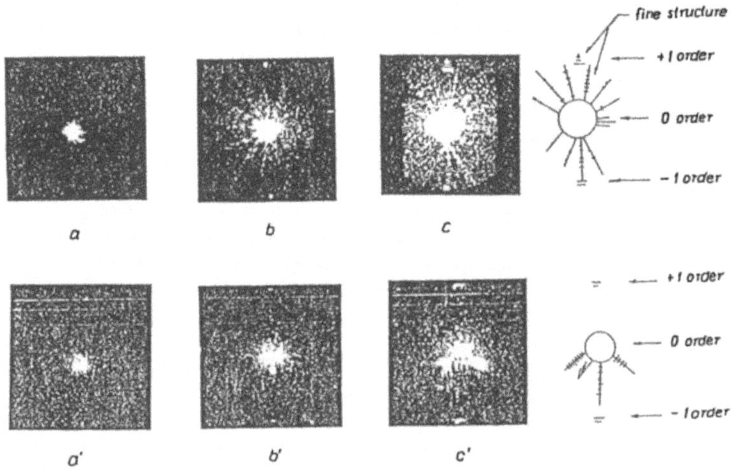

Fig. 13. Examples of A.O. diffraction patterns in nematic crystals observed in ordinary beams (after [25]).

the wave length $\Lambda_1 = \frac{c}{f}$). The pictures in the third column (c, c') correspond to the state above the intensity treshold (turbulent streaming) when the second diffraction grating (of the constant equal Λ_2) formed by vorteces is overlapped with the ultrasonic grating. In the Fig. 13 (as well as in the Fig. 14) in the last column the schematic positions of diffraction orders are drawn, too, for auxiliary interpretation (see below § 3.3). At the same time there is a clear asymmmetry (Fig. 13c) between the fine structure in "+" and "-" orders. The fine structure does not manifest itself in the ±1 orders during the observations in the polarized light in the polarization plane of the extraordinary beam (Fig. 14) which represents patterns for the same frequencies and states of the medium (different intensities) as in Fig. 13 obtained for diffraction of the ordinary light beam. The diffraction patterns in the Fig. 14c' formed for polarized light in the polarization plane of the extraordinary beam resemble a diffraction on two mutually perpendicular structures and turned by a 45° angle to the direction of the ultrasonic wave.

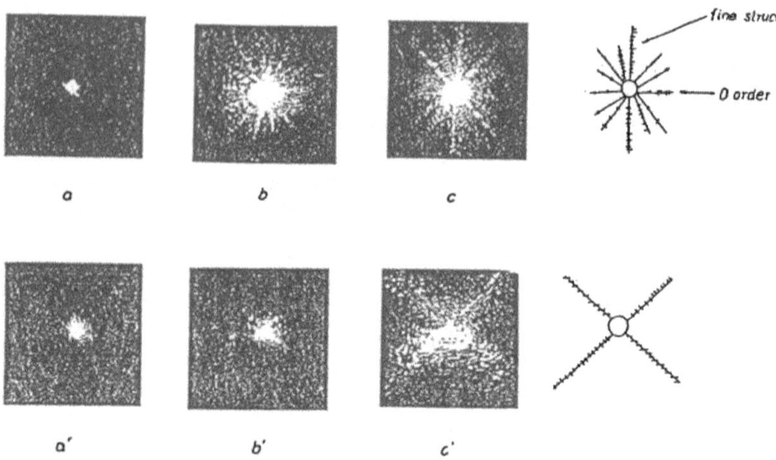

Fig. 14. Examples of A.O. diffraction patterns in nematic liquid crystal observed in extraordinary beams (after [25]).

3.3. Auxiliary method for diffraction pattern analysis.

The auxiliary drawings in the last columns in Fig. 13 and 14 present a simplification of the light intensity spots distribution in diffraction patterns for better interpretation of the correspondance between the pattern and periodic structures and irregularities of the original object.

It is known [27] from the Fraunhofer principle of diffraction (Fig. 15) that to every point of periodic object corresponds

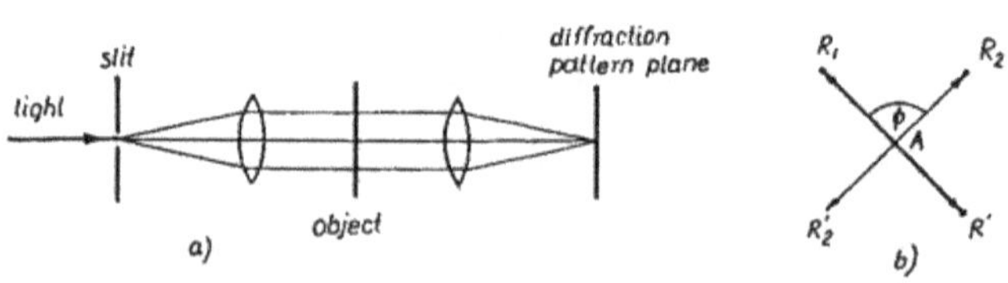

Fig. 15. Scheme of light diffraction by the object and formation of the diffraction pattern (a) and schematic presentation of points in diffraction image (b) correlated with space frequency components.

a determined component of the space frequency correlated with a determined point R_1 in the diffraction image plane (Fig. 15b). The component turned over an angle Φ gives the point R_2. Symetrically in respect to the zero component (A) the points R_1' and R_2' appear. The distances R_1A and AR' and R_2A and AR_2', respectively, are in opposite relation to the space frequencies. The stucture of the image is perpendicular to the structure of the object. The space frequency is determined by the equation

$$d = \frac{1}{kxV} \tag{19}$$

where d is the distance of the periodic structure (e.g. wavelength Λ for the UW), k is the instrumental constant, x - distance of a diffraction spot of the 1-st order in the diffraction pattern, V- magnification on the way between the object and the diffraction image plane.

The light intensity distribution in diffraction pattern allows to recognize periodic structures of the object and characteristic directions of the structures distribution. Many practical cases of comparison between structures and diffraction patterns are presented in [27] to recognize distributions of hols, rows etc. on the surface of the metal sample and one could easiely learn pictures adequate to different forms. The Fig. 16 gives an example of the polished suface of pearlite in a steel sample, its optical Fourier spectrum and the auxiliary drawing of spots distribution in the diffraction pattern.

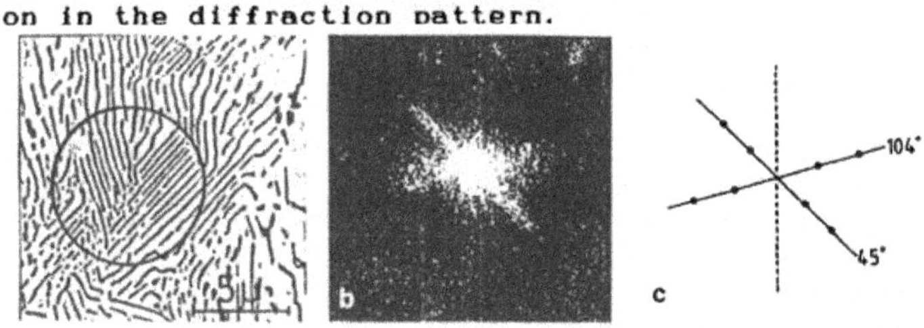

Fig. 16. Polished surface of pearlite in a steel sample having periodic structure (a) and the adequate diffraction pattern (b) and the auxiliary drawing of spots distrubeted along distinct directions under 45° and 104° degree (after 26).

In the case of the ULD phenomena which we have above considered the light intensity in diffraction pattern carries all information about the ultrasonic field and its deformation due to the regular anizotropy as well as to irregularities appearing in the medium. The diffraction patterns in liquid crystals examined by us are good examples of that situation. As we saw the primitive axiliary drawings in the Fig. 13 and 14 can be the first step to study the situation. The further step will be described below.

3.4. Investigations with the orthoscopic system.

The threshold values of the ultrasonic wave intensities at which one can observe a formation of the fine structure in the diffraction patterns of the investigated nematic crystal were determined from the ULD patterns [25] and proved with the imdependent method using the orthoscopic system [23,26] which enabled observing interference images resulting from the ordinary and ex-

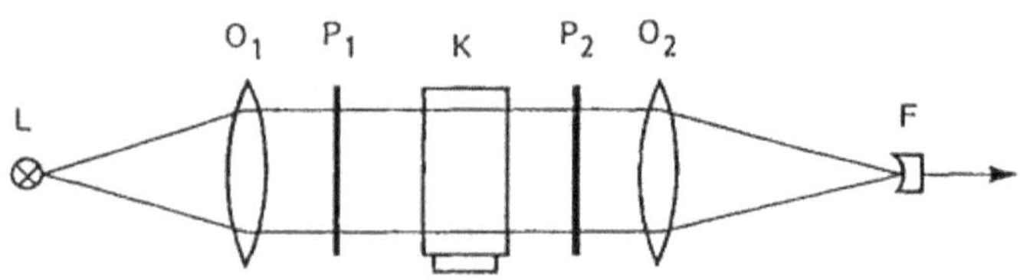

Fig. 17. Setup for an orthoscopic method: L-source of light, O_1, O_2 - objectives, P_1, P_2 - polarizers, K - ultrasonic cell with liquid crystal, F - photodetector or a camera.

traordinary beams interaction. The Fig. 17 presents such an arrangement which corresponds to the scheme of the Fig. 7. The examples of such interference fringes are shown in Fig. 18. The pictures were taken for different ultrasonic intensities at the frequency of 6.7 MHz. The influence of the turbulent streaming of the molecules is gradually (against intensity) visible (b and c).

Fig. 18. Interference patterns obtained in the orthoscopic system in nematic liquid crystal induced by ultrasonics of three values of intensity: a) $0.18 \cdot 10^4$ W/m^2, b) $0.51 \cdot 10^4$ W/m^2, c) $0.74 \cdot 10^4$ W/m^2. (after [25]).

To determine the threshold values a special procedure of interference fringes was applied. The angles α and β presented in the Fig. 12 can be related to the inclination of the interference fringes and the variation of the distance between them, respectively. The angles can be determined for different places of the interference image which correspond to local positions of the light beam cross - section in the sample. The coordinations of those positions taken for calculations are presented in the Fig. 19.

Measuring the inclinations of the interference fringes corresponding to the coordination net, the local changes of the direction of the optical axis for the average thickness of the sam-

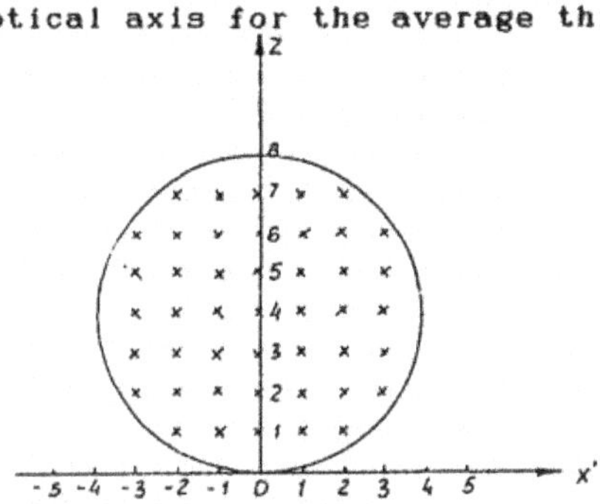

Fig. 19. Coordinate system for the light beam cross-section for local points determination.

ple (variation in the angle $\Delta\beta$) were found in the function of ultrasonic wave intensity for chosen places of the sample.

The Fig. 20 presents an example of the dependence for positions determined by x = -2 = const, and for three values of z (2, 4, 6) in the Fig. 19.

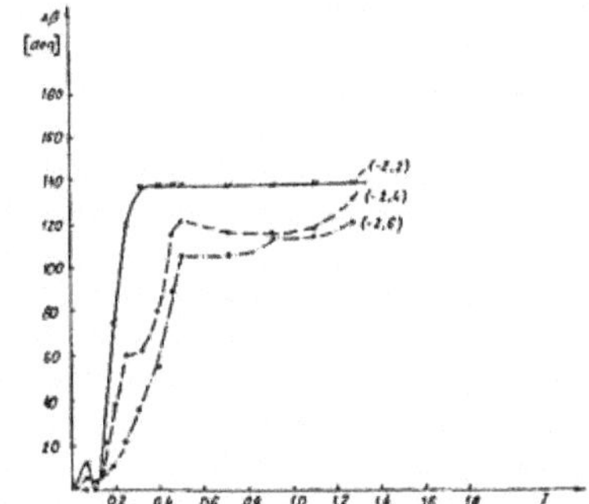

Fig. 20. Avarage inclination $\Delta\beta$ of the optical liquid crystal axes in a given local position in the sample. (after[25])

In the case presented in the Fig.21 an interacting inversion of the optical axis at a distance from the ultrasonic beam axis is evidently manifested. It must be related to the change of the direction of the molecule streaming on the opposite one at the given place of the cross - section of the sample. Such local change of the direction may confirm appearing of a vortex in the

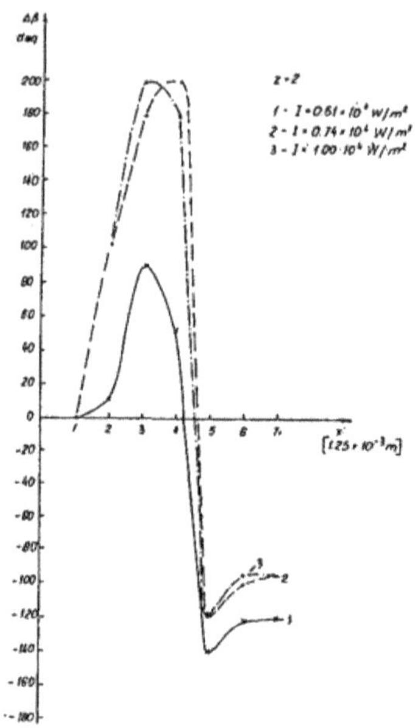

Fig.21. Example of optical axis inversion induced by ultrasonics in a given local position in the liquid crystal sample. (after [25]).

area and the inversion corresponds the passage through a izocline layer between two local streamings of opposite directions.

3.5. Automatic scaning and a.o. signal processing.

The next step of improvement in the investigation described here is an automatic scaning with a photodetector prob throughout the diffraction image plane (see Fig.11) or recording of the last one with a photodetector's matrix array. Then, the fotoelectric signal carrying all required information may be processed in any useful way. The exact reconstruction process of the ultrasonic field from the light intensities distribution in the ULD pattern is possible in a way described in the literature. The a.o. tomography method applied by R.Reibold [28,29] to study ultrasonic fields may be used to study anisotropic media, too.

A.O. methods are very useful to study not only stationary processes but also ultrasonic pulses and their deformation during propagation due to the nonlinearity of the medium. Nonlinear problems are important in anisotropic media, too and they are studied succesfully with a.o. methods [16,30].

4. Conclusions

The ULD phenomena in anisotropic media provide many posibilities of applications to examine their properties. For such applications a theoretical description of these phenomena is necessary. Fortunatelly, there exist theories which determine exact relations between the light diffrected by ultrasonics and anisotropic properties of the medium. For A.O examination the proper relations between ultrasonic frequency and light polarization states as well as geometrical conditions must be satisfied to get a defined type of ULD i.e. Raman-Nath or Bragg diffraction. In general case the situation may be complicated. It is reasonably simple for uniaxial systems. The nematic liquid crystals present such a systems in which the anisotropy may be changed by external agents like magnetic and ultrasonic fields. In the paper some results of our examinations of such media were used to examplify possible application of the A.O. methods to study anisotropic media. The diffraction patterns obtained in examination can be analized by different ways and correlated with the structure. Some ways have been described. We have not discuss wider the present day possibilities for automatic probing and processing of information in the case of A.O. methods, however, everybody may realize that such techniqes are similar in all measuring problems and can be succcessfully adapted for A.O. needs, too.

5. References

1. M.V.Berry, The diffraction of light by ultrasound, Ac. Press, New York, 1966

2. R.W.Damon, W.T.Maloney, D.H. Mc Mahon, in Physical Acoustics, Ed. W.P.Mason, Ac. Press, New York-London, 1970, vol.7, 273.

3. A.Korpel, Acousto-optics, in Applied Solid State Science, vol.3 ed. by R.Wolfe, Acad., New York, 1972, p.71-180.

4. J.Sapriel, L'acousto-optique, Masson, Paris, 1976.

5. V.I.Balakshi, V.N.Parygin, L.E.Chirkov, Fizicheskiye Osnovi Akustooptiki, Radyo i Zvyaz, 1985, Moskva, in Russian.

6. A.Korpel, Acousto-optics - a review of fundamentals, Proc. IEEE, 69, 1981, 48.

7. A.Korpel, Acoustooptics, Marcel Dekker Inc., New York, 1988.

8. A.Śliwiński, Acoustooptics and its perspectives in research and applications, Ultrasonics, 28 (1990) 195-213.

9. R.Mertens, Fifty years of acousto-optics, Proc. 4-th ICA Paris,

10. Proc. 1-st School on Acousto-optics and Applications Gdańsk-Wieżyca, 1980, Ed.A.Śliwiński, University of Gdańsk, 8, 1983, 101.

11. Proc. 2-nd School on Acousto-optics and Applications Gdańsk-Wieżyca, 1983, Ed.A.Śliwiński, University of Gdańsk.

12. Proc. 3-rd School on Acousto-optics and Applications Gdańsk-Wieżyca, 1980, Ed.A.Śliwiński, University of Gdańsk.

13. Proc. 4-th School on Acousto-optics and Applications Gdańsk-Sobieszewo, 1989, Eds. A.Śliwiński, P.Kwiek, A.Markiewicz, World Scientific, Singapore, New Yersey, London, 1990 Hong Kong.

14. Proc. of Symposium on Acoustooptics, Leningrad, 1990.

15. A.Śliwiński, Perspectives of acoustooptics and photoacoustics - in [12], p.9-14.

16. W.G.Mayer, Anisotropic material orientation with ultrasonics, in Ultrasonics methods in evaluation of inhomogen materials, Ed. by A.Alippi and W.G.Mayer, NATO ser. M.Nijhoff Publ. 1987, Dordrecht, Boston, Lancaster, p.3-11.

17. R.W.Dixon, Acoustic diffraction of light in anisotropic media, IEEE, 1967, QE3, 85-93.

18. L.L.Hope, Brillouin scattering in birefringent media, Phys. Rev. 166, 1968, 883-892.

19. V.B.Voloshinov, V.N.Parygin, Raman-Nathovskaya difraktsiya sveta na ultrazvuke v opticheski anisotropnykh sredakh, Piśma (Letts) Zdr. Tkhn. Fiz. 1981, 7, 145-148 (in Russian).

20. V.B.Voloshinov, V.N.Parygin, L.E.Chirkov, Nekotoriye osobiennosti anisotropnoi diffraktsyi Bragga, Vestn. Moskovskovo Univ., ser.3, Fiz. Astr., 17, 1976, 305-312 (in Russian).

21. L.Bergmann, Der Ultraschall, Springer Vlg, Zürich, 1954.

22. V.N.Parygin, Special regimes of Bragg anisotropic interaction, in [13] p.193-206.

23. A.Śliwiński, Acoustooptics in anisotropic media, in [12] p.224-262.

24. K.Myano, V.Shen, Phys. Rev. A, 15 (1977) 2471.

25. M.Kosmol, S.Kowalewski, A.Śliwiński, I.Wojciechowska, Investigation of light diffraction on an ultrasonic wave in liquid nematic crystals, Arch. Acoust. 16 (1991) (in press).

26. M.Witkowska-Borysewicz, A.Śliwiński, Optically detected variations of nematic liquid crystal orientation induced by ultrasound, J. Phys. 44 (1983) 411-420.

27. W.Eikhorst, S.Boseck, Eine Einfürung in the Methoden der quantitativen Erfassung metalographisch entwickelter Werkstoffgefuge durch bildanalytische Kenngrossen auf der Basis der licht optischen Diffraction (LOD), Mikroskopie, 42 (1985) 11-25.

28. R.Reibold, Light diffraction tomography, a powerful tool for ultrasonic field investigations, in [11], p.113-124

29. R.Reibold, Field reconstruction using tomographic and Fourier -optical techniques, Proc. of the 3-rd Intern. School on Phys. Acoustics, Erice, Italy, 1988.

30. W.Mayer, T.H.Neighbors, Optical probing of finite amplitude ultrasonic pulses; Proc. 10 Intern. Symp. on Nonlinear Acoustics, KOBE, 1984.

II. CONTRIBUTED PAPERS

A. Acoustic field evaluation

HETERODYNE LASER PROBES : PRINCIPLE AND APPLICATIONS TO ACOUSTIC WAVE MAPPING AND THERMOELASTIC NONDESTRUCTIVE EVALUATION

Bernard CRETIN*
Laboratoire de Physique et Métrologie des Oscillateurs du C.N.R.S.
associé à l'Université de Franche-Comté-Besançon
32 avenue de l'Observatoire - 25000 Besançon - France

ABSTRACT

Heterodyne laser probes are very sensitive instruments which do not contact the sample. They provide wide bandwidth and high resolution sensing. This paper discusses the main features of interferometers before examining applications. At low frequencies coherent electronics have been designed to improve the signal/noise ratio. An analysis of quartz resonator vibrations is presented (electromechanical transfer function and displacement mapping).

Coupling optical excitation and detection leads to thermoelastic nondestructive evaluation using an intensity-modulated power laser beam. Generated thermoelastic displacements are detected by the laser probe at the excitation point. Thermoelastic properties of solid surfaces such as thermal diffusivity and elastic constants are largely influenced by cristalline orientation or macroscopic defects. Thermoelastic microscopy is particularly suited for imaging metallic sample grains, subsurface defects, semiconductor dopants or composite structures. Results on samples with defects are presented in the form of amplitude and phase reconstructions of the thermoelastic image.

Heating conductor or a semiconductor by a periodic electrical current induces thermoelastic displacements. These displacements can be detected by the laser probe. In that case thermoelastic imaging can detect defects such as cracks or resistive weldings.

Applications to electronic components are shown and analysed.

1. Introduction

Optical detection techniques are usually used for nondestructive testing,[1-4] bulk or surface acoustic waves studies[5,6] and transducer characterization.[7,8] Several laser probes have been developed for these applications. They use various principles that can be classified as either noninterferometric techniques or interferometric techniques. The latter have

* Assistant Professor at ENSMM, Besançon

been the object of active studies since the development of all-optical inspection systems using laser generation of thermal wave and ultrasound. Thermally-generated displacements are usually detected with interferometric laser probes. Important industrial applications are viewed in on-line measurement of materials parameters (grain size in metals, film thickness or delamination, dopant concentration, ...). These optical inspection techniques need wide bandwidth and high sensitivity interferometers. Furthermore the probes can operate in industrial conditions (low frequency resonant vibrations, moving objects or surface roughness). All these requirements need specific solutions.[9]

In the case of polished objects, interferometric heterodyne laser probes are very useful. Their sensitivity is about 10^{-4} Å/$\sqrt{Hz}$ and the bandwidth can be extended to about 1 GHz when focussing the probe beam, but they are usually applied to measurements only at a selected frequency to obtain a better signal to noise ratio (measuring a displacement amplitude of 1 Å in a 100 MHz bandwidth requires a sensitivity better than 10^{-4} Å/$\sqrt{Hz}$). Heterodyne laser probes have been successfully applied to acoustic vibration measurement and mapping. They have been also used as displacement sensors in thermoelastic microscopy.

2. Laser probes description

2.1. Laser probe choice

The laser probe choice depends on the following principal parameters :

- object surface roughness
- displacement frequency
- displacement amplitude.

If the surface roughness is larger than the optical wavelength λ_0, differential interferometry or time-delay interferometry are preferable (time-delay interferometers give the velocity of the sample surface). Such sensors are not described in this paper.

At high frequencies (f ≥ 10 MHz), two simple schemes can be used to detect ultrasonic vibrations on polished surfaces :

★ surface grating techniques

This technique[10] has been applied to the detection of surface acoustic waves (SAW). The laser beam is diffracted by the grating that is generated by SAW. Intensity of the first-order diffracted beam is related to the SAW displacement amplitude (the measurement is possible only if the probe beam wavelength is larger than the acoustic wavelength).

★ knife-edge technique

This principle uses the deflection of the probe beam due to the ripple of the sample surface. The beam deviation is usually measured with a photodetector set behind a knife edge. The surface gradient induces a beam deflection and a voltage signal on the photodetector. This technique has been used to detect and measure SAW.[11, 12] It requires a beam diameter smaller than the acoustic wavelength and is sensitive only to the gradient of the vibrating surface.

2.2. Interferometric probes

a) homodyne interferometers

The most well-known homodyne interferometer is the Michelson interferometer. In such an interferometer without phase-locked control, the light intensity received by the photodetector can be written as follows :

$$I_D = I_R + I_S + 2\sqrt{I_R I_S}\cos\left[2k_o\,\delta(t) - \phi(t)\right] \qquad (1)$$

where I_R is the reference arm power incident on the photodetector, I_S the signal arm power (after reflection on the sample), k_0 the optical k-vector, $\delta(t)$ the sample surface displacement and $\Phi(t)$ the global phase depending on the path difference.

If the path difference is adjusted in such a way that

$$\phi = \pm\frac{\pi}{2} + 2n\pi\ (n\ \text{integer}) \qquad (2)$$

the photodetected signal is proportional to the out-of-plane surface displacement. That condition is obtained by mounting the reference mirror on a piezoelectric translator. Electronic control circuitry is used to stabilized the interferometer (Fig. 1).

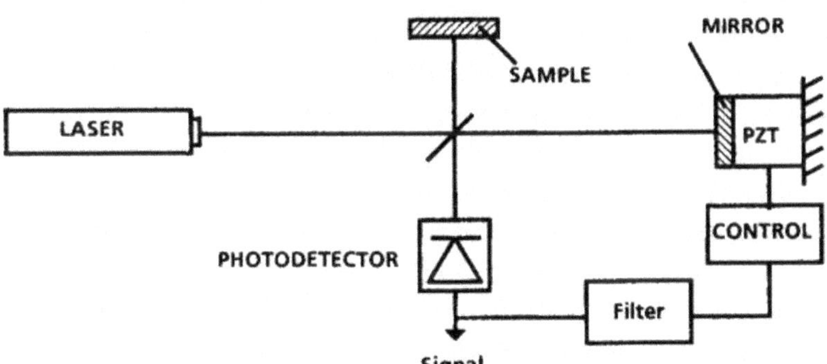

Fig. 1 : Diagram of a stabilized Michelson interferometer

Other interferometers, based on two-wave interference (Mach-Zehnder, ...), are similar.

b) heterodyne interferometers

If a frequency shifter is inserted in one interferometer arm, the probe is called a heterodyne interferometer. The optical frequency is generally shifted by a Bragg cell (acousto-optic cell in Bragg incidence).

Using a Bragg cell as a beam splitter leads to a $2f_B$ frequency shift (f_B is the Bragg cell excitation frequency). Such a setup has been developed for SAW and BAW studies.[13]

A schematic diagram of this probe is given by Fig. 2. The laser beam (frequency f_L) is split by the Bragg-cell driven at f_B. The reference beam (frequency f_L+f_B) is reflected on the mirror and goes toward the photodetector. The signal beam is phase modulated by the sample surface displacements (frequency fa) and then down shifted to (f_L-f_B) in the acousto-optic cell. The mixing of the two beams on the photodetector produces a current whose carrier frequency is $2f_B$. The information about the acoustic waves is available in sidebands at $2f_B \pm nf_a$ (n restricted to the value 1 if the displacement amplitude is less than 200 Å).

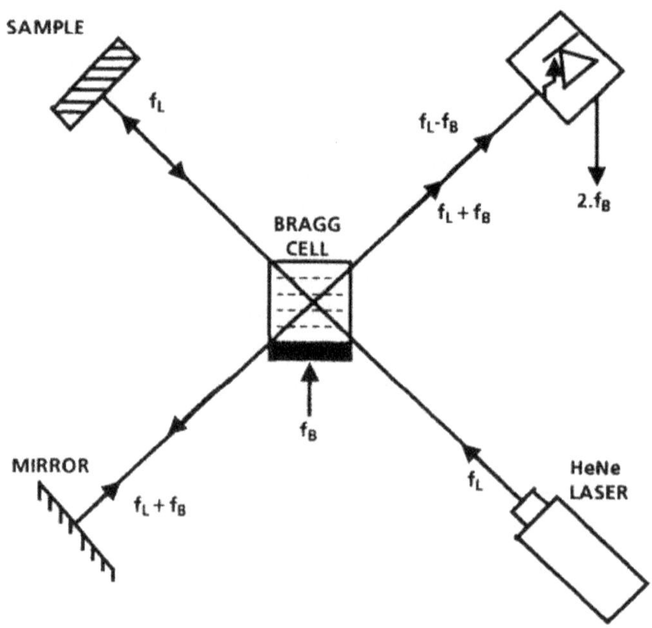

Fig. 2 : Diagram of a symmetrical heterodyne laser probe

This probe can detect a 10^{-3} Å amplitude displacement in a 1 Hz final bandwidth if the laser is isolated (with a Nicol prism or a polarizer). There are disadvantages : low frequencies instabilities, small splitting angle (10 to 20 mrad) and a long optical path (~ 1 m).

A more compact probe has been developed to shorten the optical path and to improve the stability.[14]

Fig. 3 shows its asymmetrical configuration. The HeNe laser beam is polarized horizontally. One half of the power is split by cube 1. The reference beam is directed by the Dove prism toward the photodetector. The signal beam is up-shifted ($f_L + f_B$) by the colinear input-output Bragg cell. After modulation on the sample surface, the vertically-polarized beam is reflected by cube 2. When operating, the photocurrent spectrum contains a carrier at frequency f_B and sidebands depending on the sample surface displacements.

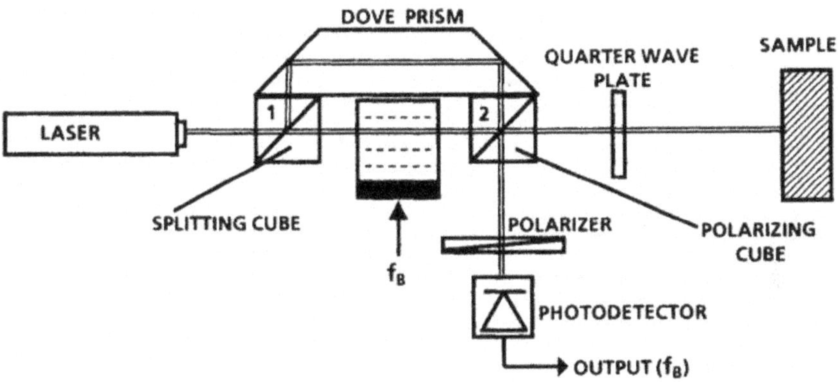

Fig. 3 : Diagram of the compact asymmetrical probe

We tested these two probes to compare their characteristics. If the probes are shot noise-limited (neglecting the thermal noise), the ultimate detected displacement d_N can be expressed, in a 1 Hz bandwidth,[15] by

$$d_N = \frac{hc}{\sqrt{2\,\eta\, e\, \pi\, P_o R_L}} \left(<V_N^2>\right)^{1/2} \tag{3}$$

where e is the electron charge, η the photodiode quantum efficiency, h is Planck's constant, c the light velocity, P_o the average optical power on the photodetector, and R_L the photodetector load, with

$$<V_N^2> = 2\,R_L^2\, F e^2\, \eta\, P_o\, \lambda_L \,/\, h c \tag{4}$$

F being the preamplifier noise figure in the power ratio.

These expressions show that $d_N \propto 1/\sqrt{P_o}$. Thus, the sensitivity is better with a high power laser. Practically, differences appear between theoretical and practical values if the source power is increased. They are attributed to the interference of the longitudinal modes in the laser cavity and spurious beams due to imperfections in optical components. Furthermore, the photodetector's acceptable power is usually limited to about 100 mW.

A comparison of theoretical and experimental values is given in Table 1. It shows that, without laser stabilization HeNe laser having a power output of 5 to 10 mW is the better choice.

Table 1 : Theoretical and experimental values of sensitivity in a 1 Hz bandwidth for heterodyne interferometers

	Theoretical* d_N (m)	Experimental d_N (m) at 50 kHz	Laser power P_L
Symmetrical probe (compact)	$0{,}8\ 10^{-14}$	$1{,}5\ 10^{-14}$	5 mW
Symmetrical probe	$1{,}7\ 10^{-15}$	$1{,}9\ 10^{-13}$	35 mW
Asymmetrical probe	$1{,}9\ 10^{-14}$	$2{,}5\ 10^{-14}$	2 mW

* effective useful laser power assumed to be $P_L/4$

In all three cases, the sources were HeNe lasers. The probe output noise was measured with a FFT analyzer after coherent electronic processing. A mirror was positioned for reflecting the signal beam.

For easy scanning and alignment, fiber optics are generally used in the signal beam arm. But all-fiber laser probes are more compact. The principle of fiber probe insensitive to external disturbances is given in Fig. 4.[16]

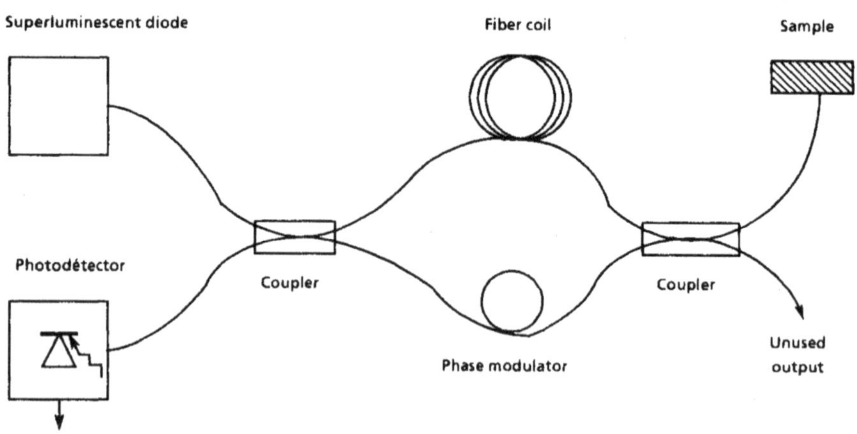

Fig. 4 : All-fiber sensor diagram

The fiber coil is used as delay loop. For small acoustic displacements, the sensor response is proportional to sin $\omega_a\tau/2$ (where ω_a is the acoustic angular frequency and τ the fiber loop delay). Thus, the fiber coil must be changed depending on the application (maximum sensitivity for $f = 0{,}5 / \tau$).

c) Heterodyne laser probe calibration

Without coherent electronic, the probe output signal is measured with a spectrum analyzer. For small displacements of frequency F_a, the resulting spectrum is composed of the carrier (the frequency F_0 is equal to f_B in the case of an asymmetrical laser probe) and two sidebands. The carrier amplitude/sideband amplitude ratio leads directly to the displacement δ

$$\frac{A_o}{A_1} = \frac{1}{k_o \delta} \simeq \frac{1\,000}{\delta} \tag{5}$$

if $\lambda_L = 0{,}633$ µm and δ is expressed in Å.

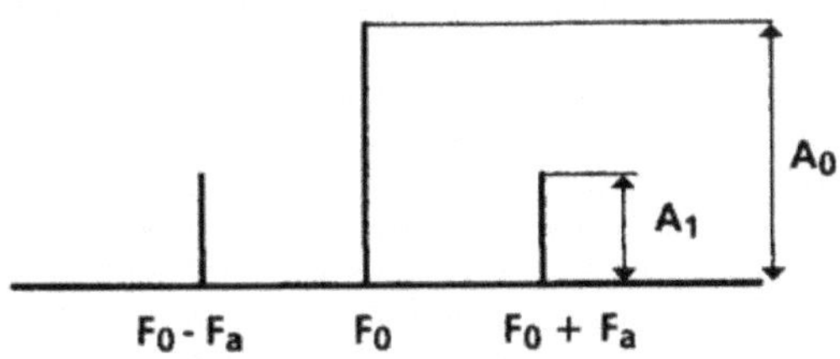

Fig. 5 : Output probe spectrum in the case of small displacements

Calibration can be controlled automatically by coherent electronics. A block diagram of a specific arrangement is shown in Fig. 6. Carrier amplitude can be considered constant by using an automatic gain control circuit. Thus sideband amplitude is proportional to displacement δ. Phase demodulation of the signal (with a phase locked loop including a VCXO, a mixer and a loop filter[17]) leads to the displacement value

$$V_s = 2\, s\, k_o\, \delta \tag{6}$$

where s is the mixer sensitivity.

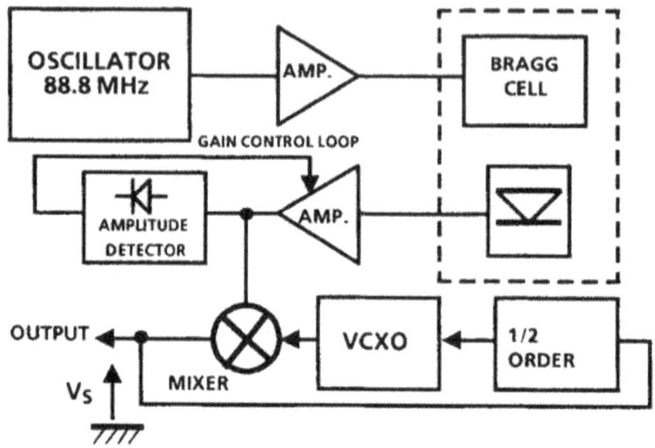

Fig. 6 : Coherent electronics used with heterodyne laser probe

Using AGC and PLL enables displacement mapping and transducer calibration (low-frequency vibration effects and reflectivity or roughness disturbances are partially suppressed).

3. Applications to vibration measurements and mapping

Most applications of laser probes depend on the transient acoustic response of excited samples. The spectral measurement provides the vibration modes. For resonators, Q-factor and displacement mapping are important information for design and fabrication.

Heterodyne interferometric laser probes are suitable for broadband detection. Thus, detection of vibration harmonics is possible at high frequencies ($f > 100$ MHz).[18, 19]

Using the coherent electronic described above yields high-sensitivity, low frequency vibration mapping. The probe output signal is processed with a lock-in amplifier for maximum frequency vibrations of about a few kilohertz.

We present here a low frequency resonator analysis. The resonator is a watch crystal. It is a 32768 Hz tuning fork obtained from leaf-type technology. A view of the resonator is given in Fig. 7. This miniature resonator is about 4 mm long and 0.125 mm thick. Focussing the laser probe provides a measured area of 15 μm diameter. Modifying the resonator orientation gives in-plane and out-of-plane components of the tuning fork arms (the laser probe is sensitive only to normal displacement).

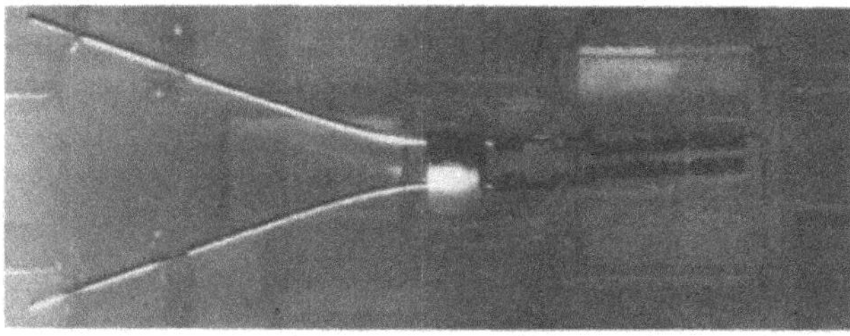

Fig. 7 : View of the quartz crystal resonator studied

Observed vibrations of tuning forks are in-plane arm flexure modes (frequency of the principal vibration is proportional to b/ℓ^2, where b is the arm width and ℓ its length). Measurements have shown that arm vibration is complicated : the out-of-plane motion component is not strictly zero. This observation is certainly related to torsional modes.

The electromechanical transfer function of the quartz crystal resonator is given in Fig. 8. The quartz crystal vibrates in air. It is continuously excited with a 0.4 V amplitude signal.

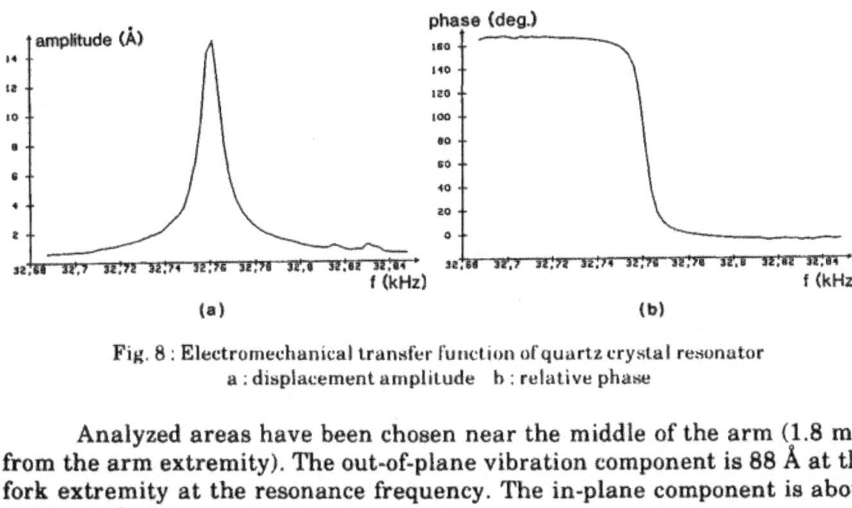

Fig. 8 : Electromechanical transfer function of quartz crystal resonator
a : displacement amplitude b : relative phase

Analyzed areas have been chosen near the middle of the arm (1.8 mm from the arm extremity). The out-of-plane vibration component is 88 Å at the fork extremity at the resonance frequency. The in-plane component is about 100 times larger.

Vibration mapping is shown in Fig. 9. The amplitude image is composed of 64 × 64 pixels. Brighter areas correspond to larger displacements (for example at the fork extremity). Important variations of displacement amplitude appear in the middle of the fork. That can be explained by a torsion movement. The phase image is noisy around the fork areas. This image clearly shows the 180° phase difference between the two resonator arms (the phase image is coded with eight bits ; 0 corresponds to -180° and 255 to a +180° phase shift).

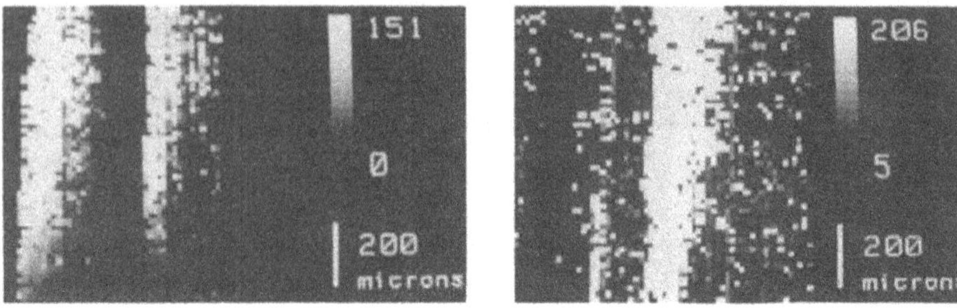

a b

Fig. 9 : Mapping of out-of-plane displacements (quartz crystal tuning fork ; f = 32755 Hz)
a : amplitude image (maximum displacement : 88 Å) b : relative phase image

4. Laser probe applications to thermoelastic microscopy

4.1. Microscope principle

a) Physical background

Heat propagation is described by a parabolic differential equation. However, in the case of periodic excitation and stationary regime, the temperature modulation propagates as a heavily damped wave having a diffusion length

$$\mu = \sqrt{\frac{2K}{\rho C \omega}} \tag{7}$$

where K is the thermal conductivity, ρ the density, C the specific heat and ω the angular modulation frequency.

The thermal waves interact with the materials. The wave front is twisted at sample boundaries or discontinuities. The medium is described by its thermal impedance

$$Z = (1 - j) / \sqrt{2 \omega \rho C K} \tag{8}$$

For a thermal wave propagating from a medium with impedance Z_1 into a medium with impedance Z_2, the reflection coefficient Γ is given by

$$\Gamma = \frac{Z_2 - Z_1}{Z_2 + Z_1} \tag{9}$$

The thermal wave reflections are detected from the resulting amplitude and phase changes that they generate.

In scanning thermoelastic microscopy, the detected displacement is proportional to the thermal expansion and depends on the elastic constants. So the technique is sensitive to the thermal and mechanical sample properties.

b) Setup

The microscope block-diagram is described in Fig. 10. The excitation laser beam (Ar) is intensity-modulated with an acousto-optic modulator (A.O.M.). The modulator is driven by a low frequency (L.F.) generator. The excitation beam is reflected on the dichroic mirror surface and is directed to the sample surface. The heterodyne laser probe detects thermoelastic displacements in reflection mode. Amplitude and phase information is obtained by demodulation electronics and a lock-in amplifier. These data are stored in a computer and then displayed on a videoscreen. The sample is scanned by two stepping motors.

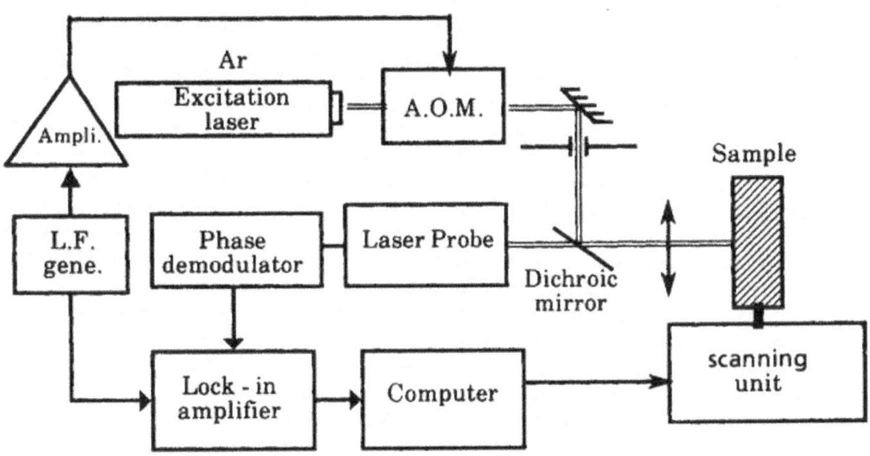

Fig. 10 : Thermoelastic microscope setup

4.2. Modeling of thermoelastic displacements

Thermoelastic displacements can be calculated theoretically in the case of an isotropic homogeneous sample. The laser beam intensity profile yields to polar coordinate analysis. A useful mathematical tool for polar coordinate calculations is the Hankel transform[20]

$$\tilde{F}(s) = \int_0^\infty r J_o(s\,r)\, F(r)\, dr \tag{10}$$

where $\tilde{F}$ is the Hankel transform of F, and J_0 the 0-order Bessel function.

For a cylindrical sample bounded by planes $z = 0$ and $z = h$ (h small compared to the sample diameter) heated with a gaussian laser beam (power P_0), the temperature (θ) can be expressed as[21]

$$\theta(r, \theta, z, t) = T(r, \theta, z)\, e^{j\omega t} \tag{11}$$

with

$$T(r, \theta, z) \simeq K' \int_0^\infty s J_o(sr) \frac{exp(-w^2 s^2/4}{a^2 - s^2 - \upsilon^2} \left[exp(-az) - \frac{a\, ch\left(\sqrt{s^2+\upsilon^2}\,(h-z)\right)}{\sqrt{s^2+\upsilon^2}\; sh\left(\sqrt{s^2+\upsilon^2}\; h\right)} \right] ds \tag{12}$$

and

$$K' = -\frac{2\,\alpha\, P_t\,(1-R)}{\pi^2 K}, \quad \upsilon^2 = j\,\omega\, a \quad et \quad a = \left(\pi\,\sigma\, f_L\, \mu_m\right)^{-1/2} \tag{13}$$

where σ is the electrical conductivity, f_L the excitation laser frequency, $1/a$ the thermal diffusivity, R the surface reflectivity, and w the beam radius at $1/e$. Numerical values of this integral are shown in Fig. 11 for different values of excitation beam radius.

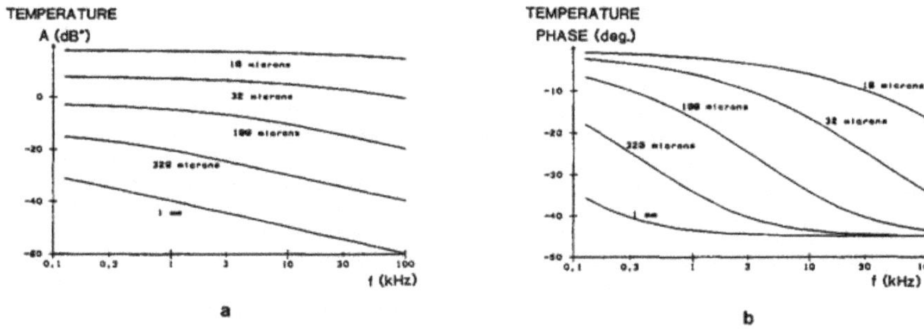

Fig. 11 : Temperature versus frequency for different values of excitation beam radius. Laser power 500 mW ; Aℓ sample
a : calculated amplitude b : calculated phase

Assuming $\sigma_{zz} = 0$ and $\sigma_{rz} = 0$ at $z = 0$ and $z = h$ (stress free surfaces), the surface displacement is calculated by using the Youngdahl stress function[22, 23] and Hankel transform (see [21] for calculation details)

$$u_z = \int_0^{\infty} s\,J_o\,(sr)\,\tilde{u}_z\;ds \tag{14}$$

with

$$\frac{\tilde{u}_z}{C_o} = s\,\frac{(3 - 4\nu)}{4\,(1 - \nu)}\,[\,Q(s) - P(s)\,] - s\,G(s) + s\,H(s)$$

$$- \frac{a\,exp\left(-w^2 s^2 / 4\right)}{a^2 - s^2 - \upsilon^2}\left[\frac{1}{a^2 - s^2} - \frac{1}{\upsilon^2}\right] \tag{15}$$

at $z = 0$, where ν is the Poisson coefficient, α_T the thermal expansion, and C_0 an excitation constant ; $P(s)$, $Q(s)$, $G(s)$ and $H(s)$ are polynomials that are given by boundary conditions.

A comparison between theoretical and experimental values of normal displacement at the excitation point (normal displacement is measured in reflection mode with an asymmetrical heterodyne laser probe) is shown by Fig. 12. Experimental conditions are : 500 mW excitation laser power, focussing with the microscope objective (N.A. 0.27) on an aluminum sample. This experiment needs adjustment of three parameters : sample reflectivity, beam radius and acousto-optic modulator delay (about 0.8 μs). The latter was not included in the model but in the curve display software.

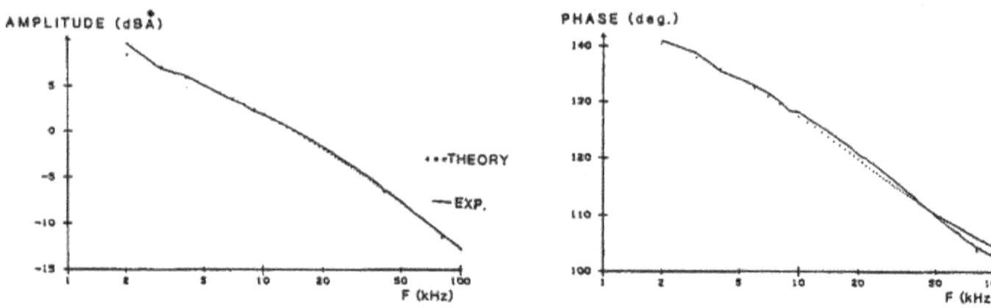

a b

Fig. 12 : Amplitude and phase (theoretical and experimental) versus frequency. Pure aluminum sample
a : amplitude response b : phase response

4.3. Thermoelastic imaging applications

a) Grain detection in metals

Thermal waves imaging yields the visualization of grain structures. In thermoelastic imaging the image contrast is related to the grain orientation.[24,25]

The sample studied is a Fe polycrystal. It has been polished first with diamond paste, then electrolytically and finally etched. Fig. 13a is a phase image obtained at 10 kHz frequency (thermal penetration depth ~ 25 μm). The lateral scale is given by the white bar and the adjacent indications. The good contrast of the interior of the grain is characteristic of this technique.

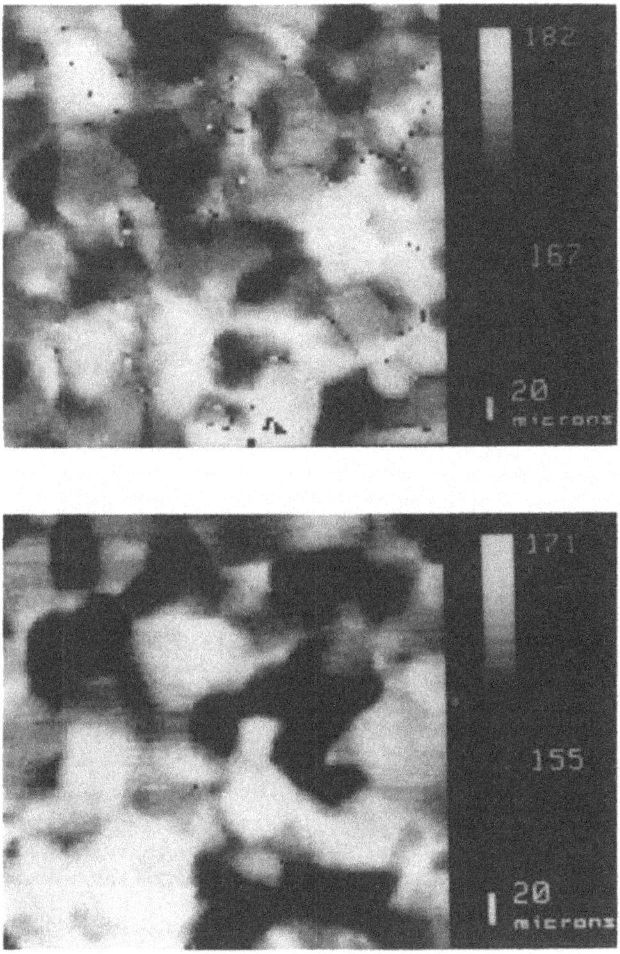

a

b

Fig. 13 : Thermoelastic phase images of a Fe polycrystal
a : sample electrolytically polished and etched
b : sample polished with diamond paste

Another part of the sample has not been electrolytically polished and etched. A thermoelastic phase image of that area is given by Fig. 13b (20 kHz excitation frequency). This image shows the grains has been flattened by diamond paste polishing.

b) Detection of subsurface thin film

A 2,000 Å thick SiO_2 strip and a 1,000 Å thick $A\ell$ layer were vapour-deposited successively onto a Ni polycrystal (Fig. 14a). The thermoelastic phase image of the sample shows the invisible SiO_2 layer (bright strip on the right) at a modulation frequency of 220 kHz (Fig. 14b). The phase scale corresponds to a range of phase of about 25 degrees.

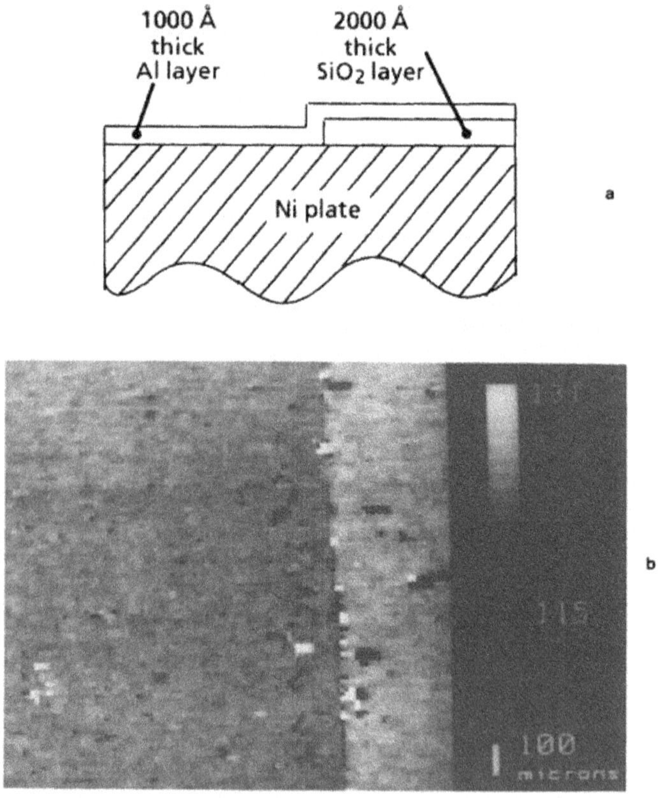

Fig. 14 : SiO_2 and $A\ell$ layers deposited onto a Ni polycrystal
a : sample diagram b : thermoelastic phase image at 220 kHz

c) Dopant imaging

Microelectronics requires a high-resolution nondestructive testing method. Thermoelastic imaging can detect low dopant concentrations ($\sim 10^{15}/cm^3$) because thermal conductivity changes with concentration.

An example of dopant visualization is shown in Fig. 15. The sample is a Si plate 370 μm thick. It has been implanted with boron ($10^{15}/cm^2$, 100 keV). Fig. 15a is a SEM (scanning electron microscope) image obtained after surface metallization (400 Å, Au). Fig. 15b presents the thermoelastic phase image of the sample (without metallization) at a 250 kHz excitation frequency.

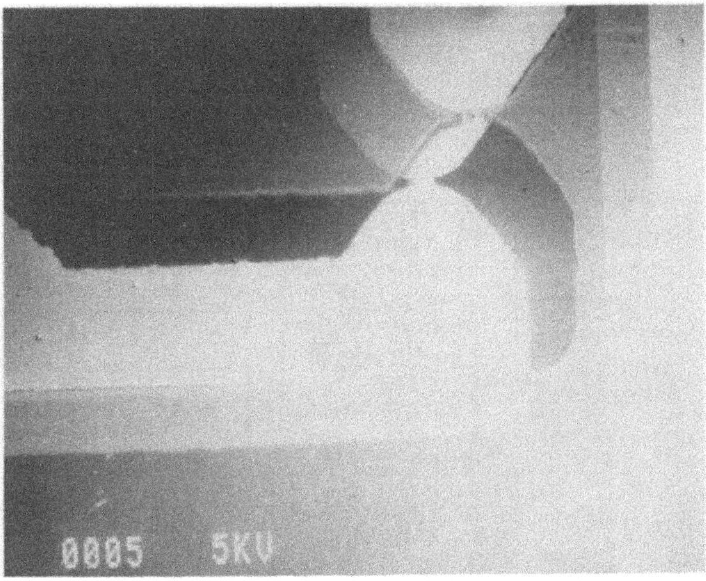

a

b

Fig. 15 : Dopant imaging in Si semiconductor
a : SEM image after metallization b : thermoelastic phase image without metallization.

5. Joule displacement microscopy

5.1. Microscope principle

Thermal excitation is generated by a periodic current (Joule effect) going through the sample. The displacement probe is a heterodyne interferometer (asymmetrical). The block-diagram of the experimental setup is shown in Fig. 16.

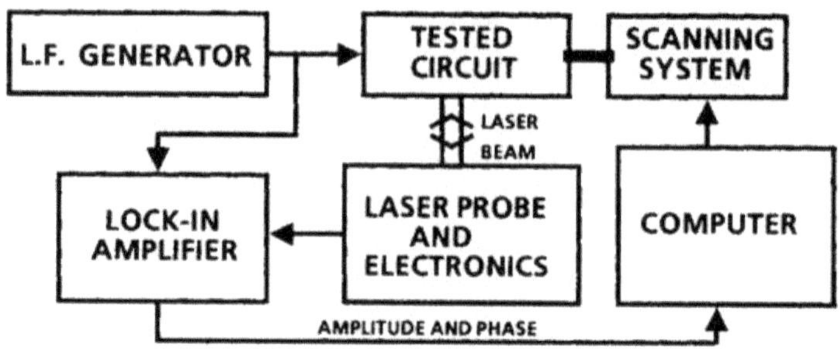

Fig. 16 : Joule displacement microscope diagram

If the thermal source is assumed to be a diffused resistor, the 1-D model correctly describes the problem if the thickness of the conductor layer and the thermal diffusion length are small compared to the doped area lateral dimensions. For an exponential doping profile, the surface displacement can be expressed as[26]

$$u(0,\ t) = \frac{I_o \alpha}{\rho C w}\ e^{j(\omega t - \pi/2)} \tag{16}$$

where I_0 is the electrical power density, α the thermal expansion, ρ the density and C the specific heat. This relation points out that the surface displacement is proportional to the electrical power and inversely proportional to the excitation frequency. The relationship has been confirmed experimentally.

5.2. Application to electronic circuit imaging

Thermoelastic microscopy can be used to image electric conductors or semiconductors (metallic layers, active discrete components or integrated circuits) heated by a periodic current and having a non-zero thermal expansion coefficient.

The application presented shows a standard transistor (2N2219). Thermoelastic images have been obtained in the following conditions : 400 × 400 micron scanned area, 4 micron scan step, 5 kHz excitation frequency and 100 mW excitation power (square wave power). Fig. 17a shows the transistor after opening the metal case. Emitter, base plating and wires are clearly visible (collector is underneath). Fig. 17b and 17c are amplitude and phase thermoelastic images of the probed chip. On the amplitude image, the bright areas correspond to the larger displacement amplitudes. The phase image shows the radial heat diffusion from the emitter area.

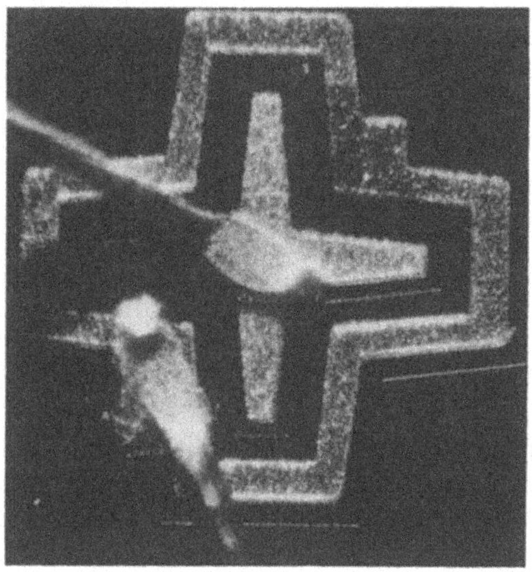

a

b

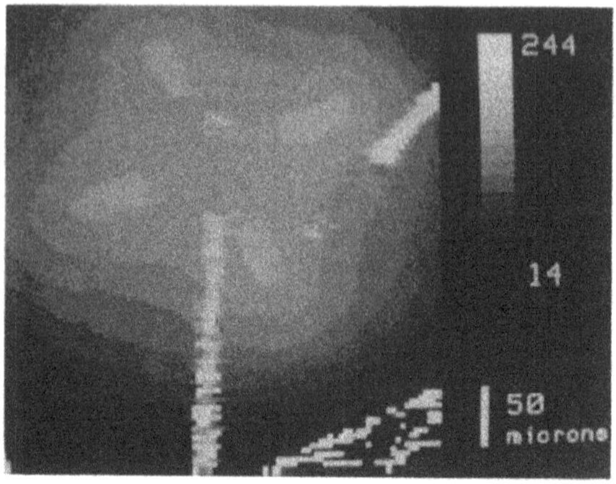

c

Fig. 17 : Thermoelastic probing of an operating transistor at $f = 5$ kHz
a : optical view of the chip   b : thermoelastic amplitude image   c : phase image

6. Conclusion

Interferometric heterodyne laser probes are useful tools for probing out-of-plane displacements. Applications include transducer calibration, SAW or BAW studies (acoustic response measurement and displacement mapping) and nondestructive evaluation. The latter is the object of recent studies using thermoelastic imaging, atomic force microscopy and near field optical microscopy.

Results presented on thermoelastic microscopy applications show that this technique can be used to probe optically-opaque specimens. The use of optical probing and excitation leads to several advantages : nondestructive, noncontact sensing, well-defined, explored volumes (lateral resolution and thermal penetration depth can be easily estimated). Optical probing of thermoelastic displacements is a method for materials analysis complementary to I.R. thermography, acoustic probing or reflection variation measurements.

The main probe limitations are surface roughness and low optical reflectivity. But in most cases solutions can be found (for example automatic intensity control in the case of varying reflectivity and phase conjugation techniques to reduce the speckle effect).

Future progress in optical components integration will certainly lead to permit more compact and more accurate laser probes (using laser diodes), resulting in lower costs and additional applications.

References

1. A.C. Boccara, D. Fournier and J. Badoz, *Appl. Phys. Lett.* **36** (1980) 129.
2. P. Cielo and C.K. Jen, *IEEE Ultrasonics Symposium Proc.* (1986), p. 515.
3. G.S. Kino, J. Fanton and B.T. Khuri-Yakub, *IEEE Ultrasonics Symposium Proc.* (1986), p. 505.
4. H.K. Wickrawasinghe, S. Ameri and C.W. See, *Electron. Lett.* **18** (1982) 973.
5. D.Hauden, H. Hassani and M. Planat, *IEEE Ultrasonics Symp. Proc.* (1984), p. 249.
6. R.L. Whitman and A. Korpel, *Applied Optics* **8** (8) (1969) 1567.
7. D. Royer and E. Dieulesaint, *Appl. Phys. Lett.* **49** (1986) 1056.
8. B.C. Moss and C.B. Scruby, *Ultrasonics* **26** (1988) 179.
9. J.P. Monchalin, *IEEE Transactions on Ultrasonics, Ferroelectrics and Frequency Control* **UFFC-33** (1986) 485.
10. A.Korpel, J.L. Laub and H.C. Sievering, *Appl. Phys. Lett.* **10** (1967) 295.
11. R. Adler, A. Korpel and P. Desmares, *IEEE Trans. on Sonics and Ultrason.* **SU-15** (1967) 157.
12. L. Noui and R.J. Dewhurst, *Appl. Phys. Lett.* **57** (1990) 551.
13. R.M. de la Rue, R.F. Humphryes, I.M. Mason and E.A. Ash, *Proc. IEEE* **119** (1972) 117.
14. D. Royer, E. Dieulesaint and Y. Martin, *IEEE Ultrasonics Symp. Proc.* (1985) p. 432.
15. B. Cretin, W.X. Xie, S. Wang and D. Hauden, Optics Com. 65 (3) (1988) 157.
16. J.E. Bowers, R.L. Jungerman and B.T. Khuri-Yakub, *IEEE Ultrasonics Symp. Proc.* (1982) p. 854.
17. B. Cretin and J.C. Mollier, *IEEE Trans. on Instr. and Meas.* **IM-34** (1985) 660.
18. A. Hassani, Thèse, Besançon (1984).
19. B. Cretin and D. Hauden, in *Ultrasonic Signal Processing* (World Scientific, 1988), p. 367.
20. H.S. Carslaw and J.C. Jaeger, *Conduction of heat in solids* (Oxford, Clarendon Press, 1959).
21. D. Hauden, B. Cretin, A. Mahmoud, J.L. Lesne, W. Xie, S. Vermersch, *Ultrasonics International* (1989).
22. C.K. Youngdahl, Int. J. Engng. Sci. 7 (1969) 61.
23. L.P. Welsh, J.A. Tuchman, I.P. Herman, *J. Appl. Phys.* **64** (11) (1988) 6274.
24. G.S. Cargill, in *Scanned Image Microscopy* (Academic Press, London, 1980) p. 319.
25. J.W. Maclachan, J.C. Murphy in *Photoacoustic and photothermal phenomena* (Springer Verlag, 1987) p. 294.
26. B. Cretin, D. Hauden in *Proc. of 4th International Conf. on Quality in electronic components* (1989) p. 114.

ACOUSTO-OPTICAL PROBING OF PERIODIC LAYERED MEDIA APPLICATION TO SUPERLATTICES.

J. SAPRIEL
Centre National d'Etudes des Télécommunications
196, Avenue H. Ravera, 92220, Bagneux, France.

ABSTRACT

The optical, acoustical and acousto-optical properties of superlattices are studied as a function of those of the constitutive layers. The propagation of acoustic and optic waves along the direction z perpendicular to the layers, as well as their acousto-optical interaction along z are particularly investigated. The problem of energy gaps for light and acoustic waves is carefully analyzed in absorbing media. A Fabry-Perot effect is evidenced theoretically and experimentally in GaAs/AlAs SL for photon energies near the stop band. Emphasize is given to light scattering from acoustic phonon for the characterization of the structure and quality of superlattices.

1. Introduction

Superperiodic structures are crystallographic arrangements where one can consider a large period D, generally superposed to a smaller one a, along the same direction z. SiC polytypes and the graphite intercalation compounds are examples of such structures. Yet the most investigated ones are the so-called "superlattices"[1] on which we'll focuse here.

A superlattice is a layered and periodic system artificially composed of alternating layers of two (or more) different materials of thickness d and d'. A new period D = d + d' is thus created in the direction perpendicular to the layers. All properties of the superlattice (SL) are transformed with respect to those of the constitutive layers. For instance, the electronic, transport and vibrational properties of these systems have been extensively studied. Many electrooptical devices have been built in the last ten years, taking advantage of their remarkable electronic behaviour based on the advent of new quantized levels. Very recently, novel acousto-optical devices have been proposed[2], based on the interaction of light with ultrasonic waves in SL. Yet, we can hardly say that SL acousto-optics have really started. This is partly due to the typical geometry of these systems wich consists of a SL, whose total thickness is at the utmost of the order of a few micrometers, epitaxialy grown on a substrate of thickness 0.5 mm. This geometry is not favorable to the interaction of light with ultrasonic waves. Nevertheless, the knowledge of optical, acoustic and photoelastic properties of SL have made much progress which is useful to review here. From the point of view of characterization techniques the difficult problem of the transducer can be avoided by using acoustic waves of thermal origin (phonons) which have the immeasurable advantage of their availability in all directions and in a very wide frequency range (a few THz). Thus the acousto-optical interaction in SL is here utilized through spectroscopic techniques, namely, the well known Raman or Brillouin scattering[1,3].

2. Propagation of acoustic and light waves along the SL axis.

The direction z perpendicular to the layer is also the direction of epitaxial growth; z is obviously a direction of translational symmetry in the SL. We'll first focuse on the dispersion relation of acoustic waves propagating without attenuation along z in the constitutive layers of the SL.

Let us consider a superlattice (Fig 1) made of two different materials of thickness d and d', density ρ and ρ' and velocity v and v'.

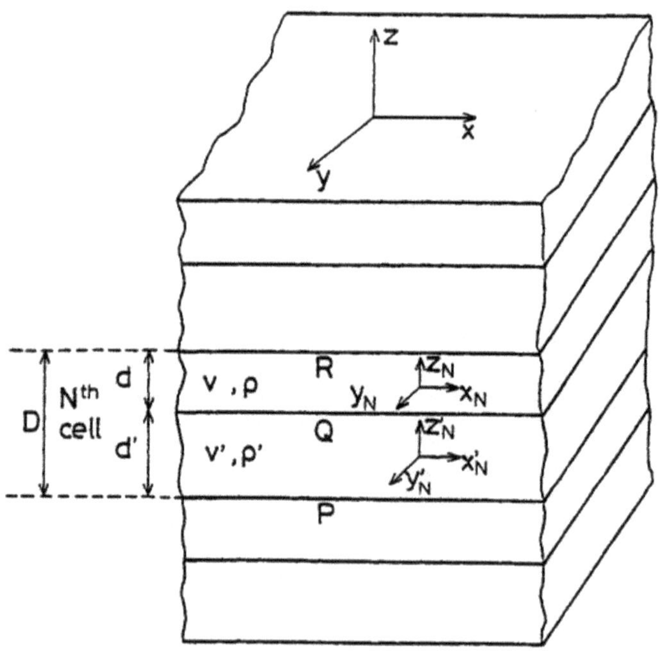

Fig 1 - Superlattice composed of two alternating layers of thickness d and d', density ρ and ρ', acoustic velocity v and v'; x_N, y_N, z_N and x'_N, y'_N, z'_N are local coordinates (origine at the middle of the layers) in the Nth cell. The general coordinates of the superlattice are x, y, z.

The displacement U_3 and the stress T_{33} are expressed in the form of Bloch functions

$$U_3 = g(z)\exp i(k_a z - \omega t) \tag{1}$$

$$T_{33} = f(z)\exp i(k_a z - \omega t) \tag{2}$$

here, $g(z+D) = g(z)$, $\quad f(z+D) = f(z) \quad$ with $\quad D = d + d'$ (3)

In the following we will omit the dependence in the time t which is trivial. U_3 and T_{33} must fulfill the continuity equations at all interfaces of the superlattice; but owing to the periodicity, the problem can be limited to two adjacent interfaces.

Consider now the Nth cell of the infinite superlattice. We define local coordinates x_N, y_N, z_N and x'_N, y'_N, z'_N with their origin at the middle of the layers (Fig 1). We thus have :

$$\begin{cases} u_{3,N} = A_N \exp\left(i\dfrac{\omega}{v} z_N\right) + B_N \exp\left(-i\dfrac{\omega}{v} z_N\right) \\ t_{33,N} = i\omega Z\left[A_N \exp\left(i\dfrac{\omega}{v} z_N\right) - B_N \exp\left(-i\dfrac{\omega}{v} z_n\right)\right] \end{cases} \quad \text{for the first layer}$$

$$\begin{cases} u'_{3,N} = A'_N \exp\left(i\dfrac{\omega}{v'} z'_N\right) + B'_N \exp\left(-i\dfrac{\omega}{v'} z'_N\right) \\ t'_{33,N} = i\omega Z'\left[A_N \exp\left(i\dfrac{\omega}{v'} z'_N\right) - B'_N \exp\left(-i\dfrac{\omega}{v'} z'_N\right)\right] \end{cases} \quad \text{for the second layer}$$

where $Z = \rho v$; $Z' = \rho' v'$ and A_N, B_N, A'_N, B'_N are constants.
The continuity relations are given by

$$u_{3,N}(-d/2) = u'_{3,N}(+d'/2) \tag{4}$$

$$t_{33,N}(-d/2) = t'_{33,N}(+d'/2) \tag{5}$$

for interface Q and

$$u_{3,N}(d/2) = u'_{3,N+1}(-d'/2) = u'_{3,N}(-d'/2)\exp(ik_a D) \tag{6}$$

$$t_{33,N}(d/2) = t'_{33,N+1}(-d'/2) = t'_{33,N}(-d'/2)\exp(ik_a D) \tag{7}$$

for interface R.

Obviously the problem of the propagation of the acoustic wave in the superlattice is completely solved by the set of Equ. (4) - (7) where only the two layers of a unit cell are involved. This is a linear and homogeneous system of 4 equations with 4 unknown variables A_N, B_N, A'_N, B'_N. The solution is obtained by solving the secular equation. One finds finally

$$\cos k_a D = \cos\frac{\omega d}{v}\cos\frac{\omega d'}{v'} - \frac{1}{2}\left(R + \frac{1}{R}\right)\sin\frac{\omega d}{v}\sin\frac{\omega d'}{v'} \tag{8}$$

where R = Z/Z' is the ratio of acoustic impedances. The Kronig-Penney Equ. (8) is an implicit relation between the frequency ω and the wavevector k_a in the SL, though ω and k_a are merely proportional in a homogeneous medium. In the following, the wave vector k_a will be written k_u when ultrasonic waves are concerned and k_p when one deals with acoustic phonons.

Likewise, in a SL with sharp interfaces, the profile of the refractive index can be described by

$$N(z) = \begin{cases} n, & \text{for } 0 < z < d \\ n', & \text{for } d < z < D \end{cases}$$

with N(z+D) = N(z)
According to the Bloch theorem, the electric field E of an electromagnetic wave propagating in the SL is given by

$$E = E(z)\exp(ikz - \omega t)$$

where $E(z + D) = E(z)$

The Kronig-Penney dispersion relation in the case of light waves gives :

$$\cos kD = \cos\frac{\omega}{c}nd\cos\frac{\omega}{c}n'd' - \frac{1}{2}\left(\frac{n}{n'} + \frac{n'}{n}\right)\sin\frac{\omega}{c}nd\sin\frac{\omega}{c}n'd' \tag{9}$$

which is quite similar to (8).

In the dispersion curve (9) in the SL has been obtained by writing the boundary conditions on the continuity of the tangential components of the electric field $\vec{E}$ and displacement field $\vec{D}$. It differs from the linear relation $k = \omega n/c$ which characterizes homogeneous media.

Consider a specific kind of SL, for instance a GaAs/AlAs SL. For a given ratio d_{AlAs}/d_{GaAs} of the layer thicknesses the acoustic and optic dispersion curves (8) and (9) can be expressed in a standard from, independent of the SL period D, provided one uses the dimensionless parameters $\Omega = \omega D$ and K = k D. The dispersion curves (see Fig 2-a and 2-b) are represented by mini-branches which are quasi-linear. Actually they show some deviations from linearity only near K = p (p integer) giving rise to small gaps (or stop bands) which depend on Z/Z' or n/n' and p. One can notice that K = p correspond to the successive boundaries of the Brillouin zones in the z directions.

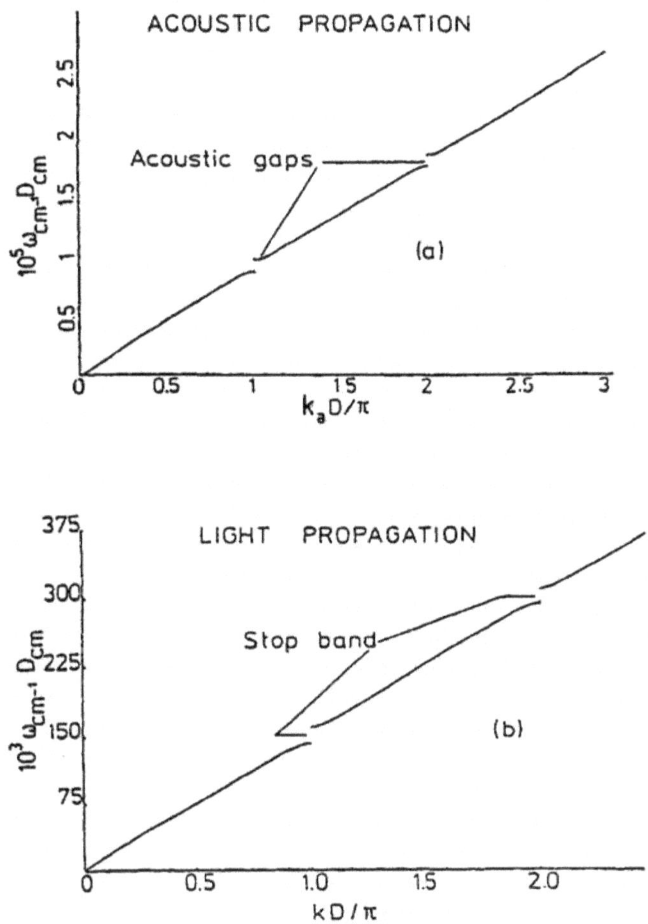

Fig 2 - Dispersion curves of : a - acoustic modes in GaAs/AlAs superlattices ($d_{AlAs} = 3\ d_{GaAs}$). b - Light waves in GaAs/AlAs superlattices ($d_{AlAs}/d_{GaAs} = 2/3$) in the extended Brillouin-zone scheme.

For Z/Z' = 1 or n/n' = 1, the acoustic or optic gaps completely disappear since there is no acoustic or optic mismatch between the layers of the SL. For Z/Z' =1, the acoustic velocity V of the SL is given by D/V = d/v + d'/v' and for n = n', one returns to the trivial case of the homogeneous optical medium where $k = \omega n/c$.

These gaps or stop bands correspond to frequency ranges for which the acoustic displacement or the electric field undergo strong attenuation during their propagation along z. Actually the larger the gap width, the stronger the attenuation of the acoustic or optic wave under consideration. Clearly the gap width increases with the acoustic mismatch or optic mismatch between two consecutive layers. For ω approaching the gap edges the attenuation decreases and cancel at the limit of the gap. It is worthwhile pointing out that the wave damping at K = p is only a SL effect, due to the new translational symmetry along z. Any, even partial, breakdown of this symmetry would modify these gaps as we'll see latter when considering SL with optically or acoustically absorbing layers. One can notice that near the stop band edges the dispersion curves flatten and the group velocity $d\omega/dk$ is close to zero.

3. Optical resonance Effects in Superlattices[2,4].

3.1. Theoretical considerations.

In this lecture we particularly consider semiconductor GaAs / AlAs superlattices, since GaAs is an excellent crystal for acousto-optics, electro-optics and nonlinear optics. As the refractive index of GaAs and AlAs are quite different, the superlattice resonances discussed here are strongly reinforced.

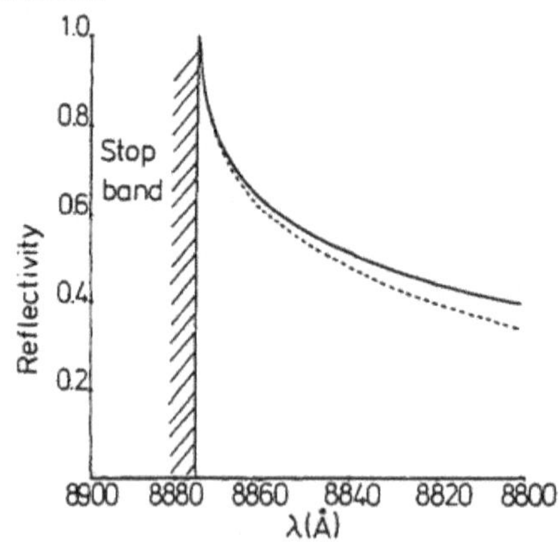

Fig 3 - Calculated reflectivity of a semi-infinite SL (dGaAs = 840 Å, dAlAs = 560 Å) as a function of the light wavelength in air. For light incident from air (solid line) and from the GaAs substrate (dashed line).

Consider now a light beam at normal incidence on the free surface of a semi-infinite SL or between the boundary between a substrate and a semi-infinite SL. In the stop band the reflectivity is equal to 1. In the vicinity of the stop band the reflectivity R is high i.e. close to 1 (see Fig 3).

Thus a finite SL limited by air from one side and by a substrate from the other side acts as a Fabry Perot cavity ended by two mirrors (the two boundaries of the SL) as far as one uses wavelengths close to the stop bands. The calculated reflectivity R of a finite SL of 50 periods of GaAs/AlAs is represented Fig 4-a near the lowest energy stop band. The reflectivity displays a succession of maxima and minima. The minima of R correspond to maxima of light energy density gain G in the SL (Fig 4-b).

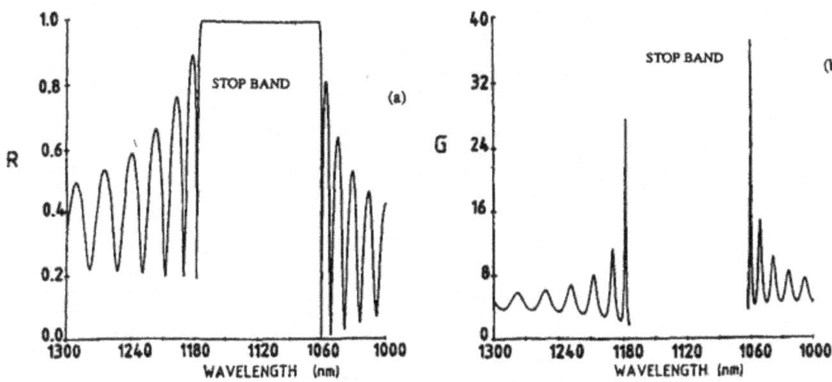

Fig 4 - Calculated reflectivity R(a) and energy density gain G(b) in a GaAs/AlAs SL; D = 1.700 Å, d_{AlAs}/d_{GaAs} = 2/3, N = 50.

The SL plays the role of a Fabry-Perot whose resonances correspond precisely to the minima of reflectivity. Such resonances take place at light wavelenghts for which the phase change of the light Bloch wave is equal to a multiple of 2π after the propagation of one cycle in the SL. One can notice that the closer the resonance peak to the limit of the stop band and the stronger the resonance. This is due to the group velocity of light in the SL, which decreases and tends to zero for light wavelength tending to the limit of the stop band. The group velocity being the velocity of the energy propagation, a low group velocity is associated to high energy densities in the SL.

We have investigated the sharpness of the main resonance in GaAs/$Ga_{1-x}Al_xAs$ SL, as a function of x and of N (the number of periods in the SL thickness) via the calculations of G,R and L (the width at half maxima of the chief density gain peak). For the calculations we have taken d_{AlAs}/d_{GaAs} = 2/3 and D is adjusted in order to obtain the chief resonance at 1.06μ.

From Fig 5, one can notice that the resonance is more pronounced for large value of the aluminum concentration x. The best situation corresponds then to the binary SL GaAs/AlAs where the refractive index mismatch between the constitutive layers is maximum.

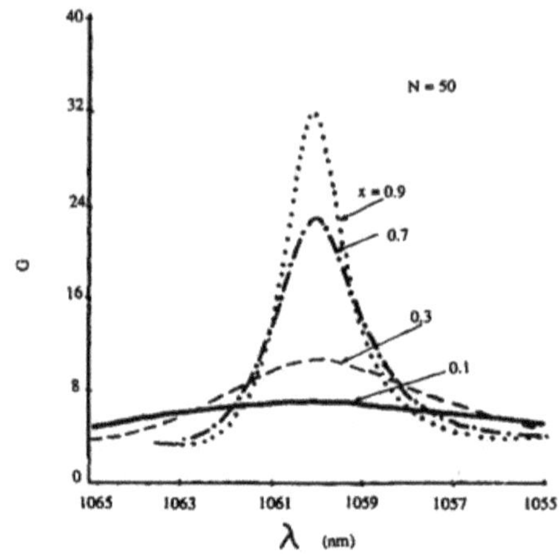

Fig 5 - Energy density gain in GaAs/$Ga_{1-x}Al_xAs$ SL as a function of x.

It is also clear, according to Fig 6, that thick superlattices corresponding to large N values are favourable to strong resonances.

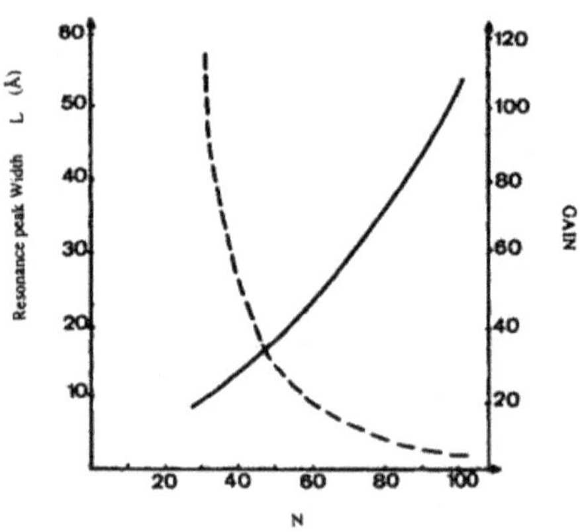

Fig 6 - Peak wodth L (dashed line) and energy density gain G (full line) as a function of N, number of unit cells in a GaAs/AlAs SL, for the main resonance peak (1.06μ).

3.2. Experiments

The only optical property of the SL discussed here, which is directly accessible to the experiment, is the reflectivity R. Actually the antiresonances of R (i.e. its minima) correspond to the resonances of G. A sample of period D~1600Å, with 30 unit cells and $d_{GaAs}/d_{AlAs} = 3/2$, grown by MBE, has been carefully studied. We have noticed inhomogeneities of period along the sample surface. The minima of R were thus shifted according to the investigated point. We measured approximately 1mev per Ångström of D variation, in good agreement with the calculations.

The best reflectivity measurements were obtained by using a Ti:Sapphir tunable laser as a source. The light beam was focused on the SL sample, the diameter of the spot being of the order of 50μm to avoid inhomogeneous broadening of the peaks. One can notice (Fig 7) that near the limit of the stop band the reflectivity undergoes approximately 100% variations on a wavelength range of 15Å. The minima of the reflectivity (nearly zero) corresponds to the resonances and are indicated by arrows. One can notice that the closer the minimum of R and the sharper the resonance.

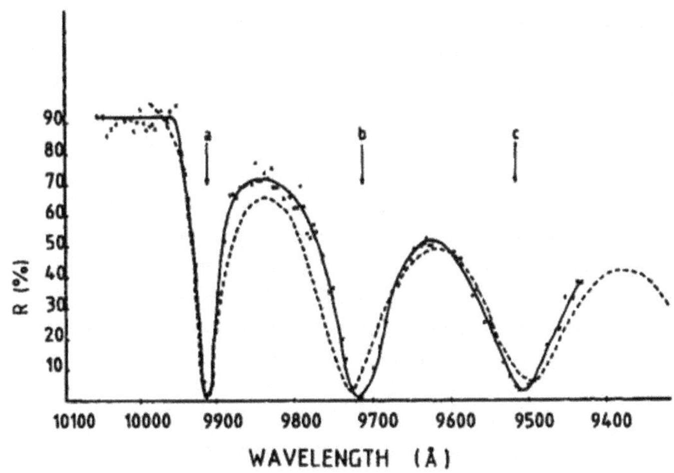

Fig 7 - Reflectivity R of a GaAs/AlAs SL; D = 1590 Å; $d_{AlAs}/D = 0.4$; N = 30; experimental (full line), theory (dashed line) ($n_{GaAs} = 3.53$, $n_{AlAs} = 2.96$).

3.3. Light Ultrasound Interaction in SL.

In a homogeneous bulk medium, the phase-matching conditions are expressed by

$$\overrightarrow{k_d} = \overrightarrow{k_i} + \overrightarrow{k_u} \tag{10}$$

Here $\vec{k_d}$, $\vec{k_i}$ and $\vec{k_u}$ are the wave vectors of the diffracted light, incident light and ultrasonic wave, respectively.

As the wave vector k_u of the ultrasonic wave is much smaller than that of light, the direction of propagation of the light should be nearly normal to that of sound (Fig 8a).

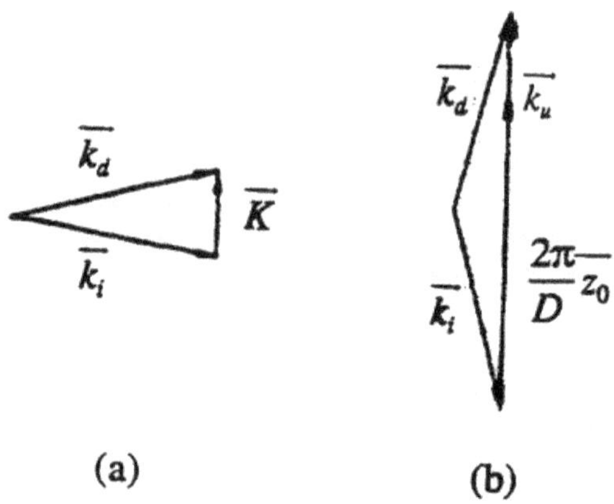

Fig 8 - Wave vector composition for acousto-optical interaction : (a) In a homogeneous bulk medium. (b) In a superlattice.

In contrast, in a superlattice, because of the periodicity along the growth axis, the interaction is assisted by a reciprocal lattice vector; for example one can have (Fig 8b)

$$\vec{k_d} = \vec{k_i} + \vec{k_u} + (2\pi/D)\vec{z_0} \tag{11}$$

Here $\vec{z_0}$ is the unit vector along the growth axis. Because of the intervention of the reciprocal lattice vector, the diffraction can take place in a backscattering configuration along the superlattice, in spite of the fact that $k_u << k_i$. This interaction then occur for light wave vector $k \sim \pi/D$, i.e. near the conditions of Bragg reflection of light on the SL interfaces. If one takes advantage of the optical resonance discussed here above, the interaction can be strongly increased. This is equivalent to a laser intracavity acousto-optical interaction, which benefits by the high energy gain localized between the two end mirors of the laser. As in the case of intracavity devices, the SL must be transparent to the light beam to minimize the optical losses.

In the case of a resonant acousto-optical interaction in SL's, the interaction length l between light and ultrasonic waves which occurs in a backscattering geometry is very small since it is equal to the SL total thickness (~ 5μm). Small l values allow a frequency broad band interaction, and short access times. A new generation of fast and compact acousto-optical devices based on superlattice structures can now be conceivable[1].

4. SL characterization by light scattering from acoustic phonons.

4.1. Frequency variations of acoustic phonons versus the scattering wave vector.

As already pointed out, the light scattering technique allows a much larger versatility, compared to the usual diffraction method from ultrasonic waves. The experiments are performed in the backscattering geometry along the SL axis and the scattering wave vector q which is defined as the change of the light wave vector in the scattering process, is related to the phonon wave vector k_p by :

$$k_p = q + 2\pi m/D \tag{12}$$

Here $q = k_s - k_i \approx 2k_i = 4\pi N/\lambda$. k_s and k_i are the wave vector of the scattered and incident light respectively of wavelength λ in the air, and N is the refractive index in the SL. For a given q, conditions of Eq. 12 is fulfilled by several phonon wave vectors; $2\pi m$ / D is a reciprocal lattice vector. The integer m can take the values m = 0, ± 1, ± 2, ± 3, etc... The frequency ω of the phonon is related to its wave vector by the implicit Eq. 8.

As far as q values are not too close to the gaps, the energies of the phonons involved in the scattering process are given by $\omega_p \approx |k_p| V$, with D / V = d / v + d / v'. The experiment consists of observing these modes which appear on the spectra and measuring their shifts ω_p with respect to the laser radiation energy. These shifts are expressed in cm^{-1} (1 cm^{-1} corresponds to a frequency shift of 30 GHz). The ω_p determination is performed by scanning a high resolution and high contrast Raman monochromator. The stray or spurious light which comes from the laser (Rayleigh line) is strongly attenuated by using a double or triple-grating monochromator. Sometimes a simple pass Fabry-Perot interferometer is added in tandem to increase the resolution and contrast. Modes as low as 1 cm^{-1} can be detected on the spectra.

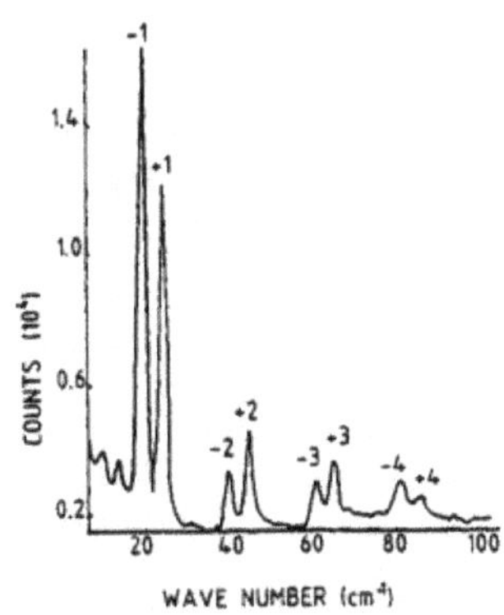

Fig 9 - First folded doublets corresponding to FLA_m modes m = ± 1, ± 2, ± 3 and ± 4 in a GaAs/AlAs SL, with a period of ~80 Å. From Ref 5.

According to symmetry considerations, in the backscattering geometry, light can only interact with longitudinal acoustic (LA) waves. The results for a SL made of GaAs (d = 56 Å) and AlAs (d' = 24 Å) are given in Fig 9. The different modes are labeled by their index m $= \pm 1, \pm 2, \pm 3$ and ± 4.

These modes are called, folded acoustic modes FLA_m because they can be considered as originating from the acoustic branch of Fig 2-a after folding in the Brillouin zone $0 < k_p < \pi/D$. m = 0 is the Brillouin mode which is the only longitudinal acoustic mode observed in homogeneous media. Normally the existence of several FLA_m in the spectrum allows a precise determination of N, V, d, and d', i.e of the composition and structure of the SL. The experiments are performed for photons in the visible. Argon-ion or Krypton lasers are used whose wavelengths range between 450 to 680 nm. In spite of the fact, that the semicondustor SL's are absorbing for these radiations, a penetration depth exists in the sample. For GaAs/AlAs SL light penetrates at a distance between a few thousands Angströms to a few microns, depending on the light wavelength. It is also clear that by changing the wavelength, one modifies the scattering wave vector amplitude $q = 4\pi N(\lambda)/\lambda$. One must keep in mind that the refractive index $N(\lambda)$ near and above the absorption edge of semiconductors, is strongly dependent on the photon energies. Likewise, in this energy range the photoelastic constants undergo strong variations.

One can also define a reduced scattering wave vector $Q = qD/\pi$ which can vary either by modifying q or by changing the SL period. In Fig 10 we have presented the frequency measurements for GaAs/AlAs SL samples corresponding to different periods, but to approximately the same GaAs/AlAs proportion (d/d' = Cte). The synthesis of all the results is given in the form of standard dispersion curves depending of dimensionless parameters, of the lowest LA and TA branches. The calculated variations of $\Omega_p = \omega_p D$ as a function of Q, drawn in Fig 10 are obtained by combining Eqs. 8 and 12.

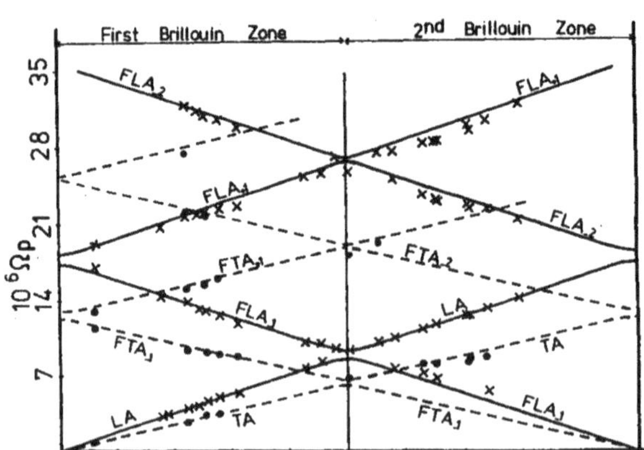

Fig 10 - Standard acoustic curves of SLs such as $d_{AlAs}/d_{GaAs} = 3$. Longitudinal (x) and transverse (.) modes (experimental). Full and dashed lines are calculated from the elastic model and correspond to longitudinal and transverse waves respectively. From Ref 6.

In the (001) oriented SL's only the LA and FLA modes can really couple with light. The diffracted light is then polarized parallel to the incident light. Yet, the TA anf FTA (folded transverse acoustic) can also be observed owing to a polarization leakage and are always very weak modes, without a definite polarization (Fig 11).

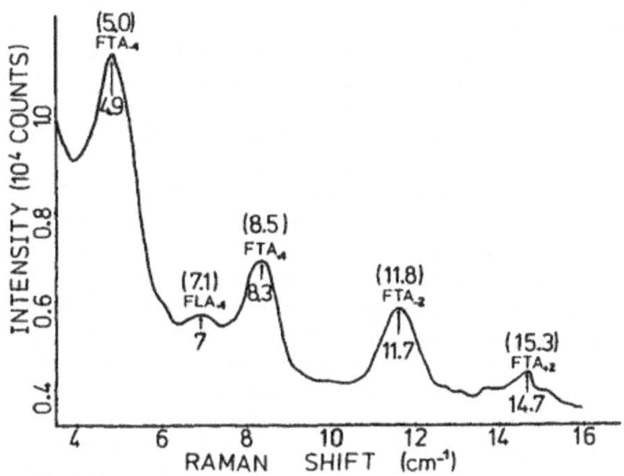

Fig 11 - Folded transverse acoustic modes obtained for crossed polarizations of the incident and scattered light. (The results of the calculations are indicated by arrows). From Ref 6.

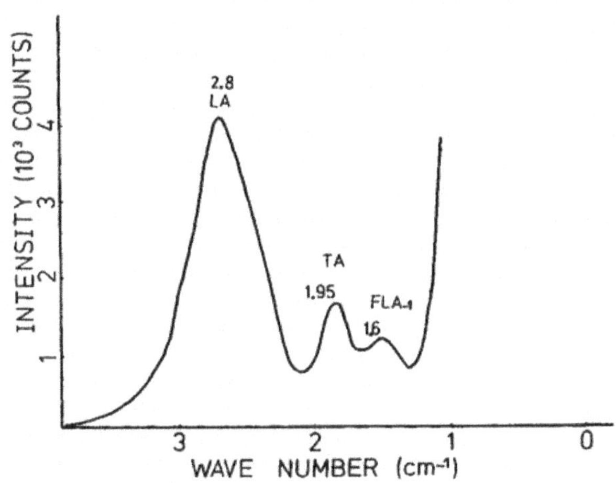

Fig 12 - Low-frequency modes in an SL in the 2th Brillouin-zone (Q > 1). See Ref 6.

In Fig 12 is represented a typical low-frequency Raman spectrum carresponding to $\pi/D < q < 2\pi/D$ (second Brillouin-zone). One can notice the presence of the Brillouin TA and of the FLA_{-1} mode below the Brillouin LA.

4.2. Intensity of the FLA peaks.

The light scattered by the folded acoustic modes, besides its dependence on the acoustic properties and the refractive index, is a function of the ratio p'/p of the constitutive layer photoelastic constants. There is also a strong dependence of the mode intensities as a function of the reduced scattering wave vector Q. A complete agreement between the calculated and measured intensities of the FLA peaks is obtained over a wide range of Q values.

As a matter of fact, the comparison between the peak intensities in the Raman spectra, allows a good determination of the ratio p/p', as emphasized in Ref. 6 and 7. The light scattering technique has been applied to $Ga_{1-x}Al_xAs$ and $Si_{1-x}Ge_x$ materials. The photoelastic constants of $Ga_{1-x}Al_xAs$ with respect to GaAs and those of $Si_{0.5}Ge_{0.5}$ with respect to Si have been measured as a function of the wavelength of the Argon ion and Krypton laser radiations. One finds that the relative photoelastic constant of $Ga_{1-x}Al_xAs$ undergoes a nonlinear variation with the aluminium concentration x at 4880 Å (Fig 13 a) and that the ratio of the photoelastic constants of $Si_{0.5}Ge_{0.5}$ and Si, varies strongly as a function of the laser wavelength (Fig 13 b).

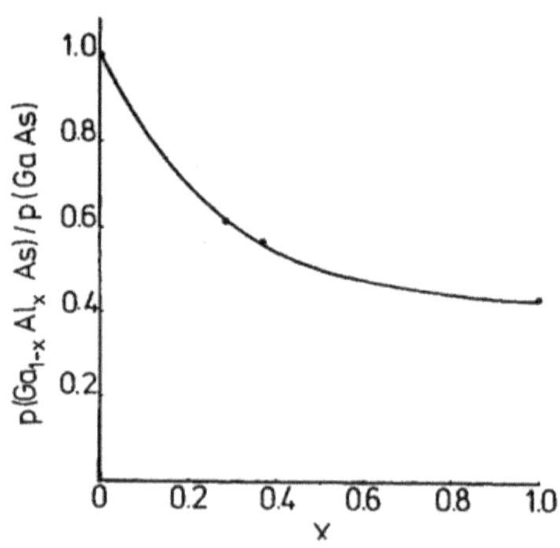

Fig 13 a - Measured photoelastic constant p_{13} of $Ga_{1-x}Al_xAs$ relative to GaAs p_{13}, as a function of x, for λ = 4880 Å. (Ref 7).

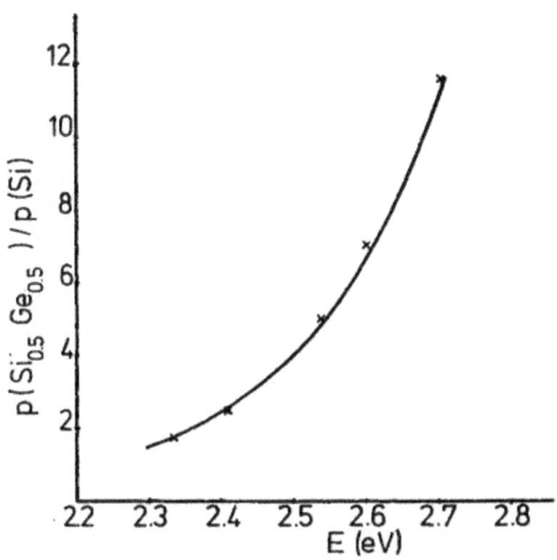

Fig 13 b - Measured photoelastic constant ratio $p(Si_{0.5}Ge_{0.5})/p(Si)$ versus the photon energy E. For photon energy between 2.3 and 2.7 eV, the variations can be approximated by an exponential (solid line). see Ref 7.

Conversely the knowledge of p/p' at a definite photon wavelength allows the determination of the relative thickness of the layers in the SL, from the measurement of the intensity ratio of the different peaks[8] (Fig 14).

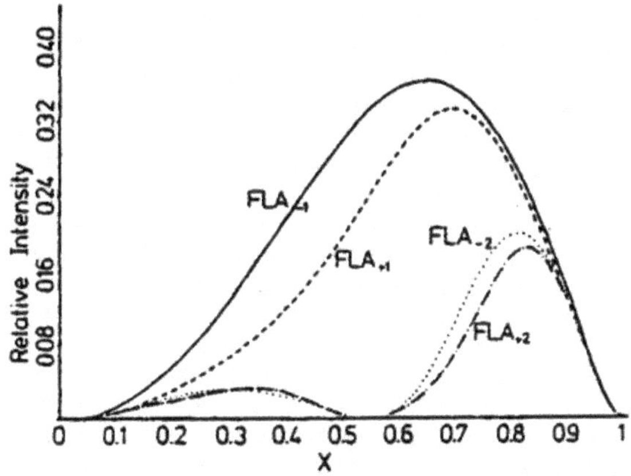

Fig 14 - Variations of the relative intensities I(FLA)m/I(LA) with the relative thickness of the layers x d'AlAs/D. The period of the SL is D = 186 Å and λ = 4880 Å (see Ref 8).

4.3. Effects of the acoustic attenuation and optical absorption.

The acoustic phonons and the optical waves undergo a damping in the constitutive layers of the SL which can be evidenced on the Raman spectra. This damping is dependent on the frequencies of the acoustic phonon and those of the photons. In this section, the result of the acoustic attenuation and optical absorption are theoretically analysed and experimentally probed in the investigated SL. Particularly striking are the damping effects for scattering wave vectors q corresponding to the limits of the successive mini-Brillouin zones. We essentially restrict our study to q values ranging around π / D, which correspond to Bragg reflection for the acoustic phonons, and near $2\pi / D$, where Bragg reflections occur for both acoustic and light waves.

Consider first the situation where the acoustic waves propagate without attenuation (i.e. $\alpha_1 = \alpha_2 = 0$ in the constitutive layers of the SL). The existence of energy gaps in the SL phonon dispersion curves is a consequence of the translational symmetry of magnitude D along the SL axis. In the energy range of the gaps the acoustic wave amplitude undergoes an attenuation, and the larger the gap the stronger the acoustic decay. If now we introduce an absorption coefficient ($\alpha_1 \neq 0$ or $\alpha_2 \neq 0$) in the layers, the wave vector of the acoustic phonons along z becomes also a complex number $K_p = k_p + i\alpha$. There is a partial loss of translational symmetry which principally affects the gaps.

The energies of the acoustic modes have been measured by light scattering at different laser wavelengths in an $Si/Si_{0.5}Ge_{0.5}$ SL ($d_{(Si)} = d_{(SiGe)} = 140$ Å) and are represented by crosses on Fig 15.

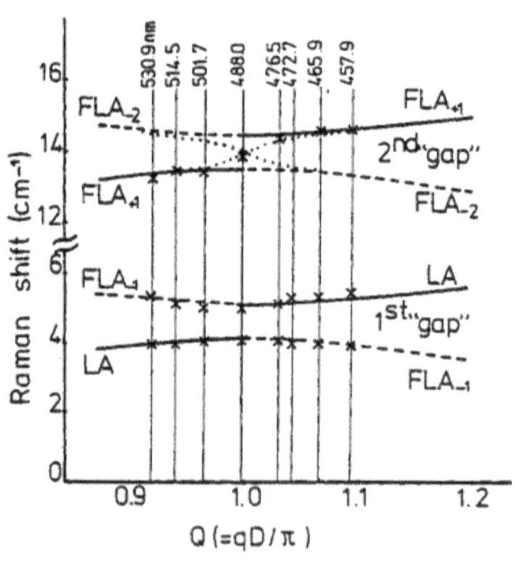

Fig 15 - Energies of the acoustic modes and the role of the acoustic attenuation on the higher energy acoustic gap in an $Si/Si_{0.5}Ge_{0.5}$ SL. from ref 9.

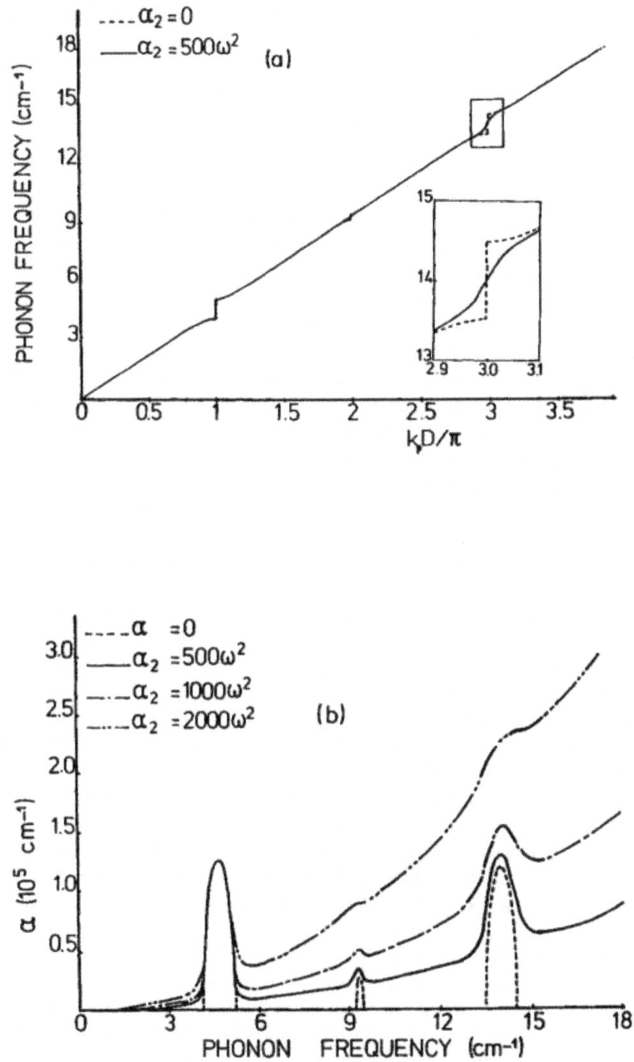

Fig 16 - Calculated dispersion curves (a) and attenuation-frequency relation of the acoustic phonons (b) in an Si/Si$_{0.5}$Ge$_{0.5}$ SL. Here the attenuation of Si is neglected and α_2 (SiGe) is taken proportionnal to the frequency square. From Ref 9.

The scattering wave vectors q are close to the value $q = \pi / D$ and are located from both sides of the limit of the Brillouin-zone, a suitable configuration to investigate the SL gaps. The solid lines correspond to the values calculated in the case of lossless media by means of the elasticity theory. At low frequencies ($\omega < 6cm^{-1}$), the cross positions are in good agreement with the solid line and a gap appears ($\sim 1cm^{-1}$ width). For $\omega > 12cm^{-1}$, we found experimental points within "the gap". Actually at these frequencies the acoustic attenuation can no longer be neglected. The acoustic damping makes the modes at the Brillouin-zone edge shift into the "forbidden gaps".

In Fig 16 are represented the calculated dispersion curve (Fig 16-a) and attenuation-frequency relation of the acoustic phonons in the considered $Si / Si_{0.5}Ge_{0.5}$ superlattice (Fig 16-b). The increase of the attenuation at high frequency changes markedly the feature of the gap at $k_p = \pi / D$ (see the inset of Fig 16 a). We have taken arbitrarily $\alpha_1 = 0$ and $\alpha_2 = A\omega^2$ (for lack of data in the litterature) for the acoustic attenuation in Si and $Si_{0.5}Ge_{0.5}$, respectively. As for the acoustic attenuation α in the SL, it increases with frequency and always displays peaks at $k_p = n\pi / D$, but the peaks are less conspicuous and almost disappear when the attenuation of the constitutive materials is very strong (Fig 16-b).

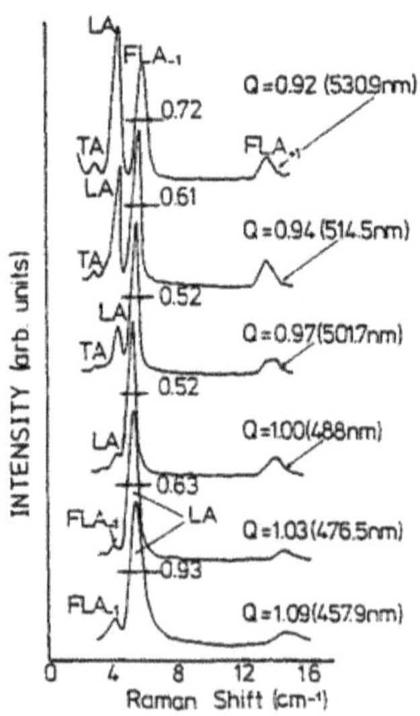

Fig 17 - Low-frequency Raman spectra of an $Si/Si_{0.5}Ge_{0.5}$ SL near Q = 1 (mini Brillouin-zone edge). See Ref 9.

The Raman peaks corresponding to the FLA modes undergo a broadening $\delta\omega$ (FWHM) that is equal to $2\alpha V_g$, $4\alpha_{op}V_g$, and $2\pi V_g/l$, due to the acoustic attenuation α, the optical absorption α_{op}, and the total thickness l of the SL, respectively; $V_g = d\omega/dk$ is the group velocity of the acoustic waves. In Fig 17 are reported the low-frequency spectra of the SL with different excitation laser wavelengths. For low-frequency modes the line width displays a minimum at Q = 1 caused by the flatness of the acoustic dispersion curves. This is particularly clear on the upper mode of the low-frequency doublet of Fig 17.

In reference 8, the relative intensities of the different modes have been calculated by taking into account the modulation of the photoelastic, acoustic, and optical properties in the SL, but the optical absorption has been neglected. A good agreement between the theory and the experiments has been obtained for a wide range of scattering wave vectors, not too close to the Brillouin-zone boundaries. However, one can expect a significant effect of the optical absorption on the relative intensities for scattering wave vectors near Q = 2, 4, 6.

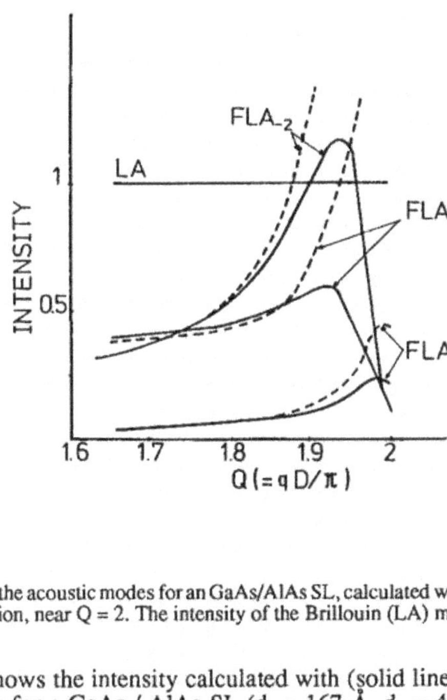

Fig 18 - Intensities of the acoustic modes for an GaAs/AlAs SL, calculated with (solid lines) or without (dashed lines) optical absorption, near Q = 2. The intensity of the Brillouin (LA) mode is normalized to 1. From Ref 9.

Figure 18 shows the intensity calculated with (solid lines) or without (dashed lines) optical absorption for a GaAs / AlAs SL (d_1 = 167 Å, d_2 = 458 Å) for a scattering wave vector near the second-Brillouin edge (Q ~ 2). When Q is very close to 2, the intensities of the folded modes with respect to the Brillouin line are much reduced by the optical absorption. The Raman spectra corresponding to this Q range are shown in Fig 19. One can see on Fig 19 that when Q tends to 2, FLA_{-2} increases, reaches a maximum and then decreases with respect to the LA (Brillouin line). This is indeed the behaviour predicted when the optical absorption is considered.

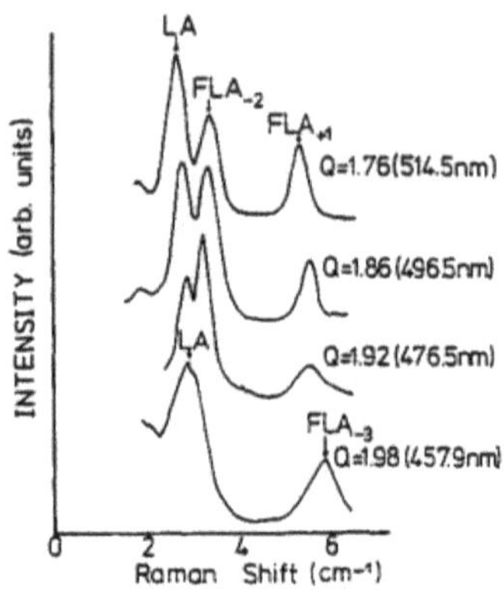

Fig 19 - Raman spectra of the SL of Fig 18 for Q values approaching 2. From Ref 9.

4.4. Disorder and intermixing effects on the acoustic phonon modes.

Till now, we have considered perfect SL's with sharp interfaces between the different layers. Special care has to be taken to obtain abrupt interfaces during the epitaxial growth. In a perfect GaAs/AlAs SL one has a succession of distinct planes Ga, Al or As, according to a well predefined sequence. In an alloy (or mixed crystal) of $Ga_{1-x}Al_xAs$ the Ga and Al planes are no more distinct. The Ga and Al atoms are then randomly distributed on the cation sublattice. In every cation plane, the proportion of Al atoms is close to x .

The alloying affects are sometimes difficult to avoid in SL, particularly in the case of very short period SL owing to the existence of a large density of interfaces. But in certain circumstances it is useful to locally provoke such effects starting from periodical SL, after inducing an intermixing (or compositional disordering) operation. For instance, to obtain lateral optical and electronic confinement in GaAs / $Ga_{1-x}Al_xAs$ multiquantum wells, different techniques have been proposed to achieve intermixing in such layered structures. The intermixing results fron the interdiffusion of Ga and Al atoms which induces the transformation of the superlattice to an homogeneous alloy medium $Ga_{1-y}Al_yAs$ with $y = xd_{Ga_{1-x}Al_xAs}/D$. Such a compositional disordering increases the bandgap energy, compared to the lowest energy transition $E_{1e\text{-}hh}$ and significantly lowers the refractive index with respect to the as-grown GaAs / $Ga_{1-x}Al_xAs$ SL. One can thus achieve a lateral optical and electronic confinement in the quantum wells which has been successfully applied to optoelectronic device fabrication. Normally, starting, from undoped SLs with defect-free interfaces one cannot obtain intermixing just by annealing since the interdiffusion coefficient is too small even at high temperatures (800 - 1100 °C). Intermixing can be achieved either by introduction of impurities by diffusion or ion implantation, or by techniques not involving impurities, as CW or pulsed laser in radiations which induce local melting followed by recrystallization.

Raman scattering can give information on the intermixing degree as the $Ga_{1-x}Al_xAs$ / GaAs SL and the corresponding $Ga_{1-y}Al_yAs$ alloy have quite different Raman signatures. This is clear on the optical mode frequencies (Fig 20 a) and also in the low-frequency FLA modes (Fig 20 b). The FLA modes are due to the existence of a periodic modulation of the acoustic, optical and photoelastic properties, and the intermixing process (here an implantation followed by an annealing) which breaks this modulation strongly diminishes the FLA mode intensity accordingly.

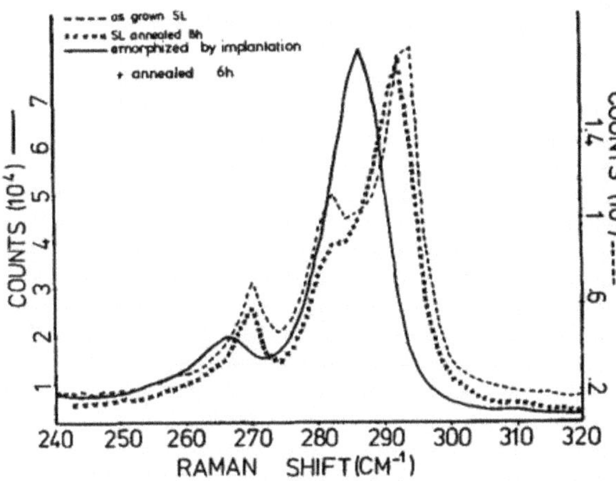

Fig 20 a - Raman spectra of a GaAs/AlAs SL under different treatements achieved to obtain an intermixing. Optical modes.

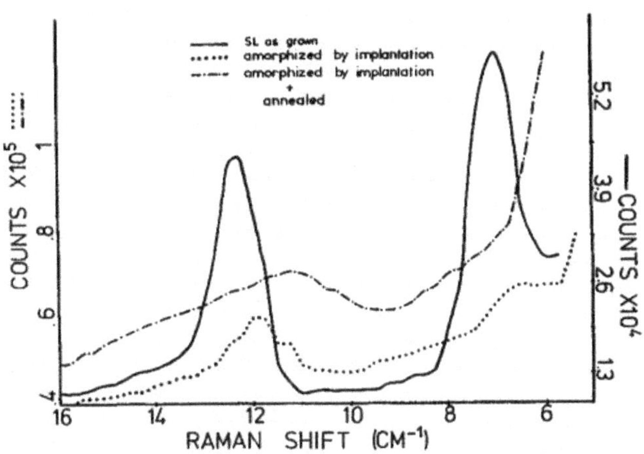

Fig 20 b - Same as - a, but for folded acoustic modes. From Ref 10.

5. Acoustic and acousto-optic properties of SL, for a general propagation of light and sound.

Table I - Expressions of the effective constants of superlattices in the long wavelength regime which have to be averaged over the constitutive layers. (See Refs 11, 12).

Dielectric tensor (all symmetries)	Elastic constants (Orthorhombic layers with principal axes parallel to x_1, x_2, x_3)	Photoelastic tensor components
$1/\varepsilon_{33}$	$1/C_{33}$	p_{3333}/C_{33}
$\varepsilon_{13}/\varepsilon_{33}$	C_{13}/C_{33}	$\varepsilon_{11}^2 p_{1133}/C_{33}$
$\varepsilon_{23}/\varepsilon_{33}$	C_{23}/C_{33}	$\varepsilon_{22}^2 p_{2233}/C_{33}$
$\varepsilon_{11}-\varepsilon_{13}^2/\varepsilon_{33}$	$C_{11}-C_{13}^2/C_{33}$	$p_{3311}-p_{3333}C_{13}/C_{33}$
$\varepsilon_{22}-\varepsilon_{23}^2/\varepsilon_{33}$	$C_{22}-C_{23}^2/C_{33}$	$\varepsilon_{11}^2(p_{1111}-p_{1133}C_{13}/C_{33})$
$\varepsilon_{12}-\varepsilon_{13}\varepsilon_{23}/\varepsilon_{33}$	$C_{12}-C_{13}C_{23}/C_{33}$	$\varepsilon_{22}^2(p_{2211}-p_{2233}C_{13}/C_{33})$
	$1/C_{44}$	$p_{3322}-p_{3333}C_{23}/C_{33}$
		$\varepsilon_{11}^2(p_{1122}-p_{1133}C_{23}/C_{33})$
	$1/C_{55}$	$\varepsilon_{22}^2(p_{2222}-p_{2233}C_{23}/C_{33})$
	C_{66}	$\varepsilon_{11}\varepsilon_{22}p_{1212}$
		$\varepsilon_{11}\varepsilon_{22}p_{1221}$
		$\varepsilon_{11}p_{1313}/C_{55}$
		$\varepsilon_{11}p_{1331}/C_{55} + \varepsilon_{11}/(\varepsilon_{33}C_{55})$
		$\varepsilon_{22}p_{2323}/C_{44}$
		$\varepsilon_{22}p_{2332}/C_{44} + \varepsilon_{22}/(\varepsilon_{33}C_{44})$

Up to now, we have only considered the propagation of light and acoustic wave along z, the direction perpendicular to the layers, which is also the direction of growth of the SL. Anyhow, due to the high refractive index of semiconductors ($n_{GaAs} \sim 3.6$) the light beam, incident from air, propagates nearly normal to the layers. One can wonder what happens for a general direction of propagation of light and sound. The problem has been solved for superlattices for which the period D is small compared to the acoustic and light wavelength (long-wavelength approximation). In this approximation the SL behaves as an homogeneous crystal whose effective properties are functions of the physical properties of the constitutive layers. Concerning our field of interest, the effective components of the rigidity, dielectric and photoelastic tensors are involved into expressions which are actually mean values, the summation being performed over the constitutive layers, each of them having a weight equal to its relative thickness. The method which is used for the obtention of the effective constants of the SL in this approximation is based on the invariance of certain components of the stress, strain, electric and displacement fields[11,12]. As to the effective density ρ^e of the SL, its expression is given by the trivial expression $\rho^e = x\rho + (1-x)\rho'$, where $x = d/D$. The elastic constants as well as the dielectric and photoelastic components of the SL are given in Table I. The effective elastic, dielectric and photoelastic constants can be obtained by equating for each of the quantity A, reported in this table its value A^e in the effective medium with its average over the two constituents of the superlattice;

$$\text{i.e:} \quad A^e = xA + (1-x)A'$$

Conclusion

Scattering of light from acoustic phonons constitutes a versatile and powerful tool to investigate optical, acoustical and acousto-optical properties of superlattices. The technique is also very useful to determine the period, the layer thicknesses and compositions, the sharpness of the interfaces, the inhomogeneities along the surface or in the direction z perpendicular to the layers.

The most investigated direction for the propagation or interaction of light and acoustic waves is along the z axis. Particularly striking is the coupling of light with the so-called "folded acoustic modes" and the creation of gaps and stop bands in this direction.

We have shown that for periods of the SL close to half the light wavelength, the SL acts as a Fabry-Perot with high energy density gain inside the structure.

Yet, concerning the interaction of light with ultrasonic waves, one needs superlattices adapted to the thickness of the superlattices. For current SL's whose thickness does not exceed a few microns, one needs ultrasonic waves of the order of a few Gigahertz, to obtain an effective coupling between light and sound since the interaction length has to be at least or the order of 4 acoustic wavelengths[13]. The expected result is a 180° diffraction of light which occurs with a very short accesstime microdevices[2]. To be lucid, one can say that scattering of light by acoustic phonons is now efficient for SL characterization and that diffraction of light by ultrasonic waves in SL is just a promising technique for light processing applications.

Acknowledgements

It is a pleasure to thank B. Djafari-Rouhani and J. He for their contributions to this study, and J. Chavignon for helping in the experiments and for the preparation of this manuscript.

References

1. J. Sapriel and B. Djafari-Rouhani, *Vibrations in Superlattices, Surface Science Reports* **10**, (1989) 189 and references therein.
2. J. He and J. Sapriel, *Appl. Phys. Lett.*, **55** (1989) 2292.
3. J. Sapriel, *Spectroscopic characterisation techniques for semiconductors technology III, SPIE,* **Vol 946**, (1988) 136.
4. J. Sapriel, J. Yao, J. He, J. Chavignon and R. Planel, *Superlattices and Microstructures*, **8** (1990) 301.
5. J. Sapriel, J. Chavignon, F. Alexandre and R. Azoulay , *Phys. Rev.* **B_34** (1986) 7118.
6. J. Sapriel, J. He, B. Djafari-Rouhani, R. Azoulay and F. Mollot, *Phys. Rev.* **B37** (1988) 4099.
7. J. He, J. Sapriel and H. Brugger, *Phys. Rev.*_**B39** (1989) 5919.
8. J. He, B. Djafari-Rouhani and J. Sapriel, *Phys. Rev.* **B37** (1988) 4086.
9. J. He, J. Sapriel and R. Azoulay, *Phys. Rev.* **B40** (1989) 1121.
10. J. Sapriel, E.V.K. Rao, F. Brillouet, J. Chavignon, P. Ossart, Y. Gao and P. Krauz, *Superlattices and Microstructures*, **4** (1988) 115.
11. B. Djafari-Rouhani and J. Sapriel, *Phys. Rev.* **B34** (1986) 7114.
12. J. Sapriel and B. Djafari-Rouhani, *Proceedings of the 3rd International School on Acousto-optics*, Wiezyca Poland (1986) 70.
13. J. Sapriel, *Acousto-optics*, (Wiley, New York, 1979).

PHASE ANOMALY OF FOCUSED SURFACE ACOUSTIC WAVE BEAMS

A. ALIPPI*, S.M. KIKKARIN+, P. VERARDI* and I.B. YAKOVKIN+

** Istituto di Acustica "O.M.Corbino", Via Cassia 1216, Rome, Italy*
+ Semiconductor Physics Institute, Lavrenteva 123, Novosibirsk, URSS

ABSTRACT

The acoustic field is analyzed, which is produced on the surface of a crystal by interdigital transducers shaped as circle sectors or according to the group velocity curve: the perturbation is obtained in terms of its Fourier spectral component and the effect of phase change is set in evidence which is produced across the focal point along the axis.

The problem of surface acoustic field produced by interdigital transducers of circular shape deposited on anisotropic substrates has been previously treated in a number of articles [1,2], where the amplitude of the generated wave has been theoretically determined and experimentally investigated. It is also well known from optics that any ray that passes through a focal point undergoes a relatively abrupt phase shift, which is equal to π in the case of a three dimensional isotropic medium.

In the present work we'll consider the phase behaviour of focused waves launched along the z axis on the y-cut surface of a $LiNbO_3$ crystal by two types of transducers: those shaped as circular arcs and those shaped according to the group velocity function of the surface waves in the material. Strong angular anisotropy of the slowness curve of the surface acoustic waves is taken into consideration and the corresponding ϑ angular dependence was calculated in the range $-180^\circ < \vartheta < +180^\circ$ around the z crystallographic axis by 10° steps and by finer 3° steps in the range $-15^\circ < \vartheta < +15^\circ$, then interpolated by cubic spline for use in the computational work.

The material parameters used in the calculations are given in Table 1.

TABLE 1

Elastic constants (10^{10} N/m^2):	c_{11}=20.3, c_{12}=5.3, c_{14}=0.9,c_{33}=24.5, c_{44}=6.0
Piezoelectric constants (C/m):	e_{15}=3.7, e_{22}=-2.5, e_{31}=0.2, e_{33}=1.3
Dielectric constants:	ε_{11}=44.3, ε_{33}=29.7
Density (g/cm^3):	ρ=4.7

In order to determine both the amplitude and phase of the scalar field of the propagating beam, the conventional procedure was used of decomposing the beam in its Fourier plane wave spectrum, then of integrating it with the proper phase factor that takes into account the propagation effect:

$$u(\mathbf{r}) = \int U(\mathbf{k})\cdot\exp(i\mathbf{Kr})d\mathbf{k} \tag{1}$$

where integral is performed on all possible **k** vectors. If one considers a monochromatic wave, one can write the integral as follows:

$$u(r,\alpha) = \int_{-\pi}^{\pi} U(\vartheta)\cdot\exp[iK(\vartheta)r\cdot\cos(\alpha-\vartheta)]K(\vartheta),d\vartheta \tag{2}$$

where $U(\vartheta)$ is the angular spectral distribution given by

$$U(\vartheta) = (2\pi)^{-1} \int_{\pi+\Delta\phi/2}^{\pi-\Delta\phi/2} u_{os}\cdot\exp[-iK(\vartheta)r\cdot\cos(\alpha-\vartheta)]\cdot Rd\phi, \tag{3}$$

where (r,α) are the polar coordinates of an arbitrarily chosen observation point, $[K(\vartheta),\vartheta]$ are the coordinates in K-vector space, u_{os} is the amplitude of the plane waves at the origin source point with coordinates$[R(\phi),\phi]$, being $R(\phi)$ the transducer radius (angle-dependent in the case of a group velocity shaped transducer), and $\Delta\phi$ is the angular aperture of the transducer. It is convenient to choose the origin of the coordinates in the center of curvature of the transducer, when using a circular transducer, and in its focal point, when using a group velocity shaped transducer.

The spatial phase behaviour of a focused beam can be described in terms of local wave vector or, equivalently, in terms of the local phase velocity, which bears the information of the local phase changing ratio. Keeping in mind that in the vicinity of a point $\mathbf{r}_0$ the field can be expanded as $u(\mathbf{r}_0+\Delta\mathbf{r})=u(\mathbf{r}_0)\cdot\exp[i\Phi(\mathbf{r}_0+\Delta\mathbf{r})]$, from the local K-vector definition $\mathbf{k}_l(\mathbf{r}_0)=\partial\Phi(\mathbf{r}_0+\Delta\mathbf{r})/\partial(\Delta\mathbf{r})$ and (1), one can obtain

$$\mathbf{k}_l(\mathbf{r}) = [\int \mathbf{k}\cdot U(\mathbf{k})\cdot\exp(i\mathbf{K}\mathbf{r})d\mathbf{k}]\cdot[\int U(\mathbf{k})\cdot\exp(i\mathbf{K}\mathbf{r})d\mathbf{k}]^{-1} \tag{4}$$

which is related to the local phase velocity through $\mathbf{v}_l=\omega/\mathbf{k}_l$, with ω the angular frequency of the wave. The group velocity being a vector, it can be conveniently expressed by its magnitude and direction.

In the experiments, we used a circular interdigital transducer, conventionally produced on the surface of a y-cut $LiNbO_3$ sample, with finger distance along the radius corresponding to a wavelength $\lambda=17$ μm, or central frequency $f=205$ MHz, radius R=2.7 mm and angular width $\Delta\phi=18^\circ$ around the crystallographic direction z of the sample.

The calculated amplitude distribution for the circular transducer is given in Fig.1a:

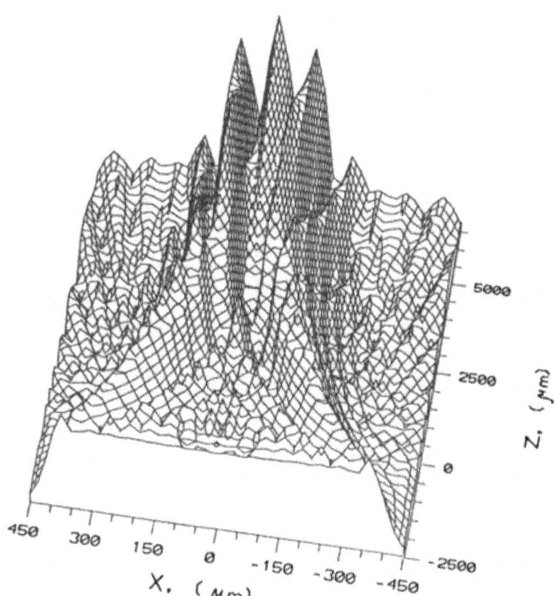

Fig.1a The spatial distribution of the focused beam amplitude for circular shaped transducer.

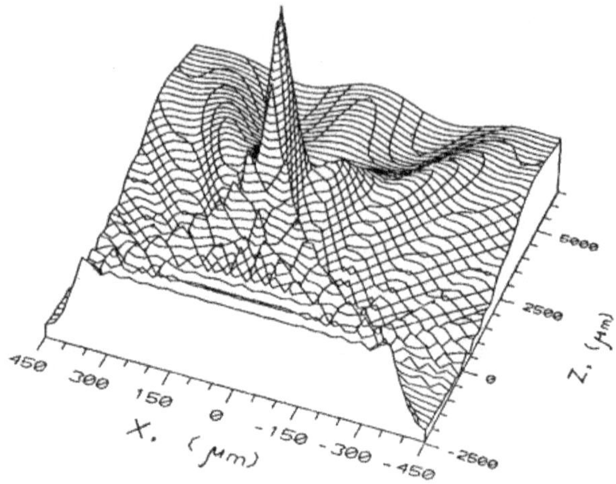

Fig.1b	The spatial distribution of the focused beam amplitude for group velocity shaped transducer.

the origin of the coordinate system corresponds to the center of the transducer, which is mainly aligned along direction x and launches waves along the crystallographic axis z, as mentioned earlier. As it is reported in literature [1], the strong anisotropy of the geometrical configuration used for the crystal does not permit good focusing from circular transducers. The field pattern has to be compared with the experimentally observed one, and good agreement is found between the data. The focal point of the transducer can be positioned at the maximum of the transversal field confinement, and it corresponds roughly to twice the radius of curvature. The case of a group velocity shaped transducer is reported in Fig.1b: in this case, there is a sharp focusing of the beam in the geometrical focal point region and the field pattern differs sensibly from the previously examined case.

The spatial distribution of the phase velocity is presented as three dimensional relief maps in Fig.2a for the case of a circular transducer, and in Fig.2b for that of the group velocity shaped one.

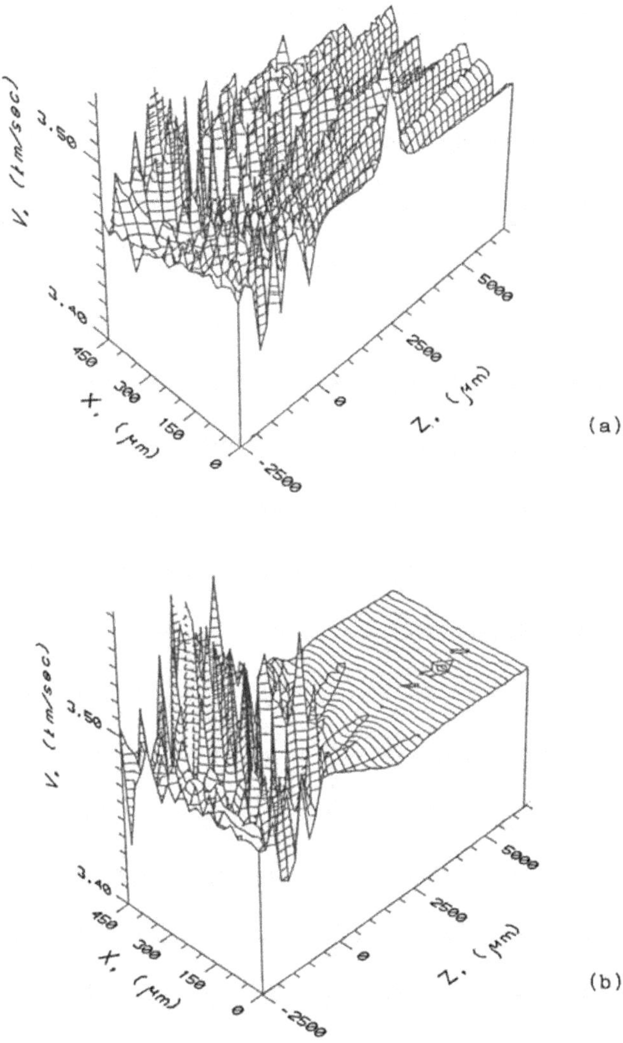

Fig.2 A half-aperture spatial distribution of the local phase velocity magnitude for (a) circular shaped, and (b) group velocity shaped transducer.

A half-aperture picture is shown in order to evidentiate the surface displacement along the z axis. Around the focal point a well developed oscillating pattern is set in evidence in both the examined cases. In the case of circular transducer there is a sharp peak of the phase velocity, which corresponds to the maximum of transversal energy confinement: an increase of the phase velocity, indeed, means a shortening of the acoustical wavelength and the accumulation of the phase shift for a ray crossing that region. According to the numerical computation, this phase shift closely approaches the value of 2π in the case of circular transducer, and $\pi/2$ in the case of the group velocity shaped one (see Fig.3).

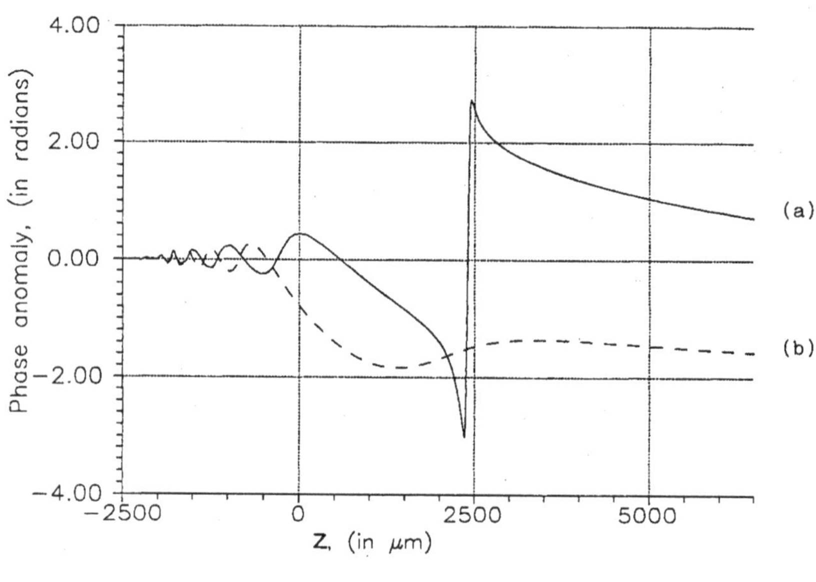

Fig.3 The phase shifting for the ray propagating along the z-axis for (a) circular shaped, and (b) group velocity shaped transducer.

This phenomenon is known as the "phase anomaly" in optics [3], where a phase shift of π is expected for a two dimensional lens focusing in an isotropic medium.

The orientation of the phase velocity vector, or of the wave vector itself, is shown in Fig.4 for both cases of the transducer shaped as a circular sector (a), or as the group velocity function (b).

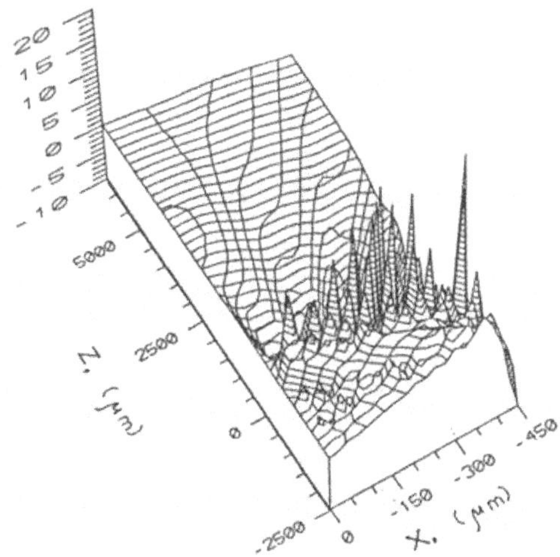

(a)

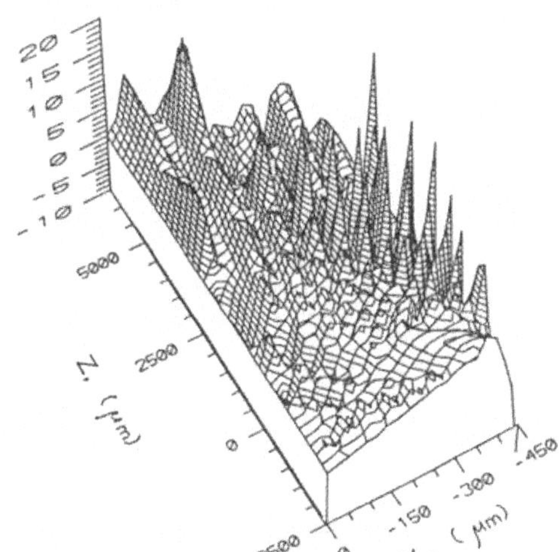

(b)

Fig.4 A half-aperture spatial distribution of the local phase velocity orientation as a deflection from the z-axis for (a) circular shaped, and (b) group velocity shaped transducer.

Data are presented as relief maps showing the angle of deflection of the local phase velocity vector from the x axis versus the coordinates x,y of the point. The rapid oscillations which are present in the "shadow" region that corresponds to the dark Airy rings in optics [3] are of no relevance, because of the vanishing amplitude of the field in that region.

In conclusions, the phase shift or "anomaly" of focused acoustic surface wave beams has been considered in terms of local phase velocity, for the case of a y-cut $LiNbO_3$ crystal sample, where waves are launched around direction z by transducers shaped as a circular sector or as the group velocity curve of the sample. The circular shaped case has been compared with experimental results. Phase shifts of 2π and $\pi/2$ have been found for rays passing through the focus for the cases previously mentioned.

REFERENCES

[1] Kharusi M. and Farnell G.W., *Proc. of the IEEE*, vol. **60**, n. 8, pp. 945-956, 1972.

[2] Borodii Yu., Grankin I. and Nepochatykh Yu., *Sov. Phys. Acoustics*, **31** (4), pp. 264-267, 1985.

[3] See, for instance, Born M. and Wolf E.: *Principles of optics*, Pergamon Press, pp. 445-449, 1975.

OPTICAL DETECTION OF VIBRATIONAL MODES IN CANTOR-LIKE COMPOSITE PLATES

F. CRACIUN, A. ALIPPI AND A. BETTUCCI
Institute of Acoustics "O.M. Corbino" C.N.R.
Via Cassia 1216 - 00189 Rome, Italy

ABSTRACT

The eigenmodes of Cantor-like piezoelectric composite plates have been measured with an acoustooptical device.Direct evidence of localized fracton modes has been obtained. It has been found that this simple fractal system presents multiple successive regimes of extended and localized modes.

1. Introduction

In recent years the gap between crystalline and disordered materials has been filled by self-similar deterministic fractals which present dilation symmetry (or scale invariance) in contrast with the translation symmetry possessed by the periodic structures. Consequently, their properties like mass and density of states (DOS) follow power laws whose exponents constitute new dimensions for the mass scaling (fractal dimension, D) and reciprocal space (spectral dimension, d_s) [1]. Real fractal systems are scale invariant over a certain range of lengths l, limited by the correlation length (beyond which the system can be considered homogeneous) and the smallest particle size a [1,2]. Correspondingly, two crossover frequencies have been identified, which separate the intermediate fractal range from the low and high frequency range, where extended phonon and particle modes are respectively expected [3]. The eigenmodes of the fractal structures in the intermediate region have been named fractons [2]: despite to the extended modes, they are strongly localized and follow the dispersion relation, $\omega \sim \lambda^{-D/d_s}$ where ω and λ are fracton frequency and wavelength, respectively [3]. The integrated DOS is given by power laws with different exponents in the phonon ($N \sim \omega^d$, d- Euclidean dimension) and fracton regimes ($N \sim \omega^{d_s}$) [3]. Experimental evidence for the different mode behaviour has been given by Courtens et al. [4] who measured the DOS by several techniques.

However, the pattern of vibration modes could not be directly probed sofar, so that the different localization of modes is usually deduced from scaling behaviour of the DOS.

In the present work we have adopted a different approach. By constructing suitable artificial one-dimensional fractal (Cantor) structures with piezoelectric ceramic and resin constituents ($a \simeq 0.8$ mm) we are able to excite vibrations over a wide frequency range (10 kHz < ω < 5 MHz) and to measure not only the frequency spectrum, but also the amplitude of the individual vibration modes over the sample surface. Our experimental mode patterns and frequency spectrum are successfully compared with the predictions of a scalar model, previously tested on several periodic samples [5].

2. Experiment and results

The samples are composite plates formed by alternating elements of piezoelectric ceramic and epoxy resin, following the iteration procedure presented in Fig.1 up to the 4th generation (a total of 31 elements). This is a simple self-similar structure, a so-called Cantor triadic sequence of fractal dimension D = ln(2)/ln(3). The size of smallest elements was chosen such that acoustic path of the ceramic and epoxy resin materials is the same. The thickness of the plate (0.3 mm) was designed in such a way that the lowest Lamb mode (S_0) propagates as an effective scalar mode to a very good approximation, as demostrated in Ref. 5 on periodic structures.

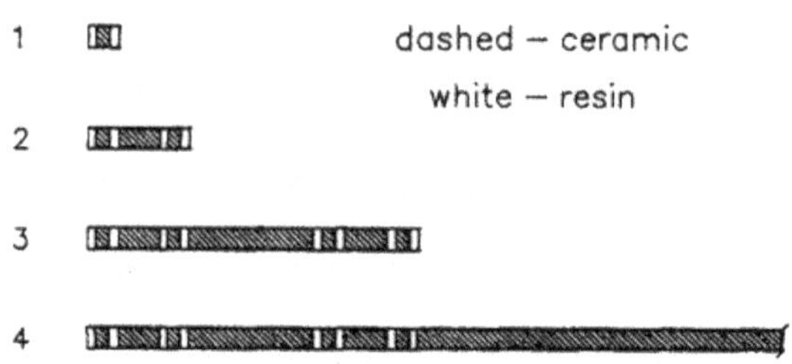

Fig. 1 First four stages in the construction of the composite Cantor triadic bar.

In order to probe the vibration spectra, highly reflecting Al electrodes were deposited on both surfaces of the plates; the admittance function was then measured in a bridge meter. Only modes symmetrical with respect to the free end of the plate are excited [5]. Once the resonant frequencies are identified, the sample is excited at each of these frequencies and the vibration amplitude is measured by a heterodyne laser probe which allows local detection of the normal component of the surface displacement down to 10^{-3}Å (Fig. 2). The coherent detection of the signal by a lock-in amplifier allows one to deduce separately the magnitude and phase of the displacement [6]. The entire sample surface was scanned by the laser beam across the different elements. Examples of vibration amplitude profiles are shown in Fig. 3a and b. The evidence of localized fracton modes and extended phonon modes is immediately apparent from their displacement. Modes 12, 13 (the number are identifying the modes in the DOS sequence; see next figure) are mostly localized on a few elements, while mode 20 essentially extends to the whole sample.

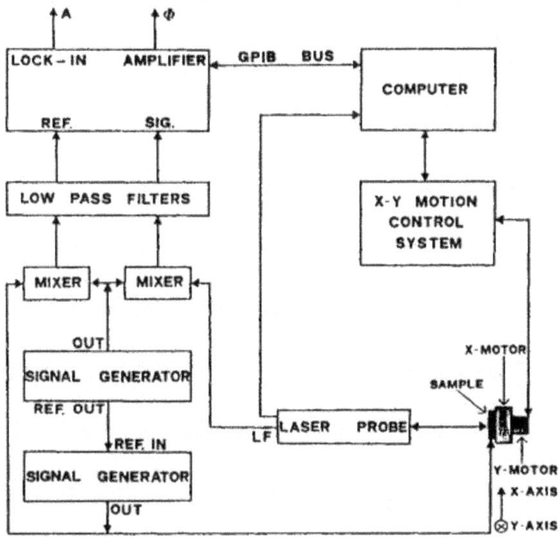

Fig. 2 Scketch of the experimental set-up for measuring surface displacement pattern of the sample.

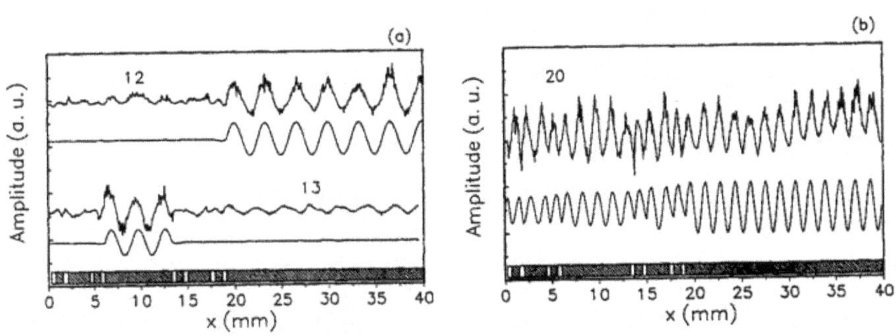

Fig. 3 Vibration amplitude profiles of localized (a) and extended (b) modes (upper curve - experimental; lower curve - calculated). The modes are numbered according to Fig. 4.

A simple scalar model for propagation [5] is used to calculate the frequency spectrum and the normal displacement amplitude profile of the plate surface. The calculated frequency spectrum (integrated DOS) is plotted on a logarithmic scale in Fig. 4a and 4b (Fig. 4b is an enlargement of the region (c) and (d) from Fig. 4a).

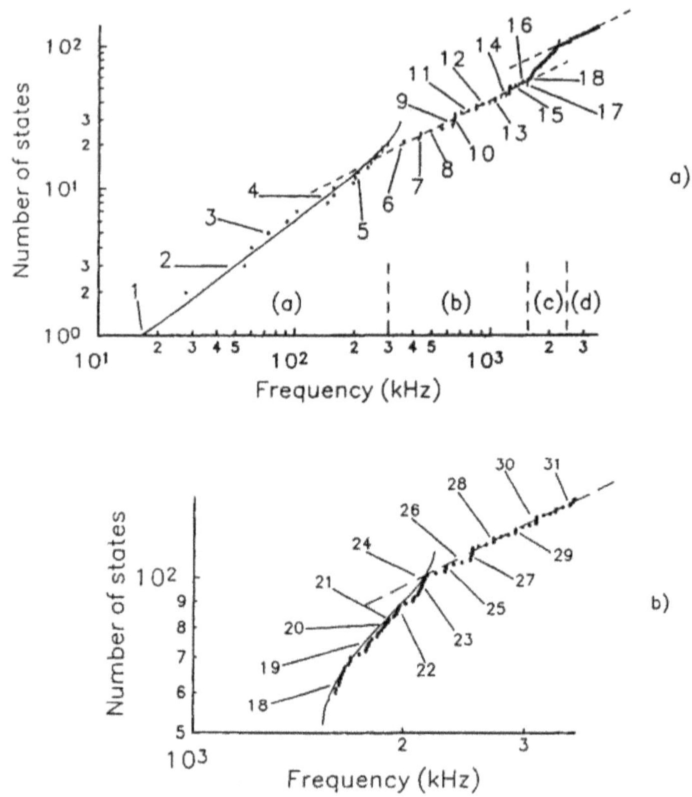

Fig. 4 a)Calculated frequency spectrum; b) Enlargement of the regions (c) and (d).

The numbers mark those modes which have been identified in the experiment on the basis of their displacements. As expected, the different localization of mode displacements closely reflects in clear crossovers in the slope of the integrated DOS: a first crossover occurs between the long-wavelength region of extended modes (a) and the subsequent region of localized modes (b), which extends up to a second crossover to the so called "particle regime" (c). However, we find that a second fracton region (d) appears at higher frequencies, also limited by a distinct crossover to region (c). This evidence is very clear both experimentally and theoretically, from our integrated DOS and from the displacement patterns. Those modes of our Cantor sample which have extended displacement profiles follow the dispersion of the corresponding periodic sample (continuous line in the regions (a) and (c)) very closely. As soon as their displacements are localized, the frequency of the modes begins to deviate from the periodic dispersion, falling in the successive gaps of the periodic structure. In brief, we are now interpreting the Cantor structure as obtained by inserting many "defects" in an otherwise periodic structure based on the smallest elements, which are in fact the most numerous. In this view we are lead to a generalization of the notion of the fracton regime, since we do expect that localized modes will appear whenever a gap is present between the extended mode bands. The dispersion of fracton modes follow a quasi-linear behaviour (in a log-log plot) which is fitted by dashed lines in regions (b) and (d). We find that the slope of these lines is very similar, and yields a spectral dimension $d_s = 0.67$ ($\simeq D$).

3. Conclusions

We have provided direct experimental evidence of the existence of multiple fracton and multiple extended-mode regimes both in the dispersion and in the displacement patterns a finite Cantor sample. We believe that our findings may be of interest to more complex fractal structures, and hope that they will stimulate further experimental and theoretical work.

4. Acknowledgements

We are grateful to M. Acciarini for expert technical assistance. Partial financial support by CNR, Progetto Finalizzato "Materiali Speciali per Tecnologie Avanzate" is acknowledged.

5. References

1. S. Alexander and R. Orbach, *J. Phys. (Paris) Lett.* **43**, L625 (1982).
2. R. Rammal and G. Toulouse, *J. Phys. (Paris) Lett.* **44**, L13 (1983).
3. A. Aharony, O. Entin - Wohlman, S. Alexander and R. Orbach, *Phil. Mag.* **B 56**, 949 (1987) and references therein.
4. E. Courtens, R. Vacher and E. Stoll, *Physica D38* **41** (1989); R. Vacher, E. Courtens, B. Coddens, A. Heidemann, Y. Tsujimi, Y. Pelous and M. Foret, *Phys. Rev. Lett.* .**65**, 1008 (1990).
5. A. Alippi, F. Craciun and E. Molinari, *Appl. Phys. Lett.* **53**, 1806 (1988).
6. A. Alippi, A. Bettucci and F. Craciun, *J. Appl. Phys.* **70**, 4004 (1991).,

ELECTRICAL, VIBRATIONAL AND ACOUSTIC FIELD TESTING OF ARRAY TRANSDUCERS FOR ULTRASONIC CLEANING

I. Gonzalez, F. Rodriguez, F.R. Montero de Espinosa and A. Ramos

UEI de Ultrasonidos, Instituto de Acústica
Serrano 144, 28006 Madrid - España

ABSTRACT

The development of new ultrasonic cleaning devices requires studies of acoustic propagation and ultrasound-medium interaction effects. The first step of these studies is the construction of an experimental set-up in which the ultrasonic transducers have an important role. The paper describes the construction and general testing of a 20 kHz array transducer designed for an experimental cleaning bath. The tests include a study of its vibration pattern by laser interferometry. Some discrepances between the simulated acoustic field and the experimental results are analyzed.

1. Introduction

The use of high intensity ultrasonic radiation in industrial applications is nowadays a common fact. Underwater cleaning[1] and cleaning baths[2] are some applications in which a liquid is used as transmission medium.

The design of new technical solutions to be applied in one of these cleaning processes involves the study of the acoustic propagation and the interaction effects produced into the bath. These studies will offer as output the acoustic frequencies and power needed to obtain the desired cleaning results. After that, the final system configuration, including the transducers, electronic excitation and bath geometry, can be defined. Normally, the frequency used in these devices lies between 20 kHz and 100 kHz. The aim of this paper is to show the design, construction and testing of ultrasonic transducer arrays made to cover the frequency range above commented. These transducers have been designed to form part of an experimental setup to measure the ultrasonic propagation parameters on an inhomogeneous medium having acoustic impedance and propagation velocity close to those of the water.

The test of the transducers must then include the measurement of their velocity amplitudes on the emitting surfaces and the study of their actual acoustic field distribution.

2. Transducer Design and Construction

The need of the study of the acoustic propagation into a so large frequency interval -20 kHz to 100 kHz- claims for the construction of a lot of different transducers to be able to cover all the range. To diminish as much as possible the number of different transducers, it is better to design transducers with large acoustic bandwidth. Nevertheless, the transducer efficiencies must be kept as high as possible to be able to produce cavitation acoustic levels in an efficient way-it is well-known that cavitation produces positive effects in the ultrasonic cleaning processes-. On the other hand, at so low frequencies, it is necessary to use transducers with large acoustic apertures if we want to perform propagation studies with acoustic fields having plane wave propagation zones and to be closer to the final design. These considerations involve the design of transducer arrays based on broad-band efficient transducers.

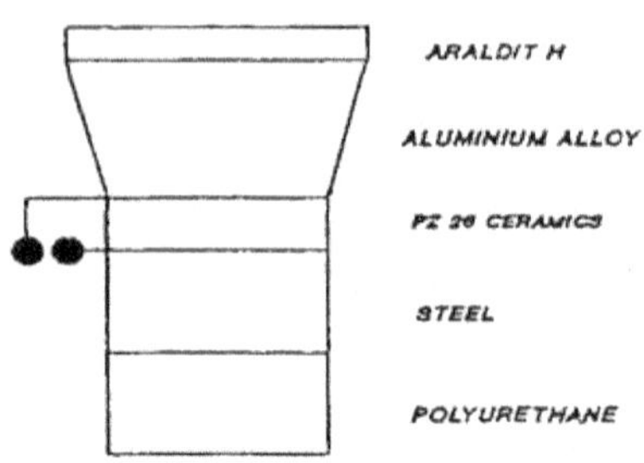

Fig. 1. Sandwich transducer scheme

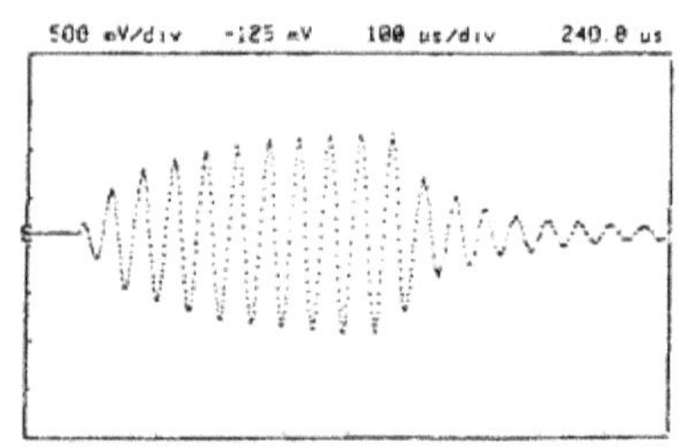

Fig. 3. Emitted acoustic signal (10 cycles)

Figure 1 shows the diagram of a 20 kHz transducer designed for efficient broad-band operation. The general structure corresponds to a typical piezoelectric sandwich transducer with a conical enlargement of the transducer head and an acoustic matching layer, to increase the efficiency and the acoustic frequency bandwidth. The transducer constituent materials are indicated in the same figure. The electrical characterization of the transducer shows the good performances of this design. Figure 2 shows the transducer conductance frequency bands in air and water loading. In figure 3 we can see the emitted acoustic signal when excited with a tone burst of 10 cycles at its resonance frequency.

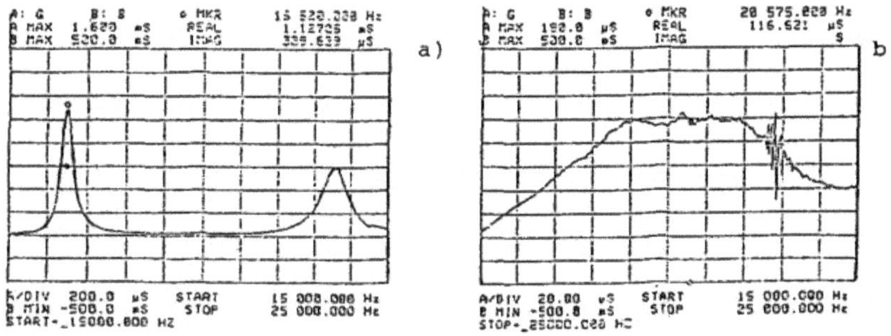

Fig. 2 . Conductance frequency bands: a) air, b) water.

The measurements were made with a HP-4194A impedance analyzer introducing the transducers in a wide water tank. If we measure the Transmission Transfer Function,a 3 dBs bandwidth of 14% is obtained. The hydrophone used is a BK 8103 type. Measuring the transducer efficiency as:

$$\eta = (G^{-1}_{max,w} - G^{-1}_{max,a}) \cdot G_{max,w}$$

where w and a refer to water and air repectively, we obtain a value $\eta = 92\%$.

Taking this transducer as elementary constituent, we have constructed a seven elements circular array placing one transducer at the center and the other six simmetrically spaced around it. The external diameter of the array is 180 mm. The moulding material is a silicone rubber.

3. Surface Velocity Amplitude Study

To ascertain the acoustic field emitted by this array, an experimental study of the transducer surface vibration was made. Two different procedures were followed: laser interferometry and piezoelectric sensors. Firstly we made scans over the array surface with the interferometric probe -Polytec OFV 2100/300- in air. The scans showed that the array surface behaves quite similar to a piston with negligible amplitude and phase variations on the transducer surfaces. The silicone rubber surface had vibration amplitude values less than the transducer ones but no more than 4 dBs down. The phase shift observed in the rubber surface was approximately 40º which can be justified thinking on how this rubber is moved by the transducers. Next we made the same

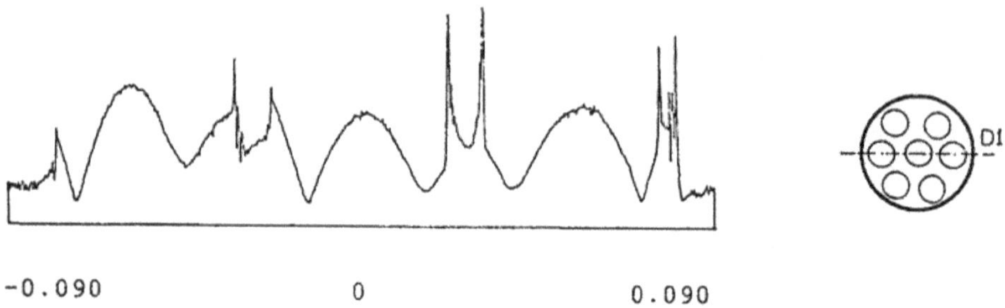

Fig. 4. Laser probe scan along the diameter D1 of the array

scans introducing the array into a water tank with perpex walls and 1x1.5x1.5 m of dimensions. Several anechoic structures were placed into the tank to diminish the acoustic reverberation. The Figure 4 shows one of these scans. The result is completely different respect to the air loading case. Each single transducer seems to vibrate with a central zone apodized in amplitude in counterphase with an external ring which presents an amplitude value 6 dBs down to the maximum of the central zone. The silicone rubber follows approximately the phase and amplitude of the external ring.

To take out the possible influence of the array structure on the vibration of the single transducers , the same scans in air and water loading were performed on one isolated sandwich transducer, showing similar behaviours than those seen in the array structure -Fig.5a-. The suspect of incontrolated vibrations on the sandwich heads led us to the measurement of the transducer surface vibrations by using piezoelectric sensors. We developed a special array vibrometer of 16 elements. This array was made following the next steps. First a piezoceramic plate was bonded over a thin cooper sheet providing electric contact between the ceramic electrode and the cooper sheet. The plate had 0.4 mm on thickness whereas the cooper was 0.07 mm thick. Next, a ceramic strip 50 mm long and 3 mm wide was cutted and then, 16 ceramic elements of 3x3 mm were obtained dicing transversally the ceramic strip. After soldering very thin cables on the free ceramic electrodes, a watertight air cavity was constructed over the upper ceramic surfaces.

Before bonding the array vibrometer over the sandwich surface with a wax mixture, the homogeneity of the 16 elements was checked. The differences on the impedance value of the elements was less than 3 dBs.

Fig. 6a shows the amplitude and phase of the vibration measured along the sandwich main diameter in air loading. The

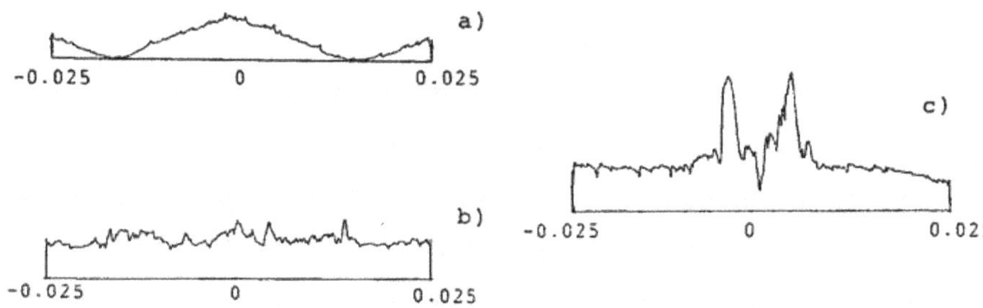

Fig. 5. Single transducer laser scans in water loading: a) Normal transducer, b) square diced and c)sector diced matching layers.

amplitude values are relative. The phase values are referred to the central vibrometer element. Figure 6b shows the same results when the sandwich was loaded with water.

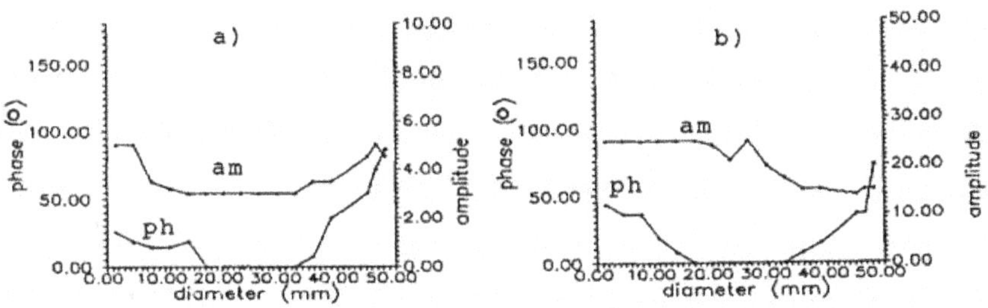

Fig. 6.- Vibration pattern of the single transducer surface measured with piezoelectric sensors: a) air and b) water

If we compare both the air and water loading results we can observe that the phase distribution remains the same whereas the amplitude pattern changes. The amplitude values are higher in the case of water loading because the transducer was acoustically optimized for these loading conditions. In the two cases, the amplitude distributions have variations less than 6 dBs and phase changes less than 90º. If we take as reference the central part of the transducer, it can be concluded that the transducer vibrates with a central part quite constant in amplitude and phase and an external ring shifted gradually in phase and with little bit higher displacements in air.

Comparing the interferometric and piezoelectric vibration measurements in air, we can deduce that the variations in amplitude and phase observed with the second technique can be originated by the interference between the longitudinal and radial vibrations on the sandwiches surfaces. The wax bonding can transmit the two kinds of vibration. Nevertheless, the fact that no phase changes were found when loading the sandwiches with water ,measuring with the piezoelectric sensors, led us to the suspect that the interferometric information presented some troubles.

Looking for a correct interpretation of the discrepances above commented other sandwich geometries were studied. Figures 5b and 5c show the surface interferometric measurements in water of two sandwiches with the same geometry and resonant frequency but with discontinuous matching layers. The first one had a matching layer with grooves forming equal sized square bars whereas the second one had the matching layer diced in circular sectors.

If we compare these figures with Fig.5a we can see that the flexural-like behaviour has disappeared. So, it should be concluded that the source of the discrepance comes from the radial vibration of the matching layer. Several authors pointed out that the changes of the refractive index of the water due to the acoustic pressure changes, can distort the interferometric measurements. In the light of the measurements here presented, we can infer that this distortion does not came from the acoustic field produced by a single piston -the diameter of the sandwiches is less than the acoustic wavelength in water- but from the acoustic field contribution of the radial vibrations -the magnitude of the radial displacements is similar to that of the longitudinal vibrations-.

The acoustic distortions of the interferometric measurements appear not only in the case of the single sandwiches but also in the array structure. In this last case, the acoustic field contribution of the silicone rubber which vibrates with a little phase shift and with non-flat emitting surfaces could justify the interference results.

The vibration measurement discrepances can be also solved measuring the acoustic field produced by the array and comparing it with the theoretical predictions using the different measured vibration patterns . This comparison is presented in the next paragraph.

4. Near and Far Acoustic Field Analysis

To assess the behaviour of the acoustic field emitted by this array a comparison between a number of measured spatial profiles and the corresponding calculated ones for an ideal piston, having the same diameter and central frequency, was performed.In all the cases the plotted magnitude was the positive peak value of the pressure signal when the transducer was excited by a tone burst of 10 cycles at its resonant frequency.The simulations were performed with free field conditions[3] using the Impulse Response Method[4].

This comparison also includes the theoretical predictions obtained when a vibration pattern as the one measured with the laser probe was supposed. In this case the impulse response of the whole array was calculated in each field point by adding the impulse responses of each of the elementary circular transducers to that corresponding to the silicone rubber radiating surface. This last response was obtained by subtraction between the responses of an aperture with the same size than the array and the circular apertures corresponding to the single transducers, assuming in this case that all of them have an uniform vibration amplitude equal to the measured average value in the silicone parts of the array.

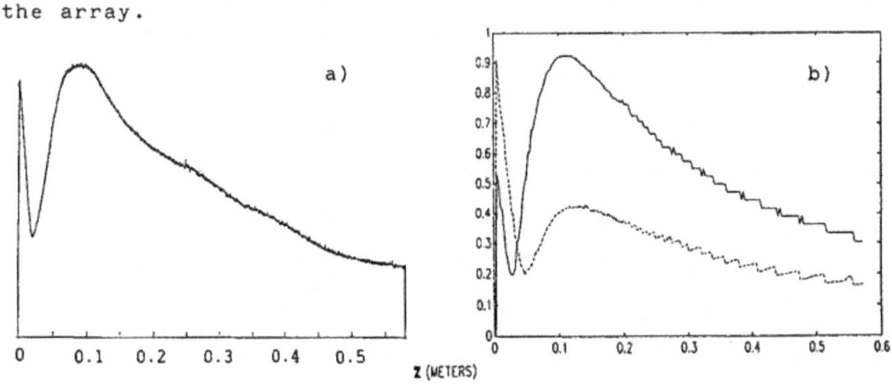

Fig. 7. a) measured array acoustic presure along the axis. b) simulated results for piston vibration assumption (——) and for the array vibration pattern of figure 4 (---).

Figure 7 gives the measured and calculated emitted acoustic pressures of the 20 kHz array along the propagation axis.It can be seen how the piston assumption agrees better with the actual array than the prediction in which a vibration pattern as that of Fig 4 was supposed.

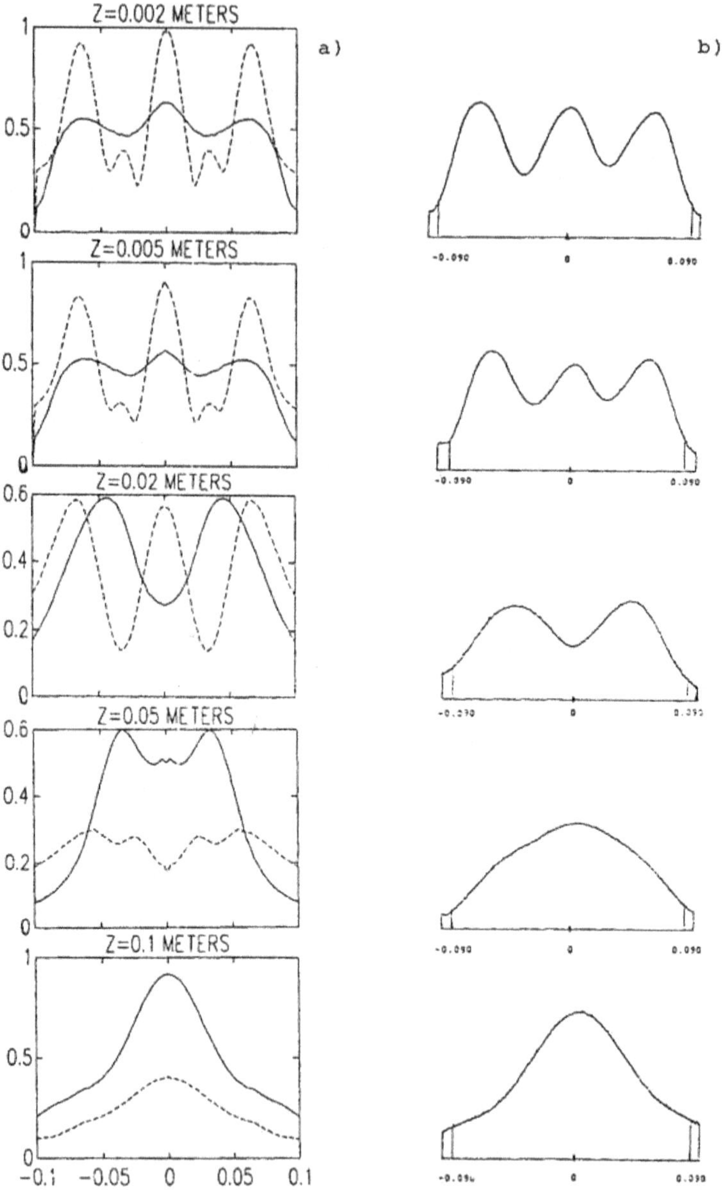

Fig. 8. a)Simulated and b) measured near field transversal profiles at different depths; —— piston vibration assumption, --- array vibration pattern of Fig 4.

This conclusion was corroborated when transversal profiles at different distances from the array surface were compared. In Figures 8 and 9 some calculated and measured transversal profiles at the near field (z=2,5,20,50,100 mm) and far field (z=200mm) zones are shown.

Whereas a reasonable agreement seems to exist between the measured profiles and the simulated piston case, strong differences, specially in the near field, appear when comparing them with the simulation results using the vibrational behaviour found with the interferometric probe.

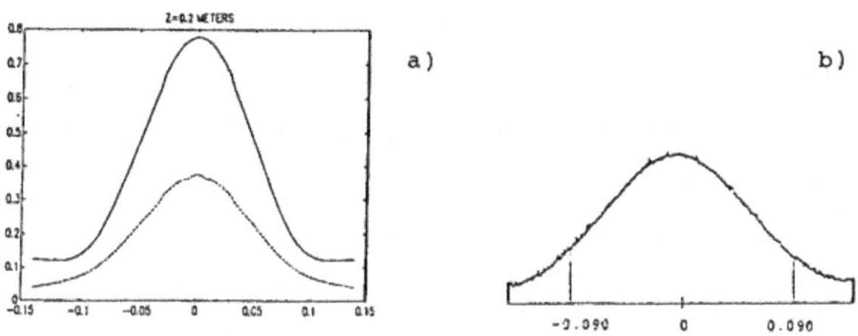

Fig. 9. Far field transversal profile, a) calculated b) measured.

5. Conclusions

Ultrasonic transducer arrays of 20 kHz of central frequency with a 3 dBs bandwidth of 14% and an efficiency of 92% were constructed.

The vibration study of the array emitting surface in water with an interferometric probe showed discrepances with the results obtained with a small array of piezoelectric sensors bonded on the emitting surface. In this last case the vibration pattern was closer to an ideal piston.

These discrepances disappeared making transducers with discontinuous matching layers, which presented a piston-like vibration with the two measurement procedures.

The comparison of the calculated acoustic field for the two different options with the experimental results confirms that the actual array presents a behaviour more coherent with a piston assumption.

The perturbations on the laser interferometry measurements could be originated by optoacoustic interferences in the water caused for the radial vibrations of the continuous matching layers and/or for the non-flat vibrational behaviour of the silicone rubber zones of the array.

6. Acknowledgements

This work continues some developments made under a BRITE contract nº BE-3009/89. The assistance of J.M. Torregrosa and M.A. García-Olias is appreciated.

7. References

1. S Trench and G. Hayward, *Proc. Ultrasonic International* (1991) (in press).
2. Rozemberg, *Physical Principles of Ultrasonic Technology* (Part III) Ed. Plenum, **vol 1,** (1973) .
3. F.J. Rodriguez, A. Ramos, J.L. San Emeterio, E. Riera *Anales de Física. Serie A.* (1991) (in press).
4. P.R. Stepanishen, *J. Acoust. Soc. Am.* **49 (5)**, (1971) p. 1629-1638

RECONSTRUCTION OF SURFACE VELOCITY DISTRIBUTIONS FROM A LIMITED NUMBER OF ACOUSTIC PRESSURE FIELD MEASUREMENTS

J.R. Blakey
Physics department, University of Surrey
Guildford, Surrey GU2 5XH, U.K.

Abstract

The accurate prediction of the acoustic field of a transducer requires a detailed knowledge of its surface velocity distribution. The normal surface velocity distribution of a piston-like plane transducer can be deduced from a spatial transformation of the pressure distribution measured over a plane parallel to the transducer surface. This method requires a large number of time-consuming precision measurements with a corresponding computational effort.

Less intense methods which can predict radiation fields with an acceptable accuracy are therefore of some interest. This contribution presents a method of reconstructing approximate surface velocity distributions by the inversion of the axial pressure distribution. Some typical results obtained are presented and discussed.

Introduction

It is well known that the radiation field of a transducer can be calculated exactly if its surface velocity distribution is known. Direct measurements of this quantity to a sufficient accuracy are extremely difficult. Thus indirect methods of surface velocity measurement offer an attractive alternative. One such method which has become increasingly popular is the reverse propagation technique [refs.1,2]. This involves the measurement of the acoustic field over a plane parallel to the transducer surface and applying a Fourier transform to this data to yield the corresponding distribution over the transducer surface.

In practice the acoustic pressure field distribution will be measured at a large number of uniformly spaced points, typically of the order of a thousand. This number of precision measurements and the corresponding computational effort is time-consuming. This fact provides an incentive for the investigation of alternative approaches which can predict radiated fields with acceptable accuracy but which require relatively few measurements.

Limited sets of measurements.

The basic feature of the limited number of measurements approach is that a transducer model is implied. One of the earliest examples of this approach is the well-known "effective radius" technique [ref.3]. The model in this case is an ideal circular piston in an infinite rigid baffle whose radius, the effective radius, differs from the physical radius. The effective radius can be found by determining the distance from the transducer of the last axial minimum. This method has been extended to spherical bowl transducers whose effective geometric parameters are found from the positions of the two final minima of the axial pressure distribution [refs.4,5].

Although the effective radius method can predict radiated fields remarkably well it only uses part of the available information, namely the positions of maxima or minima. The ideal piston would produce a series of equal amplitude maxima and zero -valued minima. The fact that this is not observed for real transducers indicates that the real distribution departs from the assumed model. This suggests that using the pressure information could improve the understanding of the radiation process. A study by the present author[refs.6,7,10] indicated that the large differences between the effective radius and the physical radius (up to 30%), which are often observed, are unlikely to be explainable by any reasonable shading distribution of surface velocity. It was found that the effective radius values could be accounted for by a phase variation over the transducer surface, in particular the phase reversal due to a nodal circle. In fact it was found that the effective radius and the nodal circle radius were identical. Similar results were also found when this approach was extended to spherical bowl transducers. It was also found that the axial pressure distribution could be reasonably well approximated using a two or perhaps three circular zone model.

It is therefore a logical step to reverse the procedure and attempt to invert the axial pressure data to find the correspond ing (approximate) surface velocity distribution. This is demonstrated in what follows.

Theory

The axial pressure distribution due to a circular piston-like transducer for which the surface distribution U(r) is a function of the radius only, can be represented by [ref.3]

$$p(z,t) = \frac{i\rho_0 ck}{2\pi} e^{i\omega t} \int_0^a \frac{U(\sigma)\exp(-ik\sqrt{z^2 + a^2}}{\sqrt{z^2 + a^2}} 2\pi\sigma d\sigma \quad (1)$$

For any arbitrary surface velocity distribution the integral is unlikely to be expressible in a convenient algebraical form. It is usual to use an appropriate finite sum approximation instead.The axial pressure can thus be expressed as

$$p(z) = \frac{ik\rho_0 ck}{2\pi} \sum_0^n u_j \frac{\exp(-ik\sqrt{z^2 + \sigma_j^2})}{(z^2 + \sigma_j^2)} .2\pi\sigma_j d\sigma_j \quad (2)$$

The axial pressure can thus be regarded as arising from a 1-d (radial) distribution of elementary Huygens ring sources with strengths proportional to u.

In order to determine the n unknown u values the form of eqn. 2 suggests that measurements of p(z) at n different values of the axial distance should be carried out. This would yield n simultaneous equations for the unknown u values. Since the pressure distribution is a complex quantity, it would be necessary in practice to determine both the pressure magnitude and phase distributions along the z-axis. Thus eqn.2 can be written more explicitly as

$$|p(z_j)|\exp[-i\phi_k] = \frac{ik\rho_0 ck}{2\pi}\sum_0^n u_j\frac{\exp(-ik\sqrt{z_k^2+\sigma_j^2})}{(z_k^2+\sigma_j^2)}.2\pi\sigma_j d\sigma_j \tag{3}$$

Taking the real and imaginary parts of eqn.2 gives two more equations:

$$|p(z_j)|\cos\phi_k = \frac{ik\rho_0 ck}{2\pi}\sum_0^n u_j\frac{\cos k(\sqrt{z_k^2+\sigma_j^2})}{(z_k^2+\sigma_j^2)}.2\pi\sigma_j d\sigma_j \tag{4}$$

$$|p(z_j)|\sin\phi_k = \frac{ik\rho_0 ck}{2\pi}\sum_0^n u_j\frac{\sin k(\sqrt{z_k^2+\sigma_j^2})}{(z_k^2+\sigma_j^2)}.2\pi\sigma_j d\sigma_j \tag{5}$$

Thus n measurements at different axial distances of the complex axial pressure will yield two sets of n simultaneous equations which are written for compactness in matrix notation.

$$\begin{aligned}[P_i^\alpha] &= [Q_{ij}^\alpha][U_j]\\ [P_i^\beta] &= [Q_{ij}^\beta][U_j]\end{aligned} \tag{6}$$

The n values of the unknown elements of [U] can now be recovered by the formal inversion of either of the equations (6) above.

$$[U_j] = [Q_{ij}^{\alpha}]^{-1}[P_i^{\alpha}]$$
$$[U_j] = [Q_{ij}^{\beta}]^{-1}[P_i^{\beta}] \tag{7}$$

SOLUTION OF THE EQUATIONS.

The sampling theorem gives an indication of the minimum number of measurements which will yield the maximum information about the surface distribution. It suggests that
$n > 4a$, where a is the transducer radius expressed in wavelengths. The use of this criterion yields two related disadvantages. For most transducers the number of points required will be large, which is contrary to the philosophy of this investigation which seeks to gain useful information from a limited number of measurements. The related disadvantage is that the solution of a large number of simultaneous equations runs the inevitable risk of ill-conditioning. The fact that many transducer source distributions are believed to vary slowly with radius implies that their spatial harmonic contents are dominated by the lowest spatial frequencies. This in turn implies that useful results might be obtained from a few measurement points.

A product-integral approach was employed for the inversion process in preference to direct matrix inversion [ref 9]. This requires that eq.1 is approximated by

$$p(z) = \frac{i\rho_0 ck}{2\pi} e^{i\omega t} \sum_{j=1}^{n} \int_{\sigma_{j-1}}^{\sigma_j} \frac{U(\sigma)\exp(-ik\sqrt{z^2+a^2})}{\sqrt{z^2+a^2}} 2\pi\sigma d\sigma \tag{8}$$

and U is taken to be constant over each interval of integration. This is seen to be equivalent to the well known annular zone approximation [ref.10]. Each integral is easily integrated to give

$$p(z_i) = \rho_0 c \sum_{j=1}^{n} u_j [\exp(-ik\sqrt{z_i^2+\sigma_j^2}) - \exp(-ik\sqrt{z_i^2+\sigma_{j-1}^2})] \tag{9}$$

The n values of the complex pressure measured at n points along the z-axisprovide n simultaneous linear equations which can now be solved.

A series of calculations have been carried out to test the principles outlined above. The variation of complex pressure with axial distance z has been calculated for a number of transducer surface velocity distribution functions using eq.1. This data was employed to perform the reconstruction via the system of simultaneous equations [9]. These equations were solved using the LU factorization technique.

A transducer diameter of 2.5 wavelengths was used in these initial simulations. In each case a 4 point reconstruction was carried out and the results compared with the original distributions. They represent a 4 zone approximation to the source using a constant zone width of 0.25a, where a is the transducer radius.

The first reconstruction attempted was for a uniform (ideal piston) surface distribution. The numerical results obtained differed from unity in the third decimal place only and are therefore not illustrated here. The results for two non-uniform surface distributions are shown in figures 1(a) and 1(b). It can be seen that the essential features of both distributions are brought out in these reconstructions.

As a further check on the method and as an indication of its likely usefulmess, the axial pressure variations due to the reconstructed source distributions have been calculated. These are also compared with the axial pressure variations due to the original source distributions and the results are shown in Figs.2(a) and 2(b). The agreement is remarkably good. The differences between the original and reconstructed distributions cannot be seen on the scale of these diagrams.

CONCLUSION

The work presented here gives support to the view that simplified models of transducer behaviour derived from a restricted number of measurements may provide predictions of the radiated field with an accuracy which is sufficient for many purposes.

The analysis presented here is valid only for circularly symmetric source distributions but real distributions may exhibit a degree of assymetry. But assymetrical distributions can be regarded as the superposition of a symmetrical component and an antisymmetrical component. The axial reconstruction method will recover the (approximate) symmetrical component, whereas the determination of the antisymmetrical component would require additional transverse measurements. The neccessary analysis is currently being carried out.

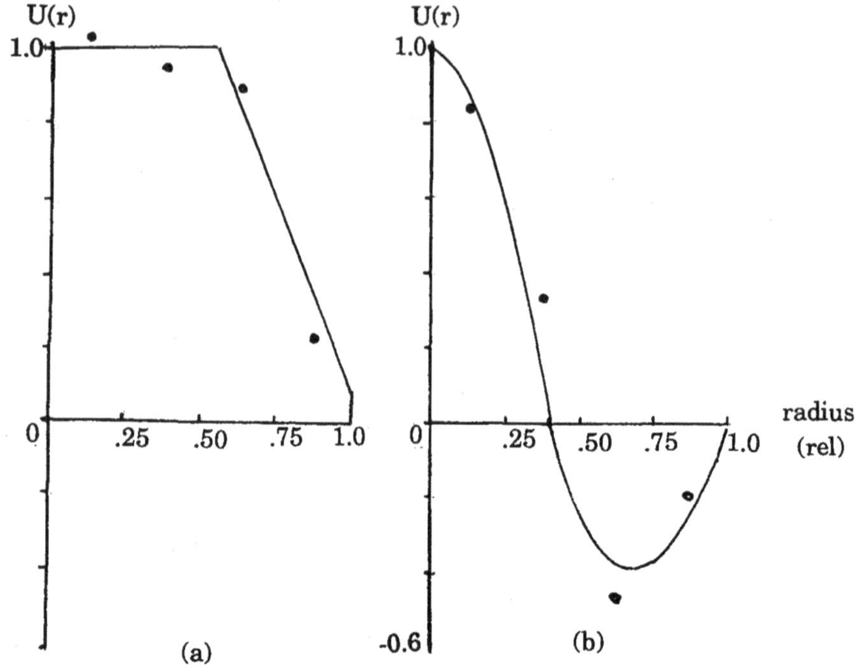

Figure 1. Comparison of source distributions and their 4-point reconstructions.
(a) Constant + linear shading.
(b) Zero-order Bessel function.

source distribution ——

reconstruction •

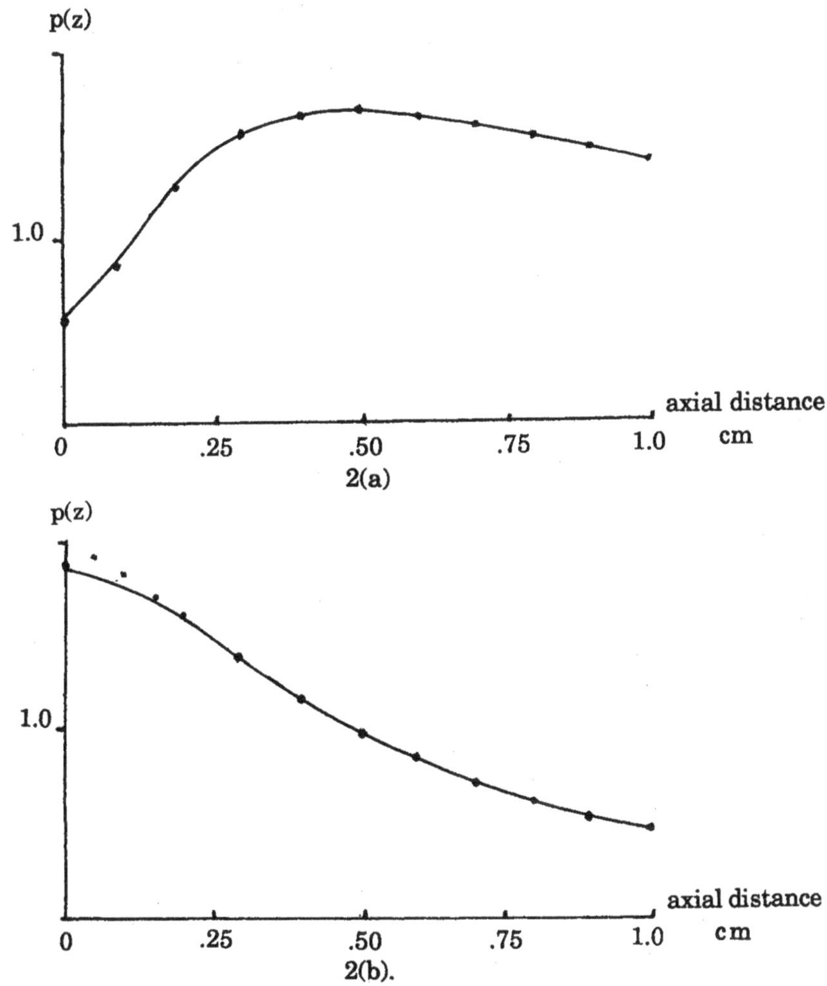

Figure 2. Comparison of axial pressure variations due to the original source distributions and the axial pressure variations due to the reconstructed sources.

(a) Constant + linear shading.
(b) Zero-order Bessel function.

REFERENCES

1. P.R.Stepanishen and K.C.Benjamin, J.Acoust.Soc.Amer. **71** (1982) 803
2 R.Reibold, Acoustica **63** (1987) 60
3. R.C.Chivers, Acoustica **70** (1990) 215
4. J.Adach and R.C.Chivers, Acoustica **62** (1986) 66
5. J.Adach and R.C.Chivers, Acoustica **70** (1990) 12,135
6. J.R.Blakey and R.C.Chivers, Acoustic Lett. **12** (1989) 217
7. J.R.Blakey and R.C.Chivers, Proc.IoA **12** (1990) 905
8. J.R.Blakey and R.C.Chivers, Acoustic Lett. **15** (1991) 26
9. I.J.D. Craig and J.C. Brown, Inverse problems in astronomy (Adam Hilger, Bristol U.K. 1986) p.58
10. R.O. Claus and P.S.Zerwekh, IEEE Trans. **SU30** (1983) 36

INFLUENCE OF THE MEDIUM INHOMOGENEITIES CORRELATION ON THE ELASTIC MEAN FREE PATH OF ACOUSTIC WAVES

E. SOCZKIEWICZ

Institute of Physics, Silesian Technical University

Gliwice, Poland

ABSTRACT

The elastic mean free path l and the transport mean free path l^t, determining character of the wave energy transport in randomly inhomogeneous media, have been calculated for various forms of the medium inhomogeneities correlation function. Values of l and l^t have been estimated in dependence on wavelength and on the mean size of the medium inhomogeneities. It has been stated that both l and l^t are greater for Gaussian form of the medium inhomogeneities correlation function than for the simple exponential form.

1. Introduction

Wave propagation in random media is encountered in a large variety of physics: acoustics, biophysics, optics, radiophysics, geophysics, atmospheric physics etc.[1,2,3]. Interest in studies of wave propagation in random media has increased in recent years in connection with their numerous applications. Using of acoustic waves in investigation of polycrystal rocks enables estimation their porosity as well as the structure of pores[4]. Measurements of ultrasonic waves attenuation are used in studies of metals structure[5,6], in investigation of tissues[7]. Intensive progress is noted in studies of acoustic waves propagation in ocean, where inhomogeneities are produced by turbulences, fluctuations in temperature and salinity[8,9]. When a wave propagates in a randomly inhomogeneous medium, energy is continually scattered out of the coherent wave into the incoherent or fluctuating wave by the inhomogeneities of the medium. These and other characteristics of wave propagation in random media have been subject of numerous investigations both theoretical and experimental[10,11]. Various of mathematical methods have been employed in studies of wave propagation in randomly inhomogeneous media[1,3], among them also the quantum field theory methods[10]. If $\lambda \ll a$, and $\lambda L \ll a^2$, where L denotes the distance traversed by the wave, λ - the wavelength, and a is the mean size of the inhomogeneities, the method of optical geometry (ray theory) can be used to calculate the mean square fluctuations of the time taken for a wave to reach a particular distant point, and also the fluctuations of phase and amplitude observed at that point[12]. The range of applicability of the method of small perturbations is determined by the inequalities: $\frac{\Delta A}{A} \ll 1$, $\Delta\varphi \ll 1$, where A denotes the amplitude of a wave, and $\Delta\varphi$ its phase fluctuations. The method of smooth perturbations (Rytov's method) has a somewhat wider range of applicability[1,11,13], as it demands only small relative changes of amplitude and phase over a distance of a wavelength. These two conditions are fulfilled if energy scattered during a wavelength of propagation distance is small in comparison to the initial energy of a wave. The method of small perturbations and of smooth perturbations have been considered in detail by Chivers[13], who has generalized Rytov's method by calculations of wave phase and amplitude fluctuations caused not only by variations in wave velocity but also in the density and the bulk modulus of the medium. The results obtained depend on the value of the parameter $D = 4L/(ka^2)$, where k is the wave number, and the other notations have been already defined above. Application of Green's function in analysis of acoustic wave propagation in random media has made multiple scattering studies feasible. If the wavelength is small in comparison to the mean scale a of the inhomogeneities, i.e. $ka \gg 1$, the parabolic equation method is used in calculations of the wave scattering coefficient[14].

The problems associated with random media may be grouped into three main categories: discrete random scatters (as in the case of emulsions), random continues and rough surfaces[3]. Because of interferences of scattered waves with the incident wave and between the scattered waves themselves, the character of wave energy propagation in randomly inhomogeneous media dependends also on the distance travelled by the wave[15]. At larger distance than the mean free path l , propagation of the acoustic field is diffusive, and when the disorder is increased the diffusion coefficient vanishes as a result of coherent interferences of scattered waves (the so called Anderson localization takes place)[15,16,17]. Sornette[17] using the method of Green's function has obtained the following formula for the wave field diffusion coefficient D dependence on the wavelength λ and the elastic mean free path l of waves:

$$D = D_0\left(1 - \frac{3\lambda^2}{\pi l^2}\right) \tag{1}$$

where $D_0 = \frac{1}{3} c(\omega)\, l(\omega)$, $c(\omega)$ denotes the effective wave velocity, $l(\omega)$ is the wave mean free path, ω - the cyclic frequency. The mean free path of waves has been defined[15] as the reciprocal of scattering cross-section of unit volume of a randomly inhomogeneous medium.

2. Statistical properties of random media

The main quantity used in discussion of acoustic waves propagation in random media is the mean square fluctuation in the acoustic refractive index $\langle \varepsilon^2 \rangle$, and autocorrelation of these fluctuations. The square of the medium refractive index we separate into mean and fluctuating parts: $n^2(\vec{r}) = 1 + \varepsilon(\vec{r})$ and the autocorrelation of fluctuations is given by: $\Psi(\vec{r}_1, \vec{r}_2) = \langle \varepsilon(\vec{r}_1)\, \varepsilon(\vec{r}_2) \rangle$ where brackets $\langle\ \rangle$ indicate an ensemble average. In the case of stationary and statistically homogeneous random media (statistical properties do not change with translation of the coordinate system) the following forms for the medium inhomogeneities correlation function are often used in theoretical considerations concerning wave propagation in random media[12] : Gaussian

$$\Psi_1(\vec{r}) = \langle \varepsilon^2 \rangle \exp(-r^2/a^2) \tag{2}$$

and simple exponential:

$$\Psi_2(\vec{r}) = \langle \varepsilon^2 \rangle \exp(-r/a) \tag{3}$$

where a denotes the radius of correlation, i.e. the mean distance over which fluctuations are correlated. Using relations: $\langle [\varepsilon(0) - \varepsilon(r)]^2 \rangle = 2[\Psi(0) - \Psi(r)]$ and $\langle (\frac{d\varepsilon}{dr})^2 \rangle = 2 \lim_{r \to 0} \frac{\Psi(0) - \Psi(r)}{r^2}$ one can demonstrate that the Gaussian form of the medium inhomogeneities correlation function $\Psi_1(r)$ corresponds to continuous changes in the medium refractive index, while the form $\Psi_2(r)$ to discontinuous changes of n . The correlation function $\Psi(r)$ determines values of scattering coefficient and the mean free path of waves in random media.

3. The acoustic scattering cross-section of unit volume of a randomly inhomogeneous medium

The differential cross-section of unit volume of a randomly inhomogeneous medium is defined as the ratio of the energy flux scattered by unit volume per unit solid angle, to the incident energy flux per unit surface[19]. In the case of acoustic waves the differential cross-section is given by the formula[3,19]:

$$\sigma(\Theta) = 2\pi k^4 \cos^2\Theta \, \varphi(2k \sin(\Theta/2)) \tag{4}$$

where Θ denotes the scattering angle, k is the wave number, while φ is the Fourier transform of the medium inhomogeneities correlation function, that in the case of a an isotropic random medium can be written, after integration over angles, in the form:

$$\varphi(k) = \frac{1}{2\pi^2 k}\int_0^\infty \sin(kr)\Psi(r) r \, dr \tag{5}$$

Using formulae given by Eq. 4 and Eq. 5, we have calculated the differential cross-section for the Gaussian form of the correlation function $\Psi_1(r)$:

$$\sigma_1(\Theta) = \frac{\langle\varepsilon^2\rangle k^4 a^3 \cos^2\Theta}{4\sqrt{\pi}} \exp(-k^2 a^2 \sin^2(\Theta/2)) \tag{6}$$

and for the simple exponential form $\Psi_2(r)$:

$$\sigma_2(\Theta) = \frac{2<\varepsilon^2>k^4a^3\cos^2\Theta}{\pi\left(1+4k^2a^2\sin^2(\Theta/2)\right)^2} \tag{7}$$

Integrating $\sigma(\Theta)$ over the solid angle 4π , one obtains the scattering cross-section σ_s :

$$\sigma_s = 2\pi\int_0^{\pi}\sigma(\Theta)\sin\Theta\, d\Theta \tag{8}$$

The obtained results are:

$$\sigma_{s1} = \frac{1}{a}\sqrt{\pi}<\varepsilon^2>k^2a^2\left[1-\frac{4}{k^2a^2}+\frac{8}{k^4a^4}-\left(1+\frac{4}{k^2a^2}+\frac{8}{k^4a^4}\right)\exp(-k^2a^2)\right] \tag{9}$$

for the Gaussian form of the correlation function $\Psi_1(r)$ and

$$\sigma_{s2} = \frac{1}{a}4<\varepsilon^2>k^2a^2\left[\frac{2k^2a^2}{1+4k^2a^2}+\frac{1}{k^2a^2}-\frac{2k^2a^2+1}{4k^4a^4}\ln(1+4k^2a^2)\right] \tag{10}$$

for the simple exponential form of correlation function $\Psi_2(r)$ given by the Eq. 3.

4. The elastic mean free path l and the transport mean free path l^t of acoustic waves in random media

The elastic mean free path of acoustic waves in randomly inhomogeneous media has been defined as reciprocal of the scattering cross-section of unit volume: $l = 1/\sigma_s$ Using Eq. 9 and Eq. 10 we have calculated $l_1 = 1/\sigma_{s1}$ and

$l_2=1/\sigma_{s2}$, respectively for the Gaussian and the simple exponential form of the medium inhomogeneities correlation function. In Fig. 1 values of $\log\frac{\langle\varepsilon^2\rangle l_1}{a}$ have been presented in dependence on ka .

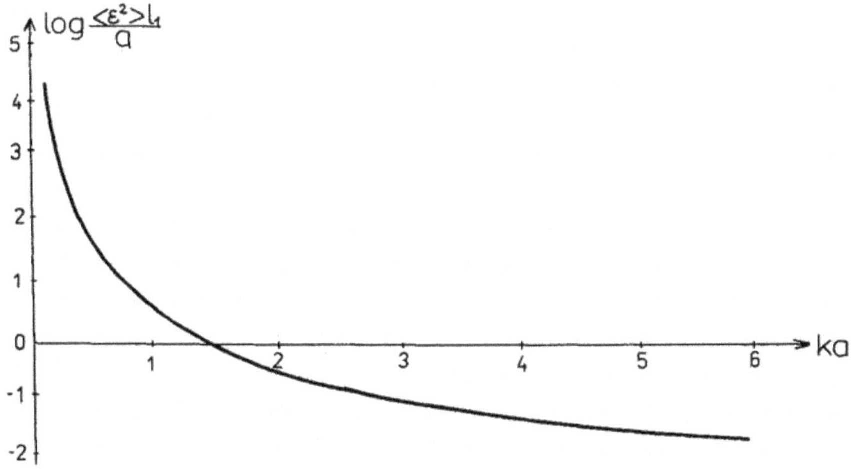

Fig. 1. Elastic mean free path l_1 of acoustic waves in random medium in dependence on ka , for the Gaussian medium inhomogeneities correlation function.

In Fig. 2 we have compared the mean free paths l_1 , and l_2 of acoustic waves in random medium respectively for the Gaussian and simple exponential medium inhomogeneities correlation functions, in dependence on ka values.
The transport of wave energy mean free path l^t , which appears in the theory of transport in the multiple scattering regime, is defined by the formula: $l^t=1/\sigma_s^t$, where σ_s^t is the transport scattering cross-section of unit volume of a random medium. σ_s^t is connected with the differential scattering cross-section $\sigma(\Theta)$ by means of the formula[15]:

$$\sigma_s^t = 2\pi \int_0^\pi \sigma(\Theta)(1-\cos\Theta)\sin\Theta \, d\Theta \tag{11}$$

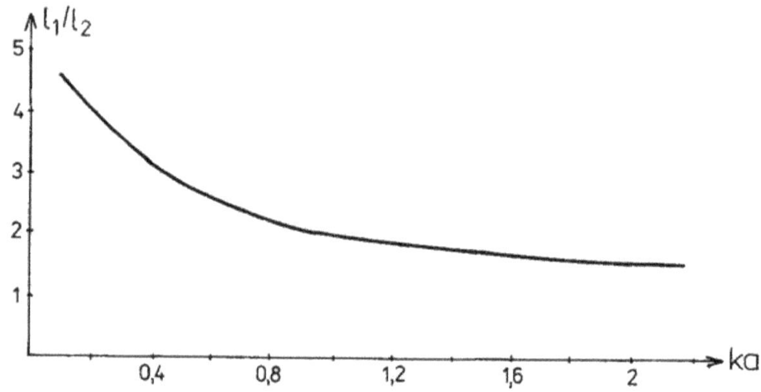

Fig. 2. Values of l_1/l_2 in dependence on ka .

Using the differential cross-sections given by Eq. 6 and Eq. 7 we have obtained the following formula for the transport cross-section in the case of the Gaussian form of the medium inhomogeneities correlation function $\Psi_1(r)$:

$$\sigma_{s1}^t = \frac{\sqrt{\pi}\langle\varepsilon^2\rangle}{a}\left[2 - \frac{16}{k^2a^2} + \frac{48}{k^4a^4} - \left(10 + 2k^2a^2 + \frac{32}{k^2a^2} + \frac{48}{k^4a^4}\right)\exp(-k^2a^2)\right] \tag{12}$$

while in the case of the simple exponential form $\Psi_2(r)$ of the correlation function:

$$\sigma_{s2}^t = \frac{\langle\varepsilon^2\rangle}{a}\left[\left(1 + \frac{2}{k^2a^2} + \frac{6}{k^4a^4}\right)\ln(1+4k^2a^2) - 2 - \frac{3}{k^2a^2} - \frac{4k^2a^2}{1+4k^2a^2}\right] \tag{13}$$

In Fig. 3 we have presented values of the transport mean free path and the elastic mean free path ratios l^t/l in dependence on ka values for medium inhomogeneities correlation functions given by Eq. 2 and Eq. 3.

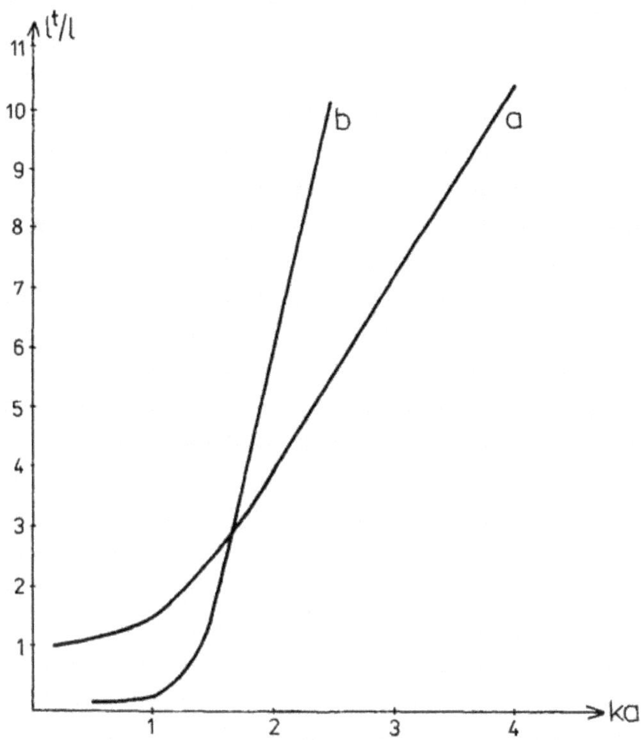

Fig. 3. Values of l^t/l in dependence on ka , a- in the case of Gaussian medium inhomogeneities correlation function, b- in the case of simple exponential form of the medium inhomogeneities correlation function.

In Fig. 4 we have compared values of transport mean free paths l_1^t and l_2^t for the Gaussian and the simple exponential forms of medium inhomogeneities correlation functions in dependence on ka values.

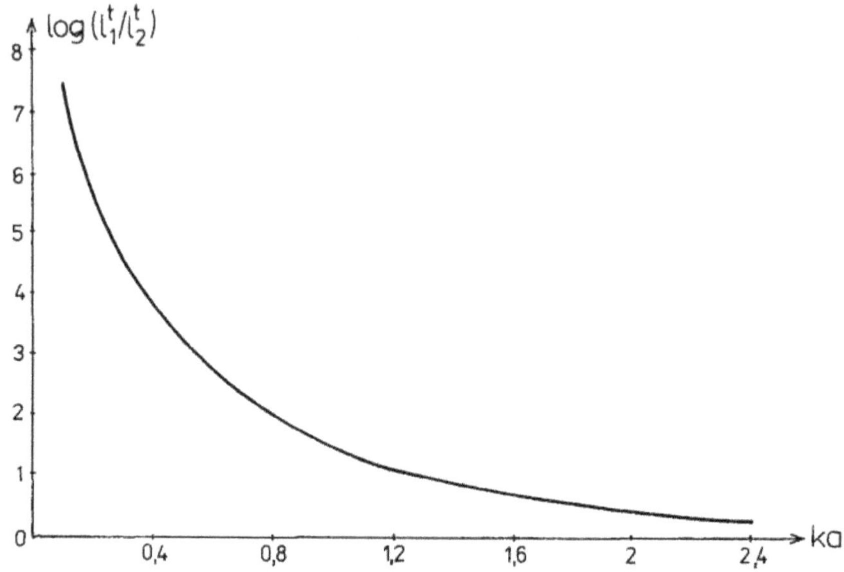

Fig. 4 Ratios of transport mean free paths l_1^t/l_2^t in dependence on ka values.

5. Conclusions

The elastic mean free path l and the transport mean free path l^t depend on the form of medium inhomogeneities correlation function $\Psi(r)$. For the Gaussian correlation function $\Psi_1(r)$ the elastic mean free path l_1 is greater than l_2 for the simple exponential correlation function $\Psi_2(r)$. Values of l_1/l_2 depend on values of wave number k and on medium inhomogeneities correlation radius a. l_1/l_2 decrease with increasing in ka values. If changes of the medium refractive index for acoustic waves are discontinuous the differential scattering cross-section $\sigma_2(\Theta)$ is greater than $\sigma_1(\Theta)$ for continuous changes in the medium acoustic refractive index. The ratio $\sigma_2(\Theta)/\sigma_1(\Theta)$ depends both on value of ka and on the scattering angle Θ . $\sigma_2(\Theta)/\sigma_1(\Theta) = 4{,}51$ for small values of ka and Θ , and it decreases with increasing in ka and Θ . In order to estimate the elastic mean free path value one needs to know the mean size a of medium inhomogeneities, and the value of the mean square fluctuations $\langle\varepsilon^2\rangle$ of the acoustic refractive index. For example in the case of acoustic waves scattering from the microstructure fluctuations in ocean, according to data given in the book of Flatte[20] : $a = 0.6\,m$, $\langle\varepsilon^2\rangle = 5\cdot 10^{-7}$, so for $ka = 5$, one obtains $l_1 = 31.2$ km.

Value of $\langle\varepsilon^2\rangle$ can be also estimated from measurement of acoustic waves effective velocity in a random medium in the case $ka \gg 1$. The method of Green's function in the Bourret approximation gives in this case the following formula for the real part of acoustic refractive index :

$$n_1 = 1 + \frac{\pi}{2}\int_0^\infty \varphi(k)\, k^2 dk \tag{14}$$

where $\varphi(k)$ is given by the Eq. 5. For the both considered in this paper forms of medium inhomogeneities correlation function, the above formula gives : $n_1 = 1 + \frac{\langle\varepsilon^2\rangle}{8}$, where $n_1 = c_0/c$, c denotes the effective wave velocity in inhomogeneous medium, and c_0 denotes the wave velocity in homogeneous medium.

6. References

1. B. J. Uscinski, *The Elements of Wave Propagation in Random Media* (McGraw-Hill, New York, 1977) .
2. S. M. Rytov, Y. A. Kravtsov and W. I. Tatarsky, *Introduction to Statistical Radiophysics* (Nauka, Moscow, 1978).
3. A. Ishimaru, *Propagation and Scattering of Waves in Random Media* (Mir, Moscow, 1981) .
4. T. J. Plona, *Acoustics of fluid saturated porous media.* 1982 *Ultrasonics Symposium*, p. 1044.

5. C. M. Sayers, *Characterization of microstructure using ultrasonics, in Ultrasonic Methods in Evaluation of Inhomogeneous Materials*, ed. A. Alippi and W. G. Mayer (Martinus Nijhoff Publishers, Dordrecht, 1987).
6. W. N. Reynolds and R. L. Smith, *Ultrasonic wave attenuation spectra in steels. J. Phys.* **D17** (1984) 109 .
7. R. C. Chivers, *Tissue characterization. Ultrasound Med. Biol.* **7** (1981) 1.
8. E. Soczkiewicz and R. C. Chivers, *Scattering of acoustic waves by turbulence. Proc. Institute of Acoustics* (G. B.) **8** (1986) 20 .
9. E. Soczkiewicz and R.C. Chivers, *Scattering of ultrasonic waves by turbulence. Acustica* **68** (1989) 33 .
10. L. A. Apresjan and Y. A. Kravtsov, *Transport Theory of Rdiation* (Nauka, Moscow, 1983).
11. E. Soczkiewicz and R. C. Chivers, *Propagation of acoustic waves in random media, a comparison of methods and results. Ultrasonics International 87 Conference Proceedings* (1987) 543 .
12. L. A. Chernov, *Waves in Random Media* (Nauka ,Moscow, 1975) .
13. R. C. Chivers, *Phase and amplitude fluctuations in propagation of acoustic waves in lossless continua with velocity and bulk modulus variations. I. F. T. R. Report* (Warsaw, 1978) .
14. E. Soczkiewicz and R. C. Chivers, *Propagation of acoustic waves in random media using the parabolic equation and Green s function methods. Acustica* **65** (.1988) 149.
15. D. Sornette, *Acoustic waves in random media,weak disorder regime. Acustica* **67** (1989) 199 .
16. D. Sornette, *Anderson localization and wave absorption. J. Stat. Physics* **56** (1989) 669 .
17. R. L. Weaver, *Anderson localization of ultrasound. Wave Motion* **12** (1990) 129 .
18. D. Sornette, *Acoustic waves in random media, coherent effects and strong disorder regime. Acustica* **67** (1989) 251 .
19. V. E. Derr and C. G. Little, *A comparison of remote sensing of the clear atmosphere by optical, radio and acoustic radar. Applied Optics* **9** (1970) 1976 .
20. S. M. Flatte, *Sound Transmission through a Fluctuating Ocean* (Cambridge University Press, Cambridge, 1979) p. 235.

SPATIAL OSCILLATIONS OF THE FRINGES IN THE DIFFRACTION OF LIGHT BY ULTRASONIC WAVES

ITALO GABRIELLI
Dipartimento di Fisica, Università di Trieste, Via A. Valerio 2
I 34127 - Trieste, Italy

ABSTRACT

After reviewing the application of the phase-vector analysis to the study of the Fraunhofer diffraction of a collimated light beam by ultrasonic waves, its application in predicting the fine structure of the diffraction fringes is presented. In the usually analyzed light-sound interaction, employing several ultrasonic waves, the diffraction lobes appear narrow, symmetric and centered on the predicted angular positions, while they become asymmetric, not centered and spatially oscillating with time, if the interacting ultrasonic waves are reduced to a very small number. We present and discuss the cases of light diffraction by one beam of traveling waves, by two adjacent antiparallel beams with frequency ratios 2/3 and 3/4 and by a beam of standing waves. The experimental verification of the predicted oscillations of the diffraction lobes is proposed.

1. Introduction

1.1. Analysis of the Experimental Arrangements

We consider the standard acoustooptic arrangements used to obtain a diffraction pattern in the Fraunhofer plane, limiting the study to the Raman-Nath (RN) range, i.e. ultrasonic (US) frequency and the RN parameter $V = 2\pi\Delta\mu L/\lambda$ must be combined avoiding to exceed such a range. Here $\Delta\mu$ is the maximum variation of the refraction index, λ is the light wavelength, L is the length of the light path in the US field. In the usual acoustooptic experiments light interacts with several US wavelengths Λ (WL's) . This research reports the predictions by phase-vector analysis (PVA) of the fine structure of the diffraction pattern, i.e. the shape of the diffraction lobes, if the light beam interacts with one or only a few (integer) number NW of US WL's. In the three considered cases, i.e. one beam of traveling waves, or two adjacent antiparallel beams of traveling waves with frequencies in the ratio of 2/3 or 3/4, or one beam of standing waves, a strong asymmetry of the diffraction lobes is found and spatial oscillations are evidenced. Antiparallel beams are defined as two parallel beams propagating in opposite directions.

The spatial and temporal modulation of light by US waves in the RN range can be analyzed by the RN method. It was applied also by our group to sophisticated arrangements. I employ here a very simple alternative method, the "phase-vector analysis" (PVA)[1,2,3], which is based on the direct application of the classical Huygens principle and presents the advantage of supplying a simple general formula for analyzing any of the usual set-ups. The principle scheme is shown in **Figures 1 and 2** taken from the Proceedings of Erice 1988 [1], and will be initially applied to a beam of traveling US waves.

1.2. The Mathematical Model

Let us first consider the screen positions corresponding to the standard angles (diffraction fringes of order n):

$$\theta_n = \arcsin n(\lambda/\Lambda). \qquad (1)$$

The light sheets of thickness dx in the x-direction, as shown in Figure 1, are considered the elemental units. As each sheet travels from the entering slit to a point of the screen defined by the angle θ_n it incurs two phase-shifts. The first one is due to the optical paths into the US field and the second is due to the geometrical path difference (see Fig, 2 b). This effect is taken into account by a wedge (of amplitude from 0 to $n2\pi$, see Fig.2 c, giving the equivalent path differences. The total phase-shift is therefore described by:

$$V\sin Kx \pm nKx, \qquad (2)$$

with $K = 2\pi/\Lambda$.

If we indicate by $\mathbf{E}_i$ the vector amplitude of the entering light beam, limited by a slit of dimensions p by q, where the width $p = \Delta x = m\Lambda$ (m integer), the vector contribution $\mathbf{dE}_n$ of said light sheet to the vector amplitude at a given screen position θ_n is then deduced, taking into account the said phase-shift:

$$\mathbf{dE}_n = \mathbf{dE}(x,V,n) = (\mathbf{E}_i/p).dx.\exp j\,\{V\sin Kx \pm nKx\}\,, \qquad (3)$$

where $\mathbf{dE}_i = (\mathbf{E}_i/p)dx$ represents the initial vector light amplitude of any single US light sheet. The total light amplitude interacting with a number m of waves is then deduced as the integral of the contributions of all light sheets included in the interval $p = m\Lambda$ (see Fig.2 d,e), namely:

$$\mathbf{E}_n\,(V) = \int_0^{m\Lambda} \mathbf{dE}_n = (\mathbf{E}_i/p).\int_0^{m\Lambda} dx.\exp j\,(V\sin Kx \pm nKx). \qquad (4)$$

It is easy to visualize the calculation of this integral, performed, of course, by a discrete sum of cosine and sine terms. **Figure 3** (from Ref.[2]) shows this for a fringe of the first order due to one wave of a single US beam. In each curve of Figure 3 the diffracting US sinusoid of the indicated RN parameter V, the dephasing wedge with amplitude from $-\pi$ to π, corresponding to the first diffraction order and the resulting diffracting grating are shown. The shifting effect $V\sin Kx+nKx-n\pi$ has been used, corresponding to Eq.2, but centered with respect to the slit p, for symmetry requirements. The "spring" or "snake" of the bottom curve represents graphically the integration, step by step, according to Eq.4, the result is shown, in amplitude and phase, by the arrow (whose large cross X is the tail). For comparison sake with the case of one WL the graphical building-up of the light vector in the case of 4 US WL is shown on the right side of the first set (marked NW 4). The resulting vector has the same angle and the relative length 0.58 of the case NW=1, as this is referred to the entering light, by dividing the vector resulting from the integral of Eq.4 by the number of WL's m We note that, for the diffraction angles defined by Eq.(1), the dephasing between two adjacent US WL is just 2π, and, as it results from Figures 2 d) and e), the relative intensity of the light is the same, independently of the (integer) number m of the US WL interacting with the light.

Both the light amplitude E_n and the light intensity I_n are measured in arbitrary units, by fixing respectively to one the amplitude of the whole entering light and to 100 the corresponding light intensity, so that: $I_n=100.E_n^2$.

Figure 4 (from Ref.[1]) shows the calculations by PVA of the light vector in case of the light diffraction by one US WL. For graphical reasons the light vectors are rotated clockwise by 90 degrees. The vertical sequences consider the temporal advancing of the US wave from the starting sinusoid position described as: $V\sin Kx$ (Eq. 2,3) to that represented by: $V\sin(Kx-\Omega t)$ where $\Omega t = \Delta x$, t increases from zero to 4/8 of the US period T and $\Omega = 2\pi/T$. The horizontal sequences are calculated not for angles fixed by Eq.1, but with dephasing wedges increasing (left side figures) or decreasing (right) from zero in steps of 1/32 of 2π. One observes that at some side points of the zero-order diffraction-lobe the vector length (computer printed) is greater than in the central position.

2. The Evolution of the Shapes of the Diffraction Lobes with Increasing Number of the Interacting Ultrasonic Waves

2.1. One traveling Wave

The calculations of the light vector amplitude and of the angle at a given screen position are used to plot the curves of **Figure 5**. It shows the spatial oscillations of the diffraction lobes obtained by the light interaction with one WL of a beam of traveling US waves in the angular interval including the orders n from -1 to 1, while the wave advances from zero to 5/8 of US period T [(+)]. The asymmetry is very strong, but the local values at the abscissae 0 and ±1, marked by the central vertical lines, are just the same as the maxima of the lobes obtained with 4 or 8 US waves interacting with the light beam (see Fig.6). Figure 5 evidences that, in case of one or very few interacting WL's, a strong mixing between the -1, zero and the +1 orders occurs and that we cannot imagine the diffraction lobe corresponding to a given order to be symmetric or quasi symmetric. This happens with optical diffraction in either amplitude or step-phase gratings.

(+) NOTE 1

In the Figures 5, 6, 8, 9, 13 and 14 the upper picture of each figure reports the relative light intensity I (from 0 to 50 % in Figures 5, 6, 8 and 9, from 0 to 100 % in Figures 13 and 14). The lower picture reports the angular behaviour from 0 to 360 (=0) degrees. The abscissa range includes a wedge increasing from -30 to 30/20 of 2π, and hence the positions ±20/20 = ±1 correspond to the order ±1, the central abscissa to the zero diffraction order. The internal vertical lines mark therefore the fringes -1, 0 and 1.

The curves shown in **Figure 6** consider the diffraction by a number NW of US WL's increasing from 1 to 8. The diffraction lobes shrink progressively and the maxima localize on the angles defined by Eq.1. Their values correspond to that calculated with one WL, as can be verified by comparing the data printed by the computer over each graph in Figures 5 and 6.

2.2. Two Adjacent Antiparallel Waves with Frequency Ratio $F_1/F_2 = 2/3$ and $3/4$, $V_1 = V_2 = 2.405$

The PVA describes such an antiparallel arrangement by writing in Eq. 4 the dephasing term due to the US waves as:

$$V_1\sin(K_1x-\Omega_1t) + V_2\sin(K_2x+\Omega_2t). \quad (5)$$

TABLE 1 reports the predictions for the light intensities of the orders from -10 to 10 in the diffraction by two adjacent antiparallel US beams of traveling waves with frequency ratio $F_1/F_2=2/3$ and $V=V_1=V_2 = 2.405$. We considered such a frequency ratio, as the simplest one not equal to 1, and the value V=2.405 because the Bessel function of zero order $J_0(2.405) = 0$. This allows experimenters to adjust the intensity of each US beam by observing the vanishing on the screen of the central diffraction fringe when only one US beam is switched on and allows theoreticians to reduce the terms to be included in the RN expressions, by dropping all products involving J_0.

In the first three columns of Table 1 the reference of the diffraction orders to the angles θ_{0n}, θ_{1n} and θ_{2n}, defined, according to Eq.1, by Λ_0, Λ_1 and Λ_2, respectively, are given. The Table has actually been calculated using only one US "fundamental" WL $\Lambda_0 = 2\Lambda_1 = 3\Lambda_2$, but the amplitude and the angle of the vector at the angular positions defined by Eq.1 are correctly predicted. The total light intensity given in the bottom line, resulting from the summation of all orders from -20 to 20 corresponds to the entering light (100 %), therefore confirming the validity of the PVA.

The theoretical prediction, which includes strong and temporally modulated interference fringes, as the first (1=3-2) and the fifth (5=3+2) orders (defined by Λ_0), not present in the light diffraction by one US wave of only WL Λ_1 or Λ_2, suggested the opportunity of an experimental test. The modulation waveforms and the temporal angular behaviour of the light vector, for the zero and the first order, are given in **Figure 7**. Also in this case only one "fundamental" WL Λ_0 has been taken into account and therefore the predicted diffraction lobes appear asymmetric and oscillate, when the two US beams are considered advancing with respect to the entering-light slit, as it is shown in **Figure 8**.

In **Figure 9** is shown the narrowing of the diffraction lobes with increasing NW.

In **Figure 10** the calculation of the light vector is graphically shown, for the cases NW = 1 and 4 units of Λ_0, for the order -1 (wedge -2π) and for a wedge of amplitude $-1.125 \times 2\pi$.

Figure 11 reports the experimental verification [4] of the modulation waveforms for frequency ratio 2/3 and orders -1, 0, and 1, with $V_1 = V_2 = 2.405$.

TABLE 2 reports the modulation waveforms predicted for the frequency ratio 3/4 for the orders from -10 to 10, and **Figure 12** reports the experimental verification [5], suggested by TABLE 2, for the diffraction orders -2, -1, 1 and 2, which also in this case are interference fringes. The total intensity, referred to any instant, as summed from the fringe -25 to 25 (last lines) shows the missing of a negligible and constant amount, which is diffracted to the higher-order fringes.

2.3. One Standing Wave

The standing wave is described as two adjacent traveling antiparallel waves of the same amplitude and frequency and hence has been simulated by using Eq.5, with:

$$V_1 = V_2 = V, \quad K_1 = K_2 = K, \quad \Omega_1 = \Omega_2 = \Omega$$

corresponding to:

$$2V\cos \Omega t \,.\sin Kx = V_s(t)\ \sin Kx, \qquad \text{with } V_s(t) = 2V\cos \Omega t\,.$$

This is also the best way to realize it experimentally, by avoiding the changing of the resonance conditions in the interferometric arrangements. Standing waves do not travel, but oscillate in amplitude. This also determines spatial oscillations of the diffraction lobes if the light beam interacts with only one or few US WL's. This is shown in **Figure 13** (see Note 1), where the time running from 0 to 5/16 of one US period T is considered for each of the US beams with RN parameter V=2. The equivalent value of the maximum amplitude $V_s(t)$ of the standing sinusoid decreasing from 4 to 0 and then again increasing from 0 to -2 is reported for each graph by VS0 = ...

In **Figure 14** is shown, for comparison sake, the narrowing, with NW=4, of the diffraction lobes. In both Figures 13 and 14 the vector angle presents abrupt phase jumps of 180 degrees in correspondence to each vanishing of the light intensity, and therefore the number of said jumps is greater for NW=4. The predicted light intensity and the angle of the light vector result in the same value independently of NW at the angles on the screen corresponding to the fringes -1, 0 and +1 (central vertical lines in the graphs). The phase jumps of the vector angle while changing the actual RN parameter V, occurring for the zero-order also with traveling waves, has been verified experimentally by light-interferometry techniques [6,7].

3. Conclusions

The study by PVA of the shape of the US diffraction lobes confirms the statement that, for the standard diffraction angles θ_n defined by Eq.1, the light intensity remains constant, independent of the lobe shape and this shape depends on the number NW of interacting US waves.

The asymmetry of the diffraction lobes and even their mutual overlapping, occurring when only a few US WL's interacts with the light beam, leave unchanged, at the angular position θ_n, both the amplitude and the angle of the light vector. This property gives us a very simple tool for the predictions, by using those positions, of the diffraction pattern and of the modulation waveforms in case of temporally modulated fringes, even for sophisticated diffracting arrangements.

On the other hand it is also confirmed that any experimental recording of the diffraction pattern or of the eventual modulation waveforms must be performed with a light beam interacting with a number NW greater than 8 or 10, to avoid the perturbing effects occurring otherwise.

The PVA has not yet been applied to the diffraction in the Bragg regime, but to predict the diffraction pattern in the RN regime for the oblique light-ultrasound interaction, where it allows overcoming the discrepancies between RN theory and experiments [3,8]. The PVA also suggests a computer program to calculate the Bessel functions, giving a physical meaning [9] to their first definition in 1824 [10], where one finds the integral of Eq. 4. In fact $E_n(V) = E_i.J_n(V)$. This method of computation is simpler than the recurrence formulae, employing the Gamma function. In **Figure 15** is given a plot of the Bessel functions obtained by a Basic program, which performs, for each point, the integration of Eq. 4, as visualized by Fig. 3 for $J_1(1.85)$.

I hope that some researcher will in addition confirm by experiments also the actual shape of the diffraction lobes here predicted for the light interaction with one or few US WL's. I believe also it is worthwhile to test experimentally the angular behaviour of the light vector for some fringes in the light diffraction by two adjacent antiparallel US beams with other frequency ratios. This could be performed by measuring the frequency variation and the phase modulation predicted by PVA in my paper presented to the Leningrad Seminar 1990 [11].

4. References

1. I.Gabrielli, Proc. *Ultrasonic Signal Processing - International School, Erice* (Oct.1988) 335-354.

2. I. Gabrielli, *Proceedings Fourth Spring School on Acoustooptics and Applications, Gdansk-Sobieszewo (Poland)* (1989) 69-91.

3. I. Gabrielli: *ibidem* 93-116.

4. I.Gabrielli, P.Kwiek, A.Markiewicz, A.Sliwinski, *Acustica* **66** (1988) 281-285.

5. I.Gabrielli, P.Kwiek, A.Markiewicz, A.Sliwinski, *Proceedings Ultrasonics International '89 Conference, Madrid* (1989) 230-235.

6. I. Gabrielli, P.Kwiek, A. Markiewicz, A.Sliwinski, "Investigation of amplitude and phase of light diffraction by an ultrasonic wave."*Proc. Ultrasonic Internat. Conf. UI 91, Le Touquet, France,* (1991), in press.

7. A. Sliwinski, P. Kwiek, A. Markiewicz: I.Gabrielli: "Selected modulation problems in light diffraction by two ultrasonic beams", *Academy of Sciences USSR,* in press.

8. F. Calligaris, P.Ciuti and I. Gabrielli, *Nuovo Cimento* **106 B** (1991) 789-798.

9. I. Gabrielli: "A physical generation of the Bessel functions." Manuscript.

10. F.G. Bessel, *Berliner Abh.* (1824) 1-52.

11. I. Gabrielli: "The angular behaviour of the light vector in the light diffraction on ultrasonic waves", *Proc. School-Seminar Acoustooptics" Leningrad* (1990) in press.

TABLE 1 - Prediction by the PVA of the light diffraction by two adjacent antiparallel US beams with frequency ratio 2/3 and $V_1 = V_2 = 2.41$ for the fringes from -10 to 10, defined by the "fundamental" WL $\Lambda_0 = 2\Lambda_1 = 3\Lambda_2$. The Table has predicted the intense interference fringes $\pm1 = \pm(3\text{-}2)$ and $\pm5 = \pm(3+2)$, experimentally verified [4]. The last line gives the total light diffracted between the fringes -20 and 20, corresponding to the entering light (100 %).

TYPE OF GRATING: SIN. - SIN.   FREQUENCY RATIO: 2/3   R-N BEAM PARAM.: 2.41 - 2.41   CALCUL. ON 192 POINTS

ORDERS REFER. TO: Λ_0	Λ_1	Λ_2	MODULAT. %	MAX. INT.	MUTUAL BEAM DEPHASING (2.PI/16) 0	1	2	3	4	5	6	7	8	9	10	11	12	13	14	15	16
-10	-5	0	51.8	4.89	2.36	2.72	3.61	4.51	4.89	4.51	3.61	2.72	2.36	2.72	3.61	4.51	4.89	4.51	3.61	2.72	2.36
-9	0	-3	16.0	1.14	0.96	1.04	1.14	1.11	1.05	1.11	1.14	1.04	0.96	1.04	1.14	1.11	1.05	1.11	1.14	1.04	0.96
-8	-4	0	39.5	6.37	6.37	5.99	5.09	4.21	3.85	4.21	5.09	5.99	6.37	5.99	5.09	4.21	3.85	4.21	5.09	5.99	6.37
-7	0	0	88.8	11.38	1.28	2.53	5.88	9.67	11.38	9.67	5.88	2.53	1.28	2.53	5.88	9.67	11.38	9.67	5.88	2.53	1.28
-6	-3	-2	43.6	0.02	0.02	0.02	0.02	0.01	0.02	0.01	0.02	0.02	0.02	0.02	0.02	0.01	0.02	0.01	0.02	0.02	0.02
-5	0	0	54.9	10.28	10.28	10.06	8.67	6.07	4.63	6.07	8.67	10.06	10.28	10.06	8.67	6.07	4.63	6.07	8.67	10.06	10.28
-4	-2	0	14.5	5.35	4.57	4.93	5.35	5.13	4.84	5.13	5.35	4.93	4.57	4.93	5.35	5.13	4.84	5.13	5.35	4.93	4.57
-3	0	-1	80.9	2.12	2.12	1.27	0.40	1.20	2.00	1.20	0.40	1.27	2.12	1.27	0.40	1.20	2.00	1.20	0.40	1.27	2.12
-2	-1	0	46.7	4.74	4.53	3.51	2.53	3.65	4.74	3.65	2.53	3.51	4.53	3.51	2.53	3.65	4.74	3.65	2.53	3.51	4.53
-1	0	0	98.4	23.30	23.30	20.36	12.65	4.12	0.37	4.12	12.65	20.36	23.30	20.36	12.65	4.12	0.37	4.12	12.65	20.36	23.30
0	0	0	100.0	2.99	0.00	1.50	2.99	1.50	0.00	1.50	2.99	1.50	0.00	1.50	2.99	1.50	0.00	1.50	2.99	1.50	0.00
1	0	0	98.4	23.30	0.37	4.12	12.65	20.36	23.30	20.36	12.65	4.12	0.37	4.12	12.65	20.36	23.30	20.36	12.65	4.12	0.37
2	1	0	46.7	4.74	4.74	3.65	2.53	3.51	4.53	3.51	2.53	3.65	4.74	3.65	2.53	3.51	4.53	3.51	2.53	3.65	4.74
3	0	1	80.9	2.12	2.00	1.20	0.40	1.27	2.12	1.27	0.40	1.20	2.00	1.20	0.40	1.27	2.12	1.27	0.40	1.20	2.00
4	2	0	14.5	5.35	4.84	5.13	5.35	4.93	4.57	4.93	5.35	5.13	4.84	5.13	5.35	4.93	4.57	4.93	5.35	5.13	4.84
5	0	0	54.9	10.28	4.63	6.07	8.67	10.06	10.28	10.06	8.67	6.07	4.63	6.07	8.67	10.06	10.28	10.06	8.67	6.07	4.63
6	3	2	43.6	0.02	0.02	0.01	0.02	0.02	0.02	0.02	0.02	0.01	0.02	0.01	0.02	0.02	0.02	0.02	0.02	0.01	0.02
7	0	0	88.8	11.38	11.38	9.67	5.88	2.53	1.28	2.53	5.88	9.67	11.38	9.67	5.88	2.53	1.28	2.53	5.88	9.67	11.38
8	4	0	39.5	6.37	3.85	4.21	5.09	5.99	6.37	5.99	5.09	4.21	3.85	4.21	5.09	5.99	6.37	5.99	5.09	4.21	3.85
9	0	3	16.0	1.14	1.05	1.11	1.14	1.04	0.96	1.04	1.14	1.11	1.05	1.11	1.14	1.04	0.96	1.04	1.14	1.11	1.05
10	5	0	51.8	4.89	4.89	4.51	3.61	2.72	2.36	2.72	3.61	4.51	4.89	4.51	3.61	2.72	2.36	2.72	3.61	4.51	4.89

TOTAL LIGHT INTENSITY FOR A GIVEN BEAM DEPHASING (SUMMATION OVER 41 ORDERS FROM -20 TO 20)

/ 100.0 100.0 100.0 100.0 100.0 100.0 100.0 100.0 100.0 100.0 100.0 100.0 100.0 100.0 100.00100. 0100.0

TABLE 2 - Prediction by the PVA of the light diffraction by two adjacent antiparallel US beams with frequency ratio 3/4 and $V_1 = V_2 = 2.41$ for the fringes from -10 to 10 referred to the "fundamental" WL $\Lambda_0 = 3\Lambda_1 = 4\Lambda_2$. The Table has predicted the intense interference fringes $\pm 1 = \pm(4\text{-}3)$, ± 2, experimentally verified [5], and $\pm 7 = \pm(4+3)$. The last two lines give the total light diffracted between the fringes -25 to 25 (99.98 % of the entering light).

TIPO RET. FREQ. 3 / 4 AMP. 2.4100 2.4100 N JU 32 N.ORD. 25 N 384 (0/ 96)

ORDERS REFER. TO: Λ_0	Λ_1	Λ_2	MODULAT. %	MAX. INT.	mutual beam dephasing JU 0	1	2	3	4	5	6	7	8	9	10	11	12	13	14	15	16
-10	0	0	41.3 /	6.44/	6.44	6.21	5.60	4.83	4.15	3.80	3.88	4.36	5.09	5.84	6.34	6.41	6.04	5.35	4.59	4.00	
				/	3.78	4.00	4.59	5.35	6.04	6.41	6.34	5.84	5.09	4.36	3.88	3.80	4.16	4.83	5.60	6.21	6.44//
-9	-3	0	2.3 /	1.09/	1.09	1.08	1.08	1.07	1.06	1.06	1.06	1.07	1.07	1.08	1.09	1.09	1.08	1.07	1.07	1.06	
				/	1.06	1.06	1.07	1.07	1.08	1.09	1.09	1.08	1.07	1.07	1.06	1.06	1.06	1.07	1.08	1.08	1.09//
-8	0	-2	18.8 /	0.12/	0.11	0.12	0.12	0.12	0.11	0.10	0.10	0.11	0.12	0.12	0.12	0.12	0.12	0.12	0.12	0.11	
				/	0.10	0.11	0.12	0.12	0.12	0.12	0.12	0.12	0.12	0.11	0.10	0.10	0.11	0.12	0.12	0.12	0.11//
-7	0	0	8.4 /	7.52/	6.89	6.95	7.12	7.30	7.45	7.52	7.50	7.41	7.24	7.06	6.92	6.90	7.00	7.18	7.36	7.48	
				/	7.52	7.48	7.36	7.18	7.00	6.90	6.92	7.06	7.24	7.41	7.50	7.52	7.45	7.30	7.12	6.95	6.89//
-6	-2	0	6.2 /	0.76/	0.72	0.73	0.75	0.76	0.76	0.76	0.76	0.76	0.76	0.74	0.72	0.72	0.73	0.76	0.76	0.76	
				/	0.76	0.76	0.76	0.76	0.73	0.72	0.72	0.74	0.76	0.76	0.76	0.76	0.76	0.76	0.75	0.73	0.72//
-5	0	0	86.6 /	10.75/	10.75	9.97	7.87	5.18	2.80	1.53	1.80	3.51	6.09	8.68	10.40	10.66	9.39	7.00	4.31	2.23	
				/	1.44	2.23	4.31	7.00	9.39	10.66	10.40	8.68	6.09	3.51	1.80	1.53	2.80	5.18	7.87	9.97	10.75//
-4	0	-1	49.8 /	0.11/	0.10	0.08	0.06	0.06	0.09	0.11	0.11	0.08	0.06	0.07	0.09	0.09	0.08	0.06	0.07	0.10	
				/	0.11	0.10	0.07	0.06	0.08	0.09	0.09	0.07	0.06	0.08	0.11	0.11	0.09	0.06	0.06	0.08	0.10//
-3	-1	0	32.7 /	0.20/	0.20	0.18	0.15	0.13	0.15	0.17	0.17	0.14	0.13	0.16	0.19	0.20	0.17	0.14	0.14	0.16	
				/	0.18	0.16	0.14	0.14	0.17	0.20	0.19	0.16	0.13	0.14	0.17	0.17	0.15	0.13	0.15	0.18	0.20//
-2	0	0	99.2 /	16.92/	0.13	1.55	5.31	10.16	14.46	16.76	16.28	13.18	8.52	3.86	0.77	0.30	2.59	6.88	11.73	15.51	
				/	16.92	15.51	11.73	6.88	2.59	0.30	0.77	3.86	8.52	13.18	16.28	16.76	14.46	10.16	5.31	1.55	0.13//
-1	0	0	73.3 /	12.52/	3.35	4.13	6.21	8.86	11.19	12.43	12.17	10.50	7.97	5.41	3.70	3.44	4.71	7.07	9.72	11.75	
				/	12.52	11.75	9.72	7.07	4.71	3.44	3.70	5.41	7.97	10.50	12.17	12.43	11.19	8.86	6.21	4.13	3.35//
0	0	0	100.0•/	0.07/	0.00	0.02	0.06	0.07	0.03	0.00	0.01	0.05	0.07	0.05	0.01	0.00	0.03	0.07	0.06	0.02	
				/	0.00	0.02	0.06	0.07	0.03	0.00	0.01	0.05	0.07	0.05	0.01	0.00	0.03	0.07	0.06	0.02	0.00//
1	0	0	73.3 /	12.52/	12.52	11.75	9.72	7.07	4.71	3.44	3.70	5.41	7.97	10.50	12.17	12.43	11.19	8.86	6.21	4.13	
				/	3.35	4.13	6.21	8.86	11.19	12.43	12.17	10.50	7.97	5.41	3.70	3.44	4.71	7.07	9.72	11.75	12.52//
2	0	0	99.2 /	16.92/	16.92	15.51	11.73	6.88	2.59	0.30	0.77	3.86	8.52	13.18	16.28	16.76	14.46	10.16	5.31	1.55	
				/	0.13	1.55	5.31	10.16	14.46	16.76	16.28	13.18	8.52	3.86	0.77	0.30	2.59	6.88	11.73	15.51	16.92//
3	1	0	32.7 /	0.20/	0.18	0.16	0.14	0.14	0.17	0.20	0.19	0.16	0.13	0.14	0.17	0.17	0.15	0.13	0.15	0.18	
				/	0.20	0.18	0.15	0.13	0.15	0.17	0.17	0.14	0.13	0.16	0.19	0.20	0.17	0.14	0.14	0.16	0.18//
4	0	1	49.8 /	0.11/	0.11	0.10	0.07	0.06	0.08	0.09	0.09	0.07	0.06	0.08	0.11	0.11	0.09	0.06	0.06	0.08	
				/	0.10	0.08	0.06	0.06	0.09	0.11	0.11	0.08	0.06	0.07	0.09	0.09	0.08	0.06	0.07	0.10	0.11//
5	0	0	86.6 /	10.75/	1.44	2.23	4.31	7.00	9.39	10.66	10.40	8.68	6.09	3.51	1.80	1.53	2.80	5.18	7.87	9.97	
				/	10.75	9.97	7.87	5.18	2.80	1.53	1.80	3.51	6.09	8.68	10.40	10.66	9.39	7.00	4.31	2.23	1.44//
6	2	0	6.2 /	0.76/	0.76	0.76	0.76	0.76	0.73	0.72	0.72	0.74	0.76	0.76	0.76	0.76	0.76	0.76	0.75	0.73	
				/	0.72	0.73	0.75	0.76	0.76	0.76	0.76	0.76	0.76	0.74	0.72	0.72	0.73	0.76	0.76	0.76	0.76//
7	0	0	8.4 /	7.52/	7.52	7.48	7.36	7.18	7.00	6.90	6.92	7.06	7.24	7.41	7.50	7.52	7.45	7.30	7.12	6.95	
				/	6.89	6.95	7.12	7.30	7.45	7.52	7.50	7.41	7.24	7.06	6.92	6.90	7.00	7.18	7.36	7.48	7.52//
8	0	2	18.8 /	0.12/	0.10	0.11	0.12	0.12	0.12	0.12	0.12	0.12	0.12	0.11	0.10	0.10	0.11	0.12	0.12	0.12	
				/	0.11	0.12	0.12	0.12	0.11	0.10	0.10	0.11	0.12	0.12	0.12	0.12	0.12	0.12	0.12	0.11	0.10//
9	3	0	2.3 /	1.09/	1.06	1.06	1.07	1.07	1.08	1.09	1.09	1.08	1.07	1.07	1.06	1.06	1.06	1.07	1.08	1.08	
				/	1.09	1.08	1.08	1.07	1.06	1.06	1.06	1.07	1.07	1.08	1.09	1.09	1.08	1.07	1.07	1.06	1.06//
10	0	0	41.3 /	6.44/	3.78	4.00	4.59	5.35	6.04	6.41	6.34	5.84	5.09	4.36	3.88	3.80	4.16	4.83	5.60	6.21	
				/	6.44	6.21	5.60	4.83	4.16	3.80	3.88	4.36	5.09	5.84	6.34	6.41	6.04	5.35	4.59	4.00	3.78//
0	0	0	0.05/	0.00/	99.98	99.98	99.98	99.98	99.98	99.98	99.98	99.98	99.98	99.98	99.98	99.98	99.98	99.98	99.98	99.98	
				/	99.98	99.98	99.98	99.98	99.98	99.98	99.98	99.98	99.98	99.98	99.98	99.98	99.98	99.98	99.98	99.98	99.98

F 3/4 LA SOMMA E' ESTESA AGLI ORD. DA -25 A + 25

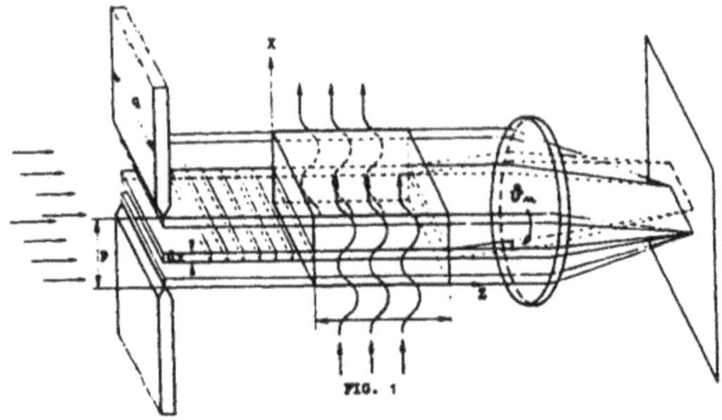

FIG. 1 - Experimental arrangement of the diffraction in the Fraunhofer plane
The sheets of light are shown.

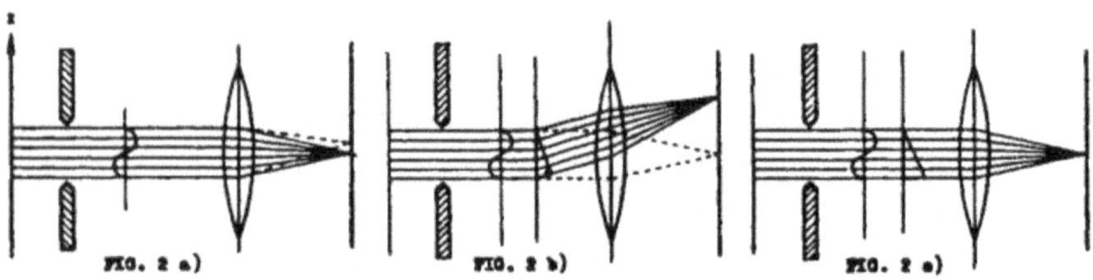

FIG. 2 a) - Schematic model of the light diffracted by one ultrasonic wave.
b) - The actual path of the light converging to a nonzero-order fringe.
c) - The equivalent path of the light rays in the model used by the PVA: definition of the additional dephasing wedge (Eq. 2).

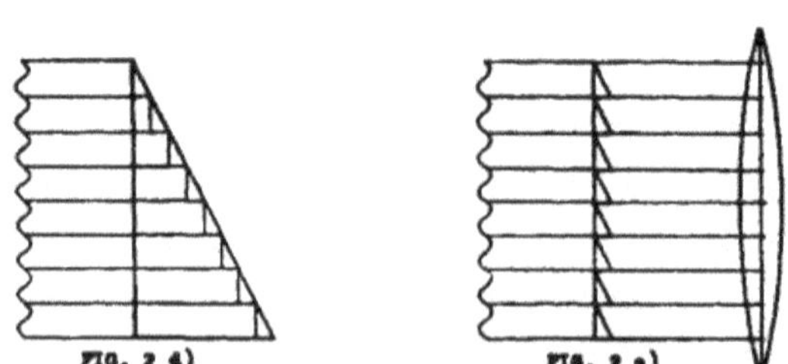

FIG. 2 d,e) - Diffraction by several (8) ultrasonic wavelengths: the dephasing wedge for the diffraction order +1.
d) - The wedge mathematically considered for the angular directions of Eq 1: the step dephasing difference is $n2\pi$.
e) - The equivalent sawtooth wedges: all the partial vectors, related to one wavelength, are equal in amplitude and phase and therefore the vectorial and scalar sums coincide.

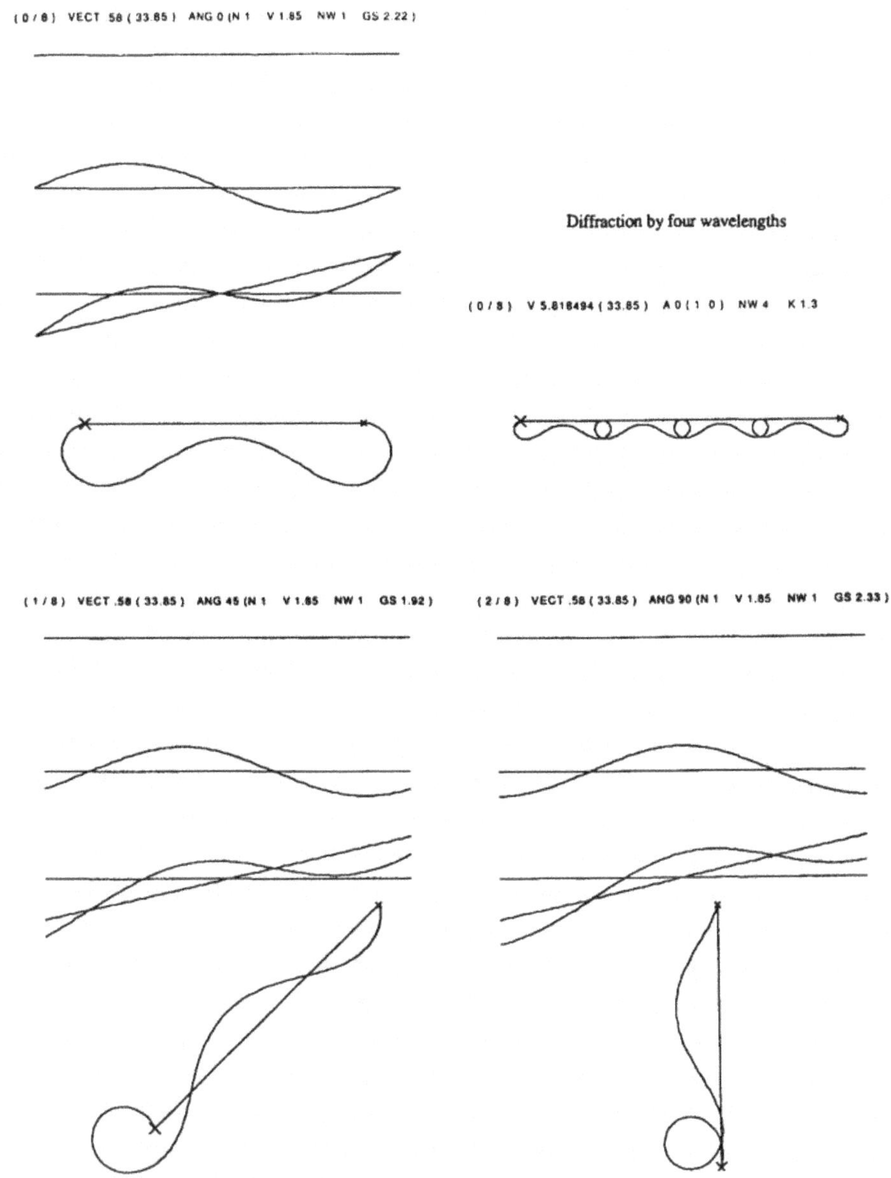

FIG. 3 - Graphical evidence of the light vector deduction, according to Eq. 4, in the light diffraction by one wavelength of one beam of waves advancing from 0 to 2/8 of T.
Four wavelengths are considered for the right top picture, giving the same relative vector length of 0.58.

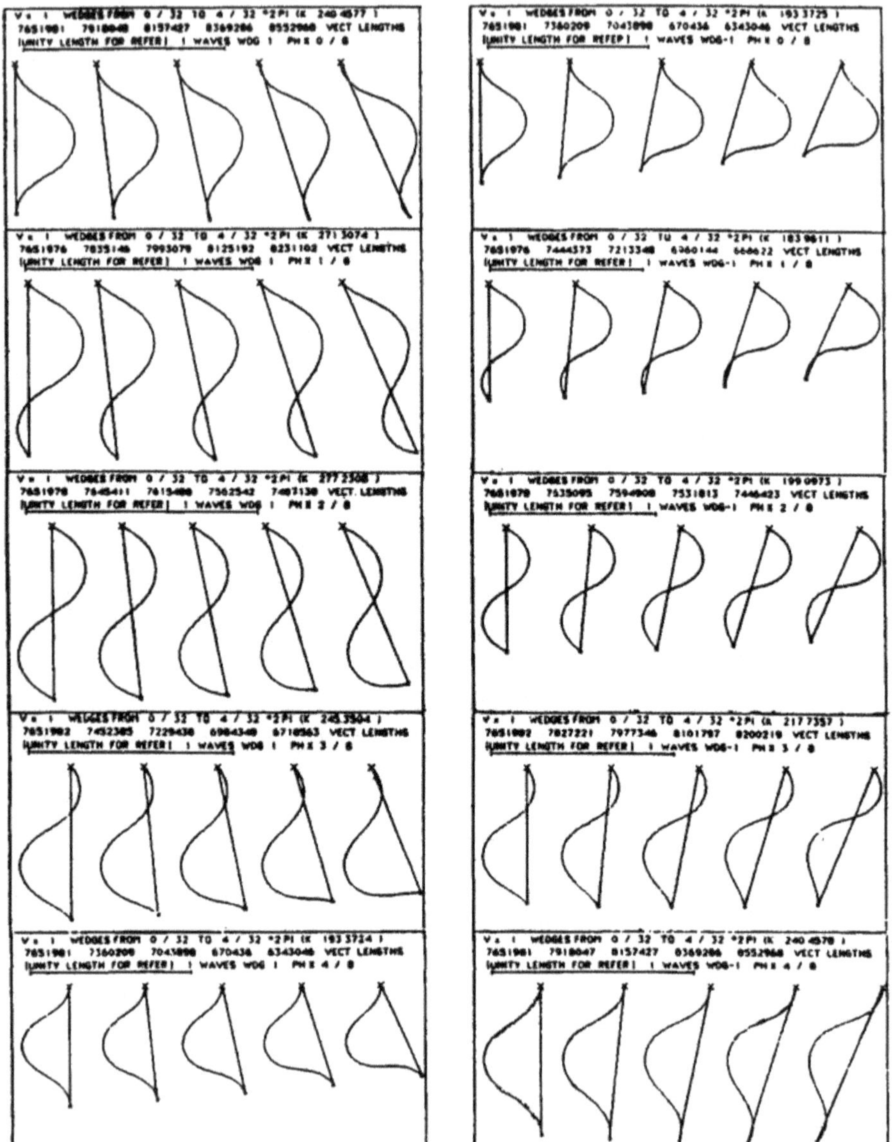

FIG. 4 - Deduction of the lobe shape for one traveling wavelength, V=1. Along each vertical line: temporal progressing of the ultrasonic sinusoid from 0 to 4/8 of Λ. Along each horizontal line: increasing of the wedge from 0 to 4/32 of 2π (positive wedge i.e. negative diffraction-orders for the left pictures - negative wedge i.e. positive orders for the right figures).

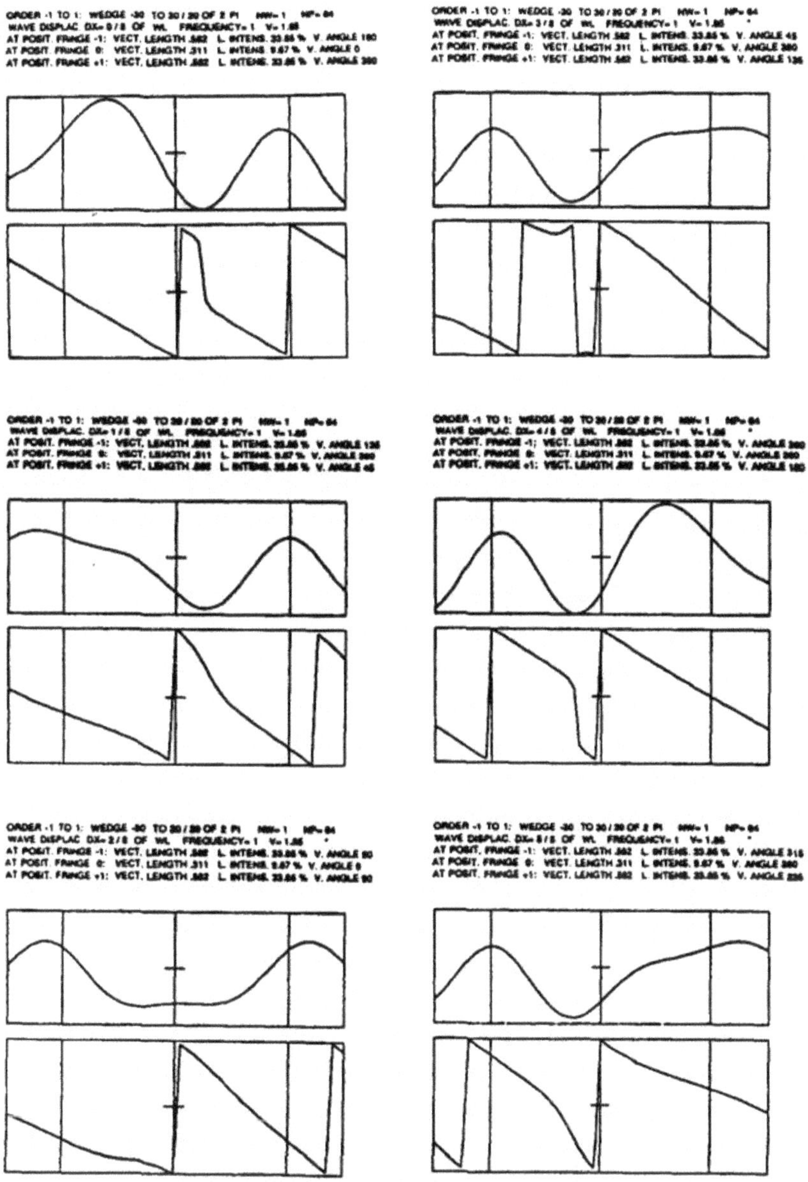

FIG. 5 - Spatilal oscillations of the diffraction lobes with one wavelength, when it advances from 0 to 5/8 of Λ. V = 1.85. See NOTE 1 in the text for the coordinate scales.

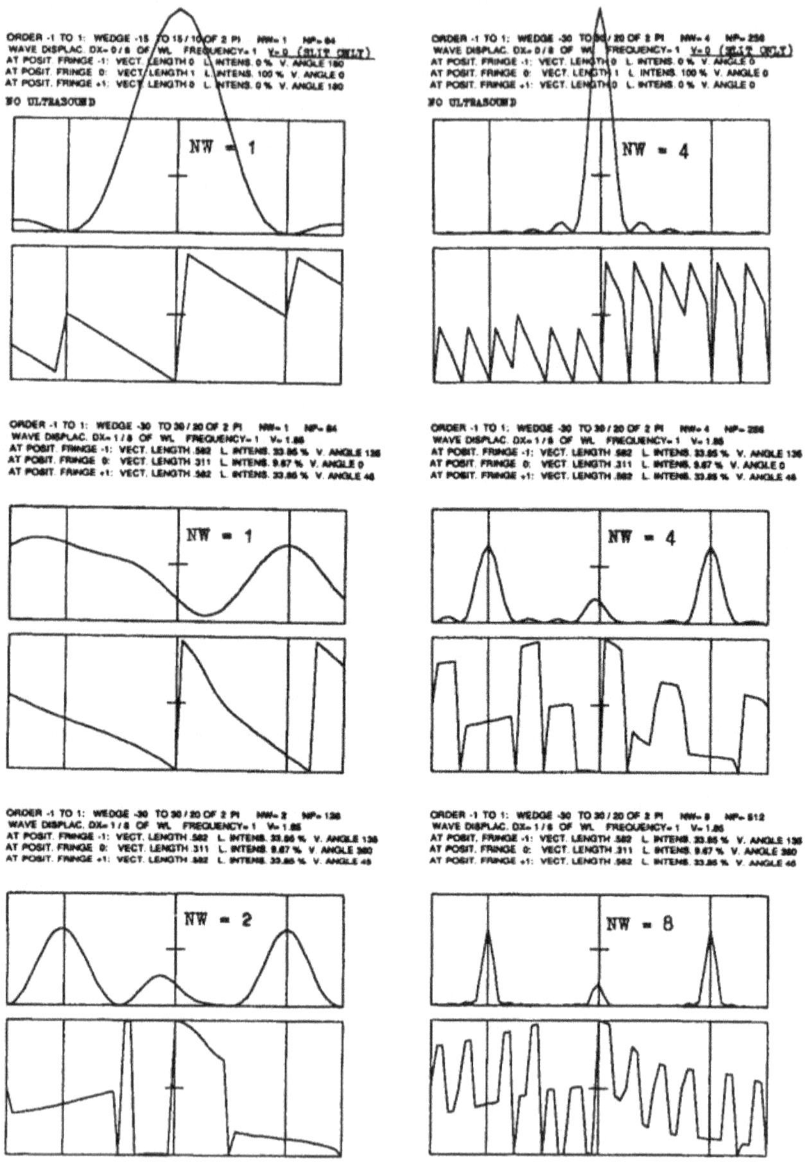

FIG. 6 - Reduction of the width of the diffraction lobes of the orders -1, 0 and 1 with increasing number NW of the interacting wavelengths. The two upper Figures show the light diffraction by only the slit of opening p=Λ or 4Λ, respectively. The sinusoid is considered displaced by 1/8 of a wavelength. See NOTE 1 in the text for the coordinate scales.

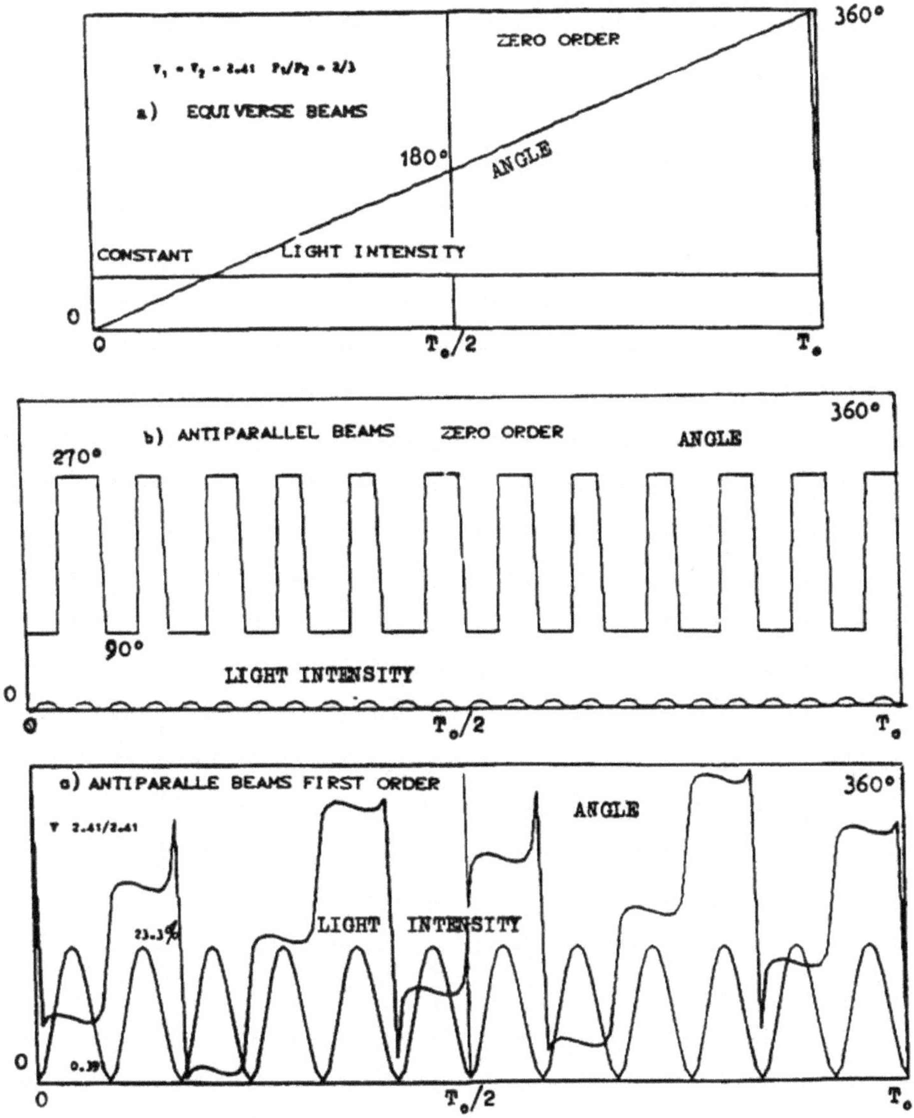

FIG. 7 - Light diffraction by two adjacent ultrasonic beams with frequency ratio 2/3 and $V_1 = V_2 = 2.41$. Each Figure reports the evolution in time of the light intensity and the angle of the light vector over one "fundamental period" T_0.
Note for the antiparallel beams the phase jumping by 180 degrees of the vector for the zero order, under b) and the frequency increase (vector rotation) with a superposed phase modulation for the order 1 under c).

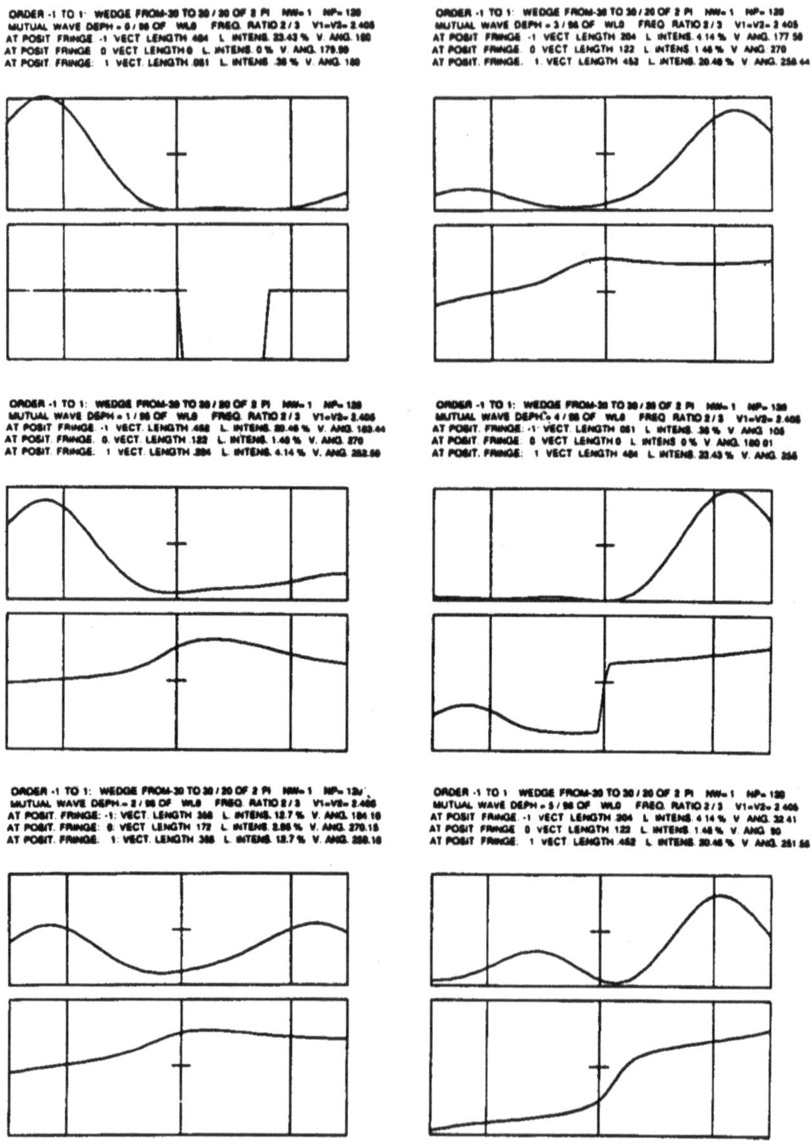

FIG. 8 - Oscillations of the lobes corresponding to the fringes -1, 0 and 1 (defined by Λ_0) in the light diffraction by two adjacent antiparallel ultrasonic beams with frequency ratio 2/3 and $V_1 = V_2 = 2.41$ (upper pictures). The lower pictures report the evolution of the vector angle. The mutual wave dephasing runs from 0 to 5/48 of Λ_0. See NOTE 1 in the text for the coordinate scales.

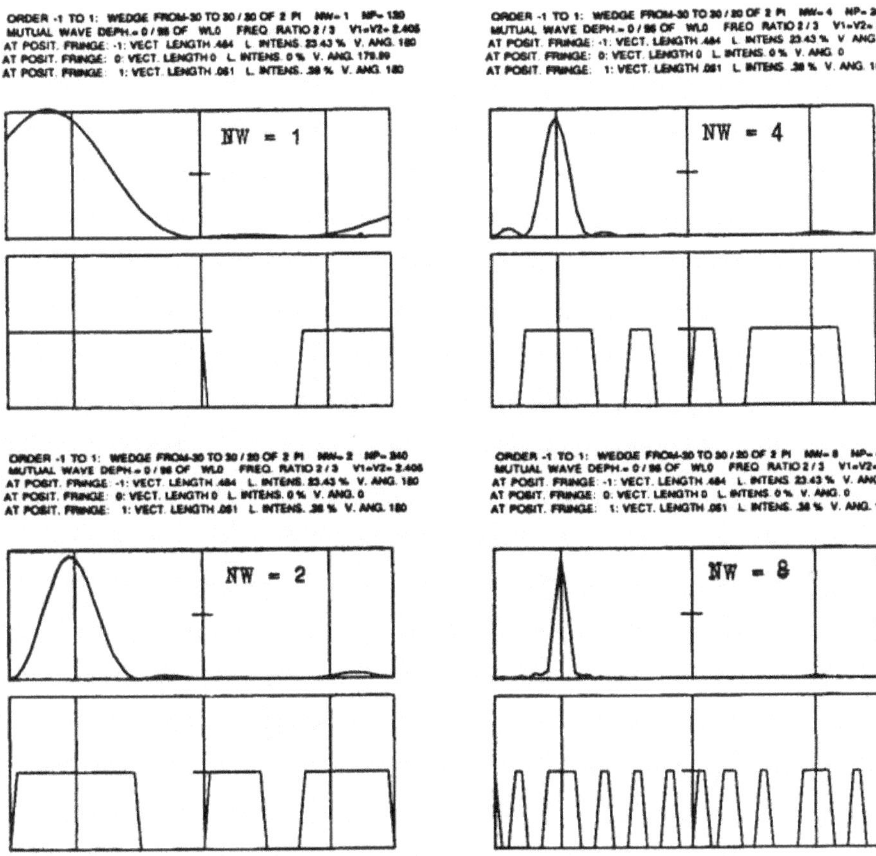

FIG. 9 - Light diffraction by two adjacent antiparallel US beams with frequency ratio 2/3 and $V_1 = V_2 = 2.41$. Reduction of the width of the lobe shapes and evolution of the behaviour of the light-vector angle at increasing number NW of the ultrasonic wavelengths interacting with the light beam. See NOTE 1 in the text for the coordinate scales.

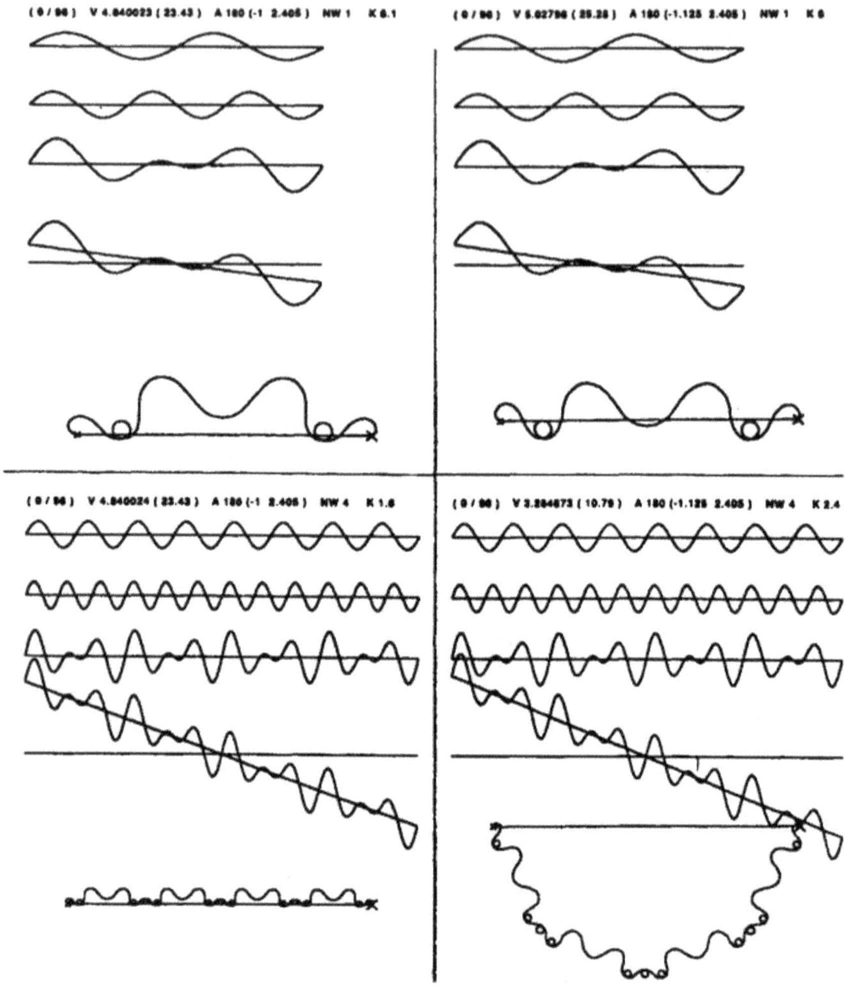

FIG. 10 - Light diffraction by two adjacent antiparallel US beams with frequency ratio 2/3 and $V_1 = V_2 =$ 2.41 . Plots of the components and of the resulting dephasing gratings used and graphical visualization of the "springs" generating the light vector, showing the effect of increasing the number NW of "fundamental" ultrasonic wavelengths Λ_0 interac-ting with the light beam:

number of wavelengths:	angular amplitude of the wedge considered:
- upper Figures: NW = 1	- left Figures: $\theta_{-1} = \text{asin}(-\lambda/\Lambda_0)$
- lower Figures: NW = 4	- right Fig.s: $\theta_{-1.125} = \text{asin}(-1.125\ \lambda/\Lambda_0)$

resulting relative length of ther light vector:

- left upper and lower: 0.484 - right upper: 0.503, right lower: 0.328

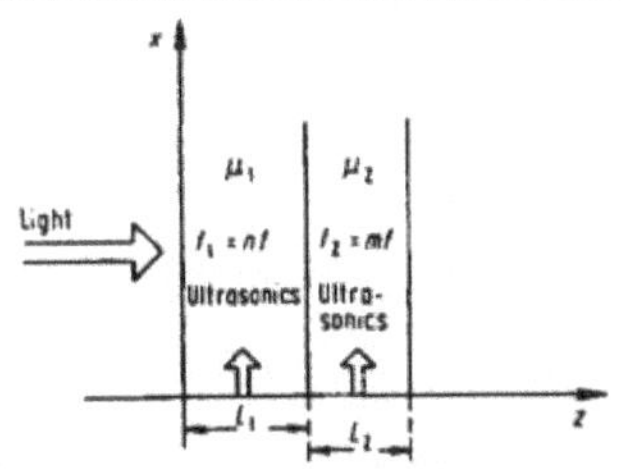

Fig. 1. The geometry of interaction of light and ultrasound.

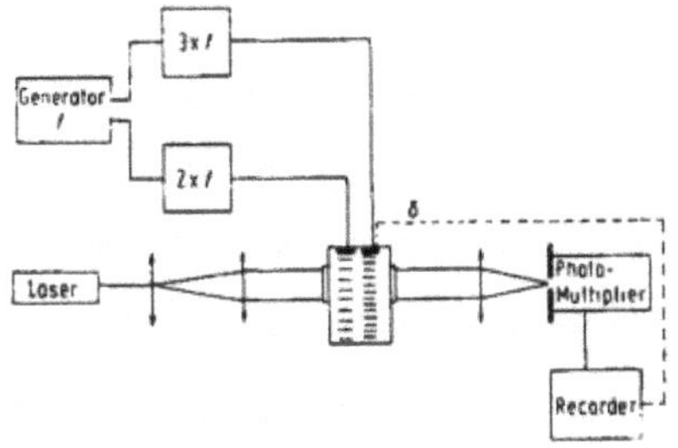

Fig. 3. Scheme of the experimental arrangement.

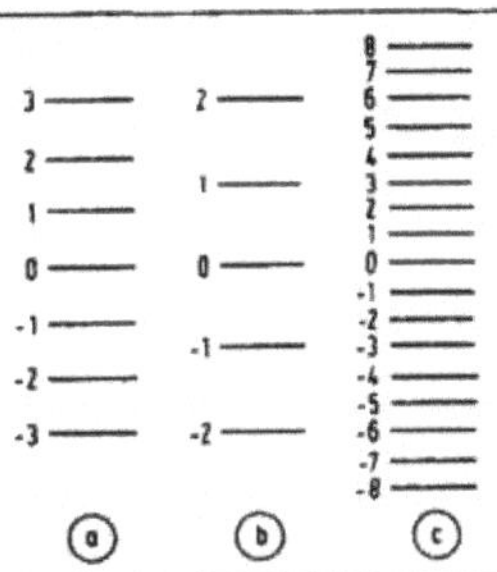

Fig. 2. Scheme of light diffraction pattern by two ultrasonic beams when their frequency ratio is $n:m = 2:3$. Diffraction pattern for:
a) the first ultrasonic beam ($2f$),
b) the second ultrasonic beam ($3f$),
c) both ultrasonic beams.

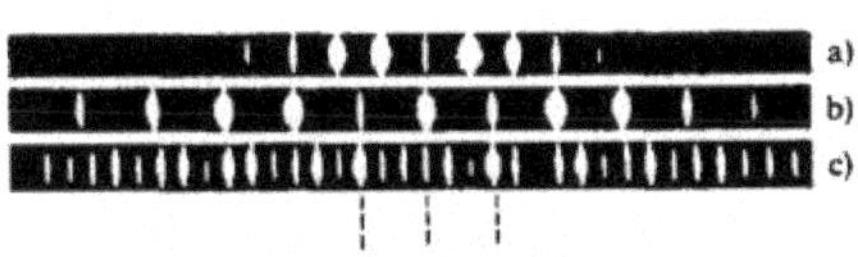

Fig. 4. Diffraction pattern observed for
a) the first ultrasonic beam of frequency $2f$,
b) the second ultrasonic beam of frequency $3f$,
c) both ultrasonic beams (frequency ratio 2:3).

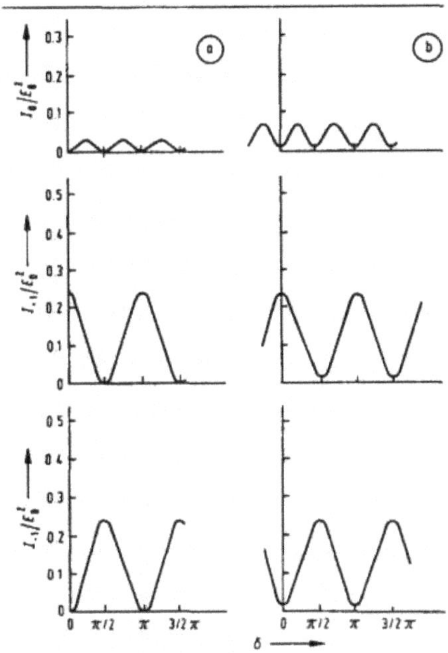

Fig. 5. The light intensity modulation as a function of δ for 0th and $\pm 1^{st}$ orders
a) experimental results,
b) theoretical results.

FIG. 11 - Light diffraction by two adjacent antiparallel US beams with frequency ratio 2/3 and $V_1 = V_2 = 2.41$. Experimental verification (Ref.4) for the modulation waveforms of the zero-order and of the "interference" diffraction fringes -1 and 1, predicted by the PVA (see FIG.7 and TABLE 1).
In the recorded diffraction pattern the fringe -5 can be seen.

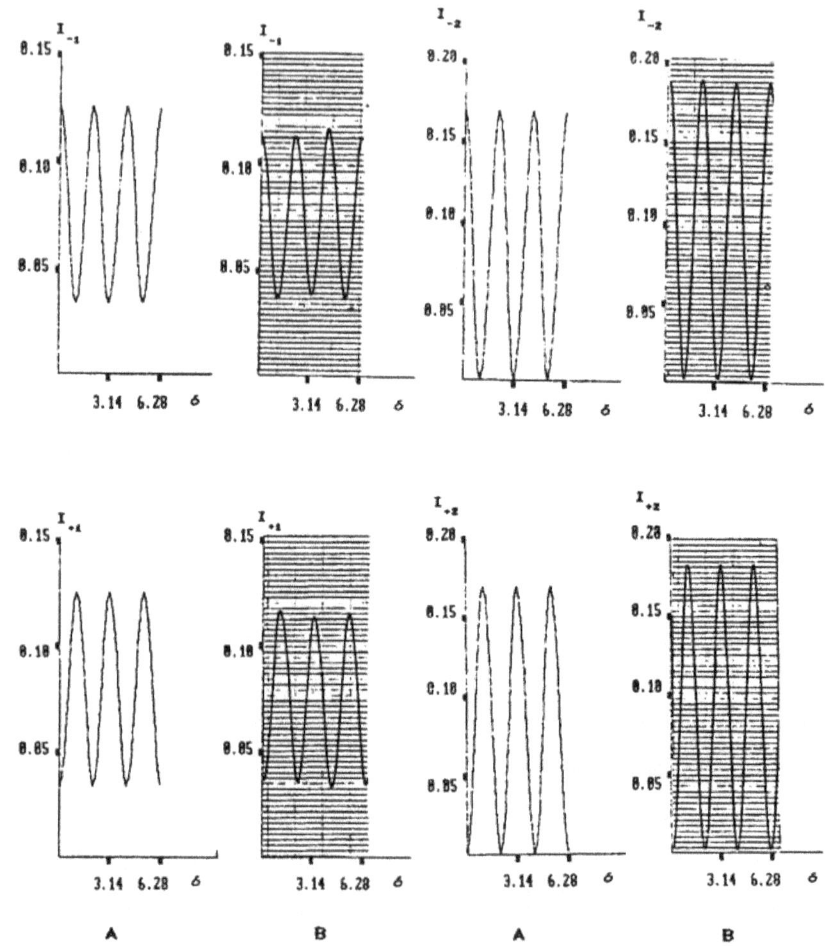

Fig. 4

The modulation of light intensity as a function of δ for $\pm 1^{st}$ orders.
A -theoretical results
B -experimental results

Fig. 5

The modulation of light intensity as a function of δ for $\pm 2^{nd}$ orders.
A -theoretical results
B -experimental results

FIG. 12 - Light diffraction by two adjacent antiparallel US beams with frequency ratio 3/4 and $V_1 = V_2 = 2.41$. Experimental verification (Ref. 5) for the "interference" diffraction fringes -2, -1, 1 and 2, predicted by the PVA (see TABLE 2).

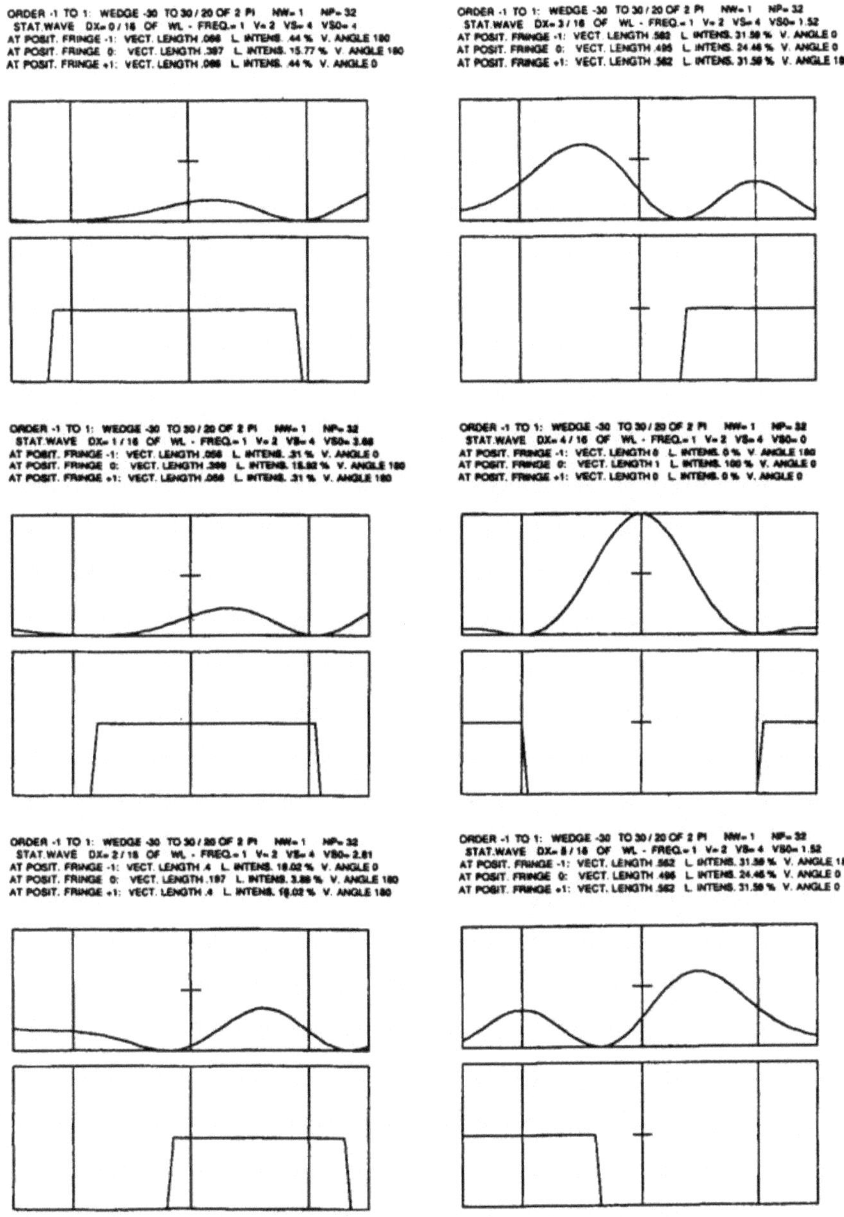

FIG. 13 - Light diffraction by one standing wave only. Oscillations of the diffraction lobes for the orders -1, 0 and 1 between t = 0 and 5/8 of one US period T. See NOTE 1 in the text for the coordinate scales.

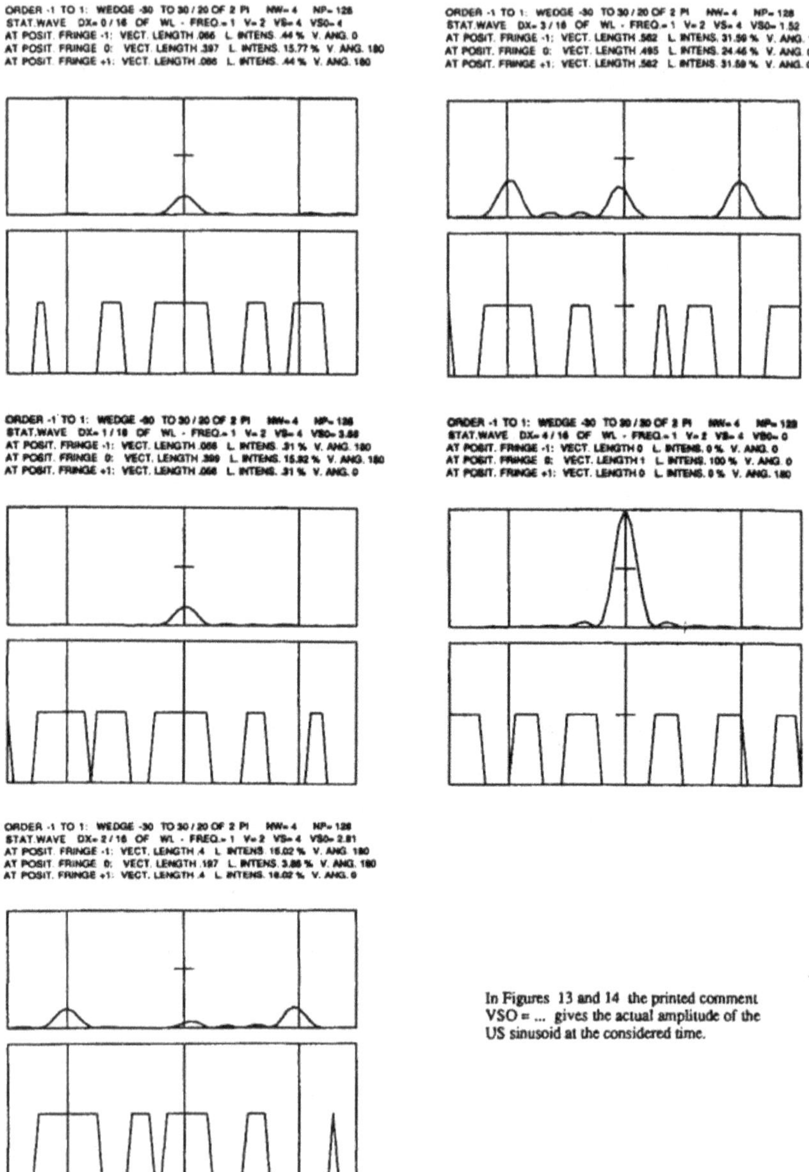

In Figures 13 and 14 the printed comment VSO = ... gives the actual amplitude of the US sinusoid at the considered time.

FIG. 14 - Light diffraction by 4 standing waves. Oscillations of the diffraction lobes for orders -1, 0 and 1 between t = 0 and 4/8 of one US period T. The reduction of the lobe width appears by comparison with Figure 13. See NOTE 1 in the text for the coordinate scales.

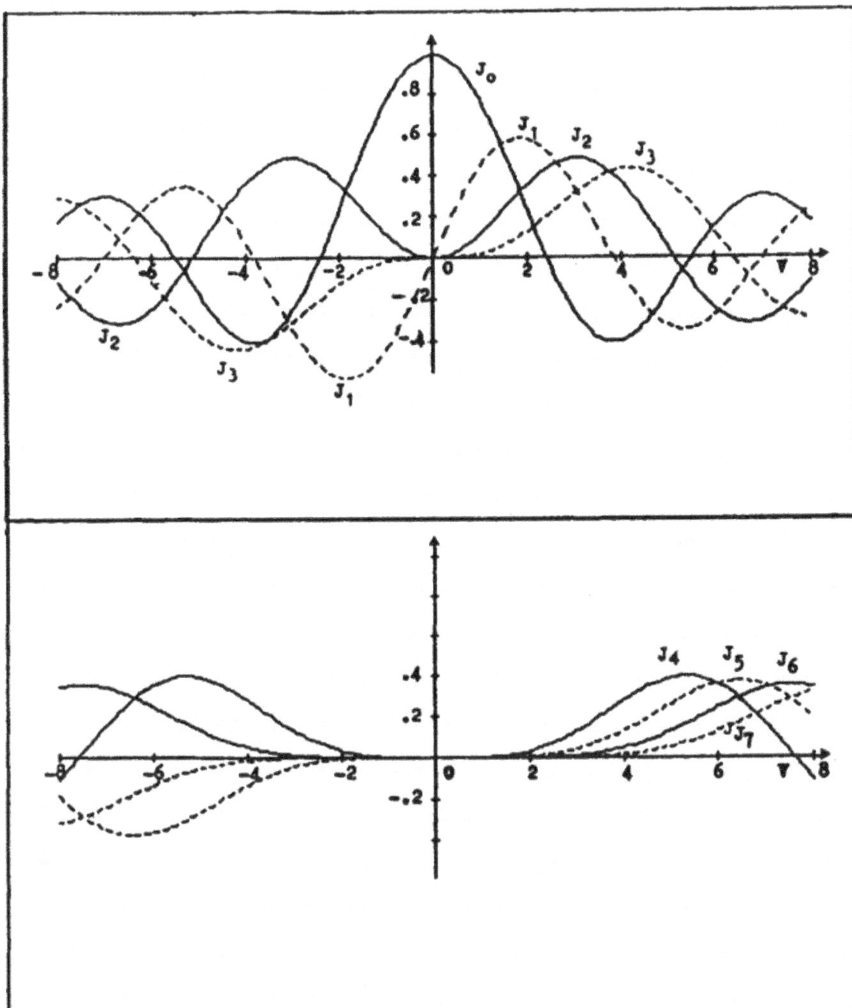

FIG. 15 - Plot of the Bessel functions from $J_0(V)$ to $J_6(V)$ for V from -8 to 8, obtained by a Basic program from Eq. 4 (See text in Conclusions).

REFLECTION AND SCATTERING OF RAYLEIGH WAVES AND GULYAEV-BLEUSTEIN WAVES ON SINGLE RESONATING ELEMENT

E.A.Garova, V.P.Plessky, N.N.Shibanova*, A.W.Simonian
Institute of Radioengineering and Electronics, USSR Academy of Sciences, Marx Avenue 18, Moscow USSR
*Moscow Institute of Physics and Technology, Institutsky per., 9, Dolgoprudny, Moscow region 141700, USSR

ABSTRACT

We have studied the characteristics of localized surface acoustic waves (SAW) reflectors made in the form classical mechanical resonator or in the form of heavy strip on the surface of elastic solid. Such structures are found to have eigen modes and as a result very strong reflection and scattering of SAW can be realized on single element in relatively broad frequency range, since the resonances have low Q-factors. Similar local surface resonances have been recently considered by A.A.Maradudin and co-workers [1,2]. This results qualitatively correlate with the experiments made by K. Yamanouchi and co-authors [3] on scattering Rayleigh waves on reflecting gratings with " large energy storage effect "

1.Introduction

In recent works by Maradudin and co-workers [1,2] the so-called local acoustic surface shape resonances have been studied. The considered resonances are essentially the eigen resonances of the ridge lying on a flat surface of elastic half-space. These oscillations attenuate with time due to emission of acoustic energy into the rigid body and have limited Q-factor.

Localized broad band reflectors for surface waves have not been found yet. We show in this paper that elements with their eigen resonances at the frequency of incident wave are able to perform this function.In SAW technology such elements may be used similar to the way mirrors are used in optics.In the present work we consider scattering of Rayleigh waves and Gulyaev -Bleustein waves on single resonating element.

2.Scattering of Rayleigh waves

For Rayleigh wave we consider two cases, firstly: a weight on a spring , secondly:a heavy strip on elastic body (Fig.1).

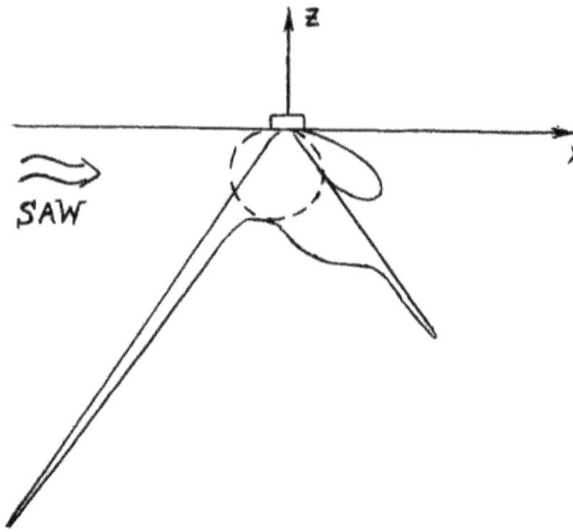

Fig.1 The strip geometry.
Scattering diagrams for longitudinal (dashed line) and transverse bulk waves.

The mass and spring are assumed to be distributed along Z axis, since our problem is two-dimensional. We consider the two cases separately. In the first case the resonator reacts only to vertical surface shifts, in the second - to the horizontal ones. In both cases the stress of the spring is assumed to be uniformly applied to the narrow strip $|x| < a << \lambda$

In our model the resonator is described by equation:

$$m\ddot{y} = -K\cdot(y - u), \tag{1}$$

where m and K - linear mass of the weight and spring elasticity, y - the weight displacement, u - surface displacement of the sound guide at point $x = 0$. Assume $u = u_0 e^{-i\omega t}$, then the weight shift amplitude y equals:

$$y_0 = \frac{\omega_0^2}{\omega_0^2 - \omega^2}\cdot u_0, \tag{2}$$

where $\omega_0 = \sqrt{K/m}$ - eigen frequency of the resonator (losses are disregarded). Here the force applied to the half -space surface is written as:

$$F = K\cdot\frac{\omega^2}{\omega_0^2 - \omega^2}\cdot u_0\cdot e^{-i\omega t}. \tag{3}$$

We'll describe the motion of isotropic rigid half-space with the aid of scalar φ and vector $\vec{\Phi}$ potentials , T.e. assuming the shift $\vec{u} = grad\,\varphi + rot\,\vec{\psi}$, where the potential satisfy the well known equations of motion :

$$\Delta\varphi + k_l^2\cdot\varphi = 0, \quad \Delta\vec{\psi} + k_t^2\cdot\vec{\psi} = 0. \tag{4}$$

Here $k_l^2 = \omega^2\rho/(\lambda + 2\mu)$, $k_t^2 = \omega^2\rho/\mu$ are wave numbers of longitudinal and transverse acoustic waves in the isotropic medium, ρ - medium density, λ and μ - elastic constants. In the considered two-dimensional geometry vector $\vec{\psi}$ has one zero component: $\vec{\psi} = (0,0,\psi)$. The general solution of wave equations (4) can be written as:

$$\varphi = \varphi_o e^{p_o y + iq_o x} + \int_{-\infty}^{+\infty} \varphi(q) e^{py+iqx}\, dq$$

$$\psi = \psi_o e^{s_o y + iq_o x} + \int_{-\infty}^{+\infty} \psi(q) e^{sy+iqx}\, dq, \tag{5}$$

where the first terms describe Rayleigh wave incident on the resonator on the left, while the second correspond to the hole spectrum of the scattered and reflected waves.

constants $p_o = \sqrt{q_o^2 - k_l^2}$, and $s_o = \sqrt{q_o^2 - k_t^2}$ describe drooping of ossilations into elastic half-space. Branches of complex square roots $p = \sqrt{q^2 - k_l^2}$, and $s = \sqrt{q^2 - k_t^2}$ ($k_{l,t} = \omega/v_{l,t}$ - wave numbers of longitudinal and transverse bulk waves, ω - frequency, v_l and v_t - corresponding frequencies) are chosen so that oscillations either droop into half space (i.e. if p,s are real, then p,s >0), or the waves are emitted from the resonator (i.e. $Im\ p,s < 0$). Values with subscript "o" satisfy Rayleigh dispersion equation: $D(q,\omega) = 4psq^2 - (q^2 + s^2)^2 = 0$. Components of deformation and stress tensors are expressed through potentials φ и ψ. In particular, due to absence of tangential stress in the case of resonator reacting to vertical surface shears it is easy to obtain for σ_{yy} component:

$$\sigma_{yy} = - \mu\cdot\int_{-\infty}^{+\infty} \frac{D(q,\omega)}{q^2 + s^2}\cdot\varphi(q) e^{iqx}\, dq \tag{6}$$

In the second case σ_{xy} component is not zero:

$$\sigma_{xy} = i\mu\cdot\int_{-\infty}^{+\infty} \frac{D(q,\omega)}{2qs}\cdot\varphi(q) e^{iqx}\, dq\ . \tag{6'}$$

Quantity σ_{yy} is determined by the force exerted by the surrounding bodies on unit surface,of the soundguide:$\sigma_{yy}= F/(2a)$ (or $\sigma_{xy} = F/(2a)$

Taking Residue in pole $q = - q_o$ it is possible to calculate reflection coefficient of Rayleigh wave from the resonators for both cases:

$$R = 2i\pi- \frac{\varepsilon p_o \Pi\left(-q_o\right) \cdot k_t^2/D'_q|_{q=-q_o}}{1 - \omega_o^2/\omega^2 + \varepsilon \cdot I_1}, \tag{7}$$

$$R = 2i\pi \frac{\varepsilon s_o \Pi\left(-q_o\right) \cdot k_t^2/D'_q|_{q=-q_o}}{1 - \omega_o^2/\omega^2 + \varepsilon \cdot I_2},$$

module $|R|$ on the incident wave frequency for several values of parameter ε

$$I_1 = \int_{-\infty}^{+\infty} \cdot k_t^2 \cdot \Pi(q)/D(q,\omega)\, dq \qquad I_2 = \int_{-\infty}^{+\infty} \cdot k_1^2 \cdot \Pi(q)/D(q,\omega)\, dq$$

$\Pi(q) = sin\ (qa)/(\pi qa)$ is Fourier transform function of mass distribution on the strip $\Pi(x) = \frac{1}{2}$ if $|x| < a$ and $\Pi(x) = 0$ at $|x| > a$. Here $\varepsilon=2\pi^2 \cdot a\rho_s\ /\rho\lambda_t^2 = \varepsilon_o k_t a$ is the basic parameter of the problem that compares linear mass of the strip with the oscillating mass of the soundguide in the region of $\lambda_t \times \lambda_t$ cross section (λ_t- wavelength of bulk transversal wave , $\varepsilon_o=\rho_o h/\rho a$).

Shown in Fig.2 is dependence of reflection and transmission coefficients on the incident wave frequency for several values ε.Here and below Poisson coefficient chosen to be $\sigma = 0,17$, (quartz glass) the width of the strip $2a = \lambda_o/8$, where λ_o -SAW wave length of the resonator frequency.

It is seen from the Fig.2 that the "light " resonator ($\varepsilon \ll 1$) strongly reflects the wave in a narrow frequency band close to ω_o , while with ε_o increasing particularly at $\varepsilon \gg 1$, the resonance frequency and resonator Q- factor greatly diminish. Here elastic body in the strip area $2a$ wide has a function of an additional spring successively coupled to the resonator , its rigidity diminishing with bandwidth $2a$.

The maximum value of reflection coefficient $|R|_{max}$ in our model depends entirely on Poisson coefficient of medium σ, that determines in the isotropic solid the relationship between longitudinal and shift components in Rayleigh wave. We determined amplitudes of the waves scattered into the volume as well.

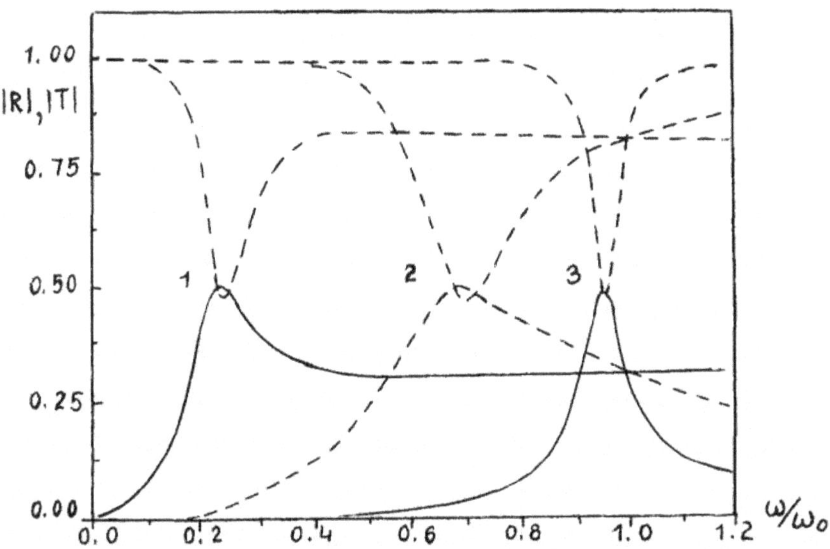

Fig. 2 The dependencies of |R| and |T| (dashed line) on frequency Curves 1 - ε = 10 curves 2 - ε = 1 curves 3 - ε = 0,1. (vertical resonator)

It is clear from above that from the heavy resonator ($\varepsilon \gg 1$) eigen frequency is determined not only by resonators parameters (m and K), but mainly by elasticity of solid and by the width of the strip $2a$. It allows us to suppose that near heavy strip immediately placed on the elastic substrate surface also can exist some localized acoustical vibration, "acoustical surface state " which can be excited by incident Rayleigh wave and , as a result, which can cause strong reflection and scattering of the incident wave.

For the heavy strip occupy some part of the surface on the isotropic elastic body in the region $x < a$. Surface density of the inertia load is $\rho_s = \rho_0 \cdot h$ where "h" is layer thickness. We'll consider that "h" and "a" sizes of the strip are small compared to the length of the incident wave: $a, h \ll \lambda$, which permits to consider the strip motion of uniform single mass. Besides, we presume that the density ρ_0 is much greater than that of the substrate material, $\rho_0 \gg \rho$, so that the band is rather heavy.The procedure of calculation gives following results for reflection of coefficient:

$$R=-2\pi i \cdot \frac{\varepsilon\, k_t^2\, \Pi\left(-q_0\right)}{D_q'\big|_{q=-q_0}} \cdot \left[\frac{p_0/k_t}{1-\varepsilon I_1} + \frac{s_0/k_t}{1-\varepsilon I_2}\right] \tag{8}$$

We have obtained frequency dependence of reflection factor $|R|$ for some values of parameter ε , .It should be noted that curve for the heavy strip has resonant character (with very small Q-factor).

Physically the resonator describes oscillation of the heavy strip on elastic half-space acting as a spring. The resonance frequancy is decreasing with the increasing of the mass of strip $\omega_0^2 \sim \mu/\rho_s a$ (μ is a shear module of soundguide material). Also note that at high frequencies the strip remains practically motionless, while $|R|_{max}$ is independent of frequency and is determined by the Poisson factor of the medium. For σ=0.17 (melted quartz) $|R|_{max}$=0.27. We have also found the scattering waves, the diagrams of the direction of the scattering waves are shown schematically on the Fig.1.

3. Scattering of Gulyaev- Bleustein waves

We have considered scattering and reflection of Gulyaev-Bleustein waves at normal incidence on the rectangular ridge with sizes comparable to the wavelength. Bleustein - Gulyaev shear waves have one component along OZ axis. We consider piezoelectric of class C_{6v} occuping half- space Y <0 with symmetry axis parallel to OZ.

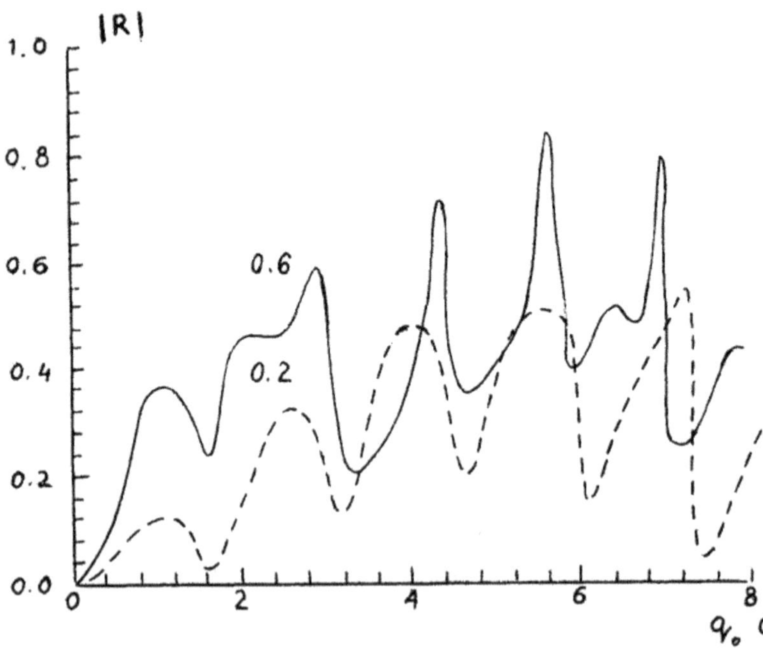

Fig.3. Amplitude-vs-frequency dependence of reflection coefficient for strong perturbations h/λ = 0.2 (curve 1) and h/λ =0.6 (curve 2)

On the surface there is a ridge (or a strip of other isotropic material) of rectanguler profile of height h and width 2a stretched along OX axis. We shall assume that BGW is incident from the left on the high ridge that have been recently considered in detail by Maradudin and co-workers . These localized in the ridge region surface acoustic resonances attenuate with time since they emit energy into piethoelectric in the form of bulk waves.
It is necessary to match together solution inside the ridge and in the field Y<0. We omit the corresponding calculation quite similar to those made in [1,2].

Shown in the Fig.3 are frequency dependence of reflection coefficients, for stronger perturbations. It is worth noting that BGW reflection coefficient in a strong piethoelectric (η= 0.55 for $LiIO_3$) from single ridge may reach R^2 = 0.60 (with $q_0 a$ = 7.3). Directive diagram of the scattered waves is determined by the ratious $h/\lambda, a/\lambda$ and by presence or absence of resonance at a given frequancy.

5.CONCLUSIONS

We have originally studied the characteristics of localized SAW reflectors made in the form classical mechanical resonator or in the form of heavy strip on the surface of elastic solid. Such structures are found to have eigen modes which can be excited by incident Rayleigh wave or Gulyaev -Bleustein wave. As a result very strong reflection and scattering of SAW can be realized on single element in relatively broad frequency range, since the resonances have low Q-factors. This results qualitatively correlate with the experiments made by K. Yamanouchi and co-authors [3] on scattering Raileigh waves on reflecting gratings with " large energy storage effect " wich is likely to be connected with excitation of localized vibrations in studied structures. However direct comparison with results of our calculation is impossible due to a more complex geometry of the elements used in their experiments.

References

1. Maradudin A.A., McGurn A.R., *Phys.Rev. B*, **39**, 12, 8732 (1989)
2. Maradudin A.A., Ruan P., McGurn A.R., *Phys.Rev.B*.**38**, 5,3068 (1988)
3. Yamanouchi K.,Megure T.and Chen Z.H., *in Ultrasonics Symposium Proceedings*, 1987 edited by B.R.Mcavoy (IEEE Publications, New York, NY, 1987),p.173

VIBRATIONAL PROPERTIES OF FRACTAL STRUCTURES

A. PETRI
Istituto di Acustica del C.N.R. "O.M.Corbino",
Via Cassia 1216, 00189 - Roma, Italy
and
Dipartimento di Fisica, Università di Roma "La Sapienza",
Piazzale A. Moro 2, 00185 - Roma, Italy

ABSTRACT

After recalling some basic concepts of the fractal geometry, we review the main properties of *fractons*, the localized modes of fractal strucures. We focus in particular on the features of the normal modes of percolating lattices in the approximation of scalar elasticity, and on their spatial amplitude fluctuations, that turn out to be multifractal. The physical implication of this last result is that self-similar fluctuations are an intrinsic characteristic of fractons, and therefore smoothing or averageing procedures when describing the spatial shape of modes can be questionable. Multifractality could also allow to clarify the ambiguities that are present in the definition of the correlation length.

1. Introduction

Fractal geometry has been found to be a good model for a large number of systems in nature showing irregularities over many length scales [1]. The leading concept of this geometry is the *scale-invariance* or *self-similarity*. A fractal object is such that a small part of its looks, exactly or statistically, similar to a larger part, or tho the whole object. Common examples are ramified structures, like trees or snowflakes, and the trajectory of a brownian motion. By magnifing a little portion of them we are not able to say at which lengthscale we are looking at.

Self-similarity is a kind of symmetry and can therefore be exploited for simplifying the determination of solutions of problems or of their properties, in a fashion similar to more conventional symmetries, like rotational or translational. At the same time, unlike the regular and smooth euclidean geometry, it is very suitable to the description of irregularities of nature, since self-similarity implies or trivial homogeneity either high singularity in the properties of an object.

In this paper we discuss some vibrational properties of random fractals, that is of fractals generated by random processes and for which self-similarity holds as an average statistical property. Deterministic fractals, the fractals generated by iterating a well determined rule, as e.g. a hierarchical Cantor set, will not be treated here, as they are discussed in other contributions to this School [2].

Section two constitutes a brief introduction to some useful concepts in fractals and to the percolation model, for those who are not very familiar with them. Section three gives a short overview on some features of the frequency and Fourier spectra of vibrational modes, paying particular attention to the the problem of the way of decaying in space of *fractons*, the localized modes of a fractal. It is an open question if their smoothed profile can be better described by an exponential function or by a super-exponential, that is by a function of the form $e^{-x^{d_\phi}}$, with $d_\phi > 1$. Finally in section four we suggest a way out of this problem by appealing to the multifractal properties of modes. After recalling the notion of multifractality, we show that this is actually a property possessed by the amplitude distribution of vibrational modes. This implies that intrinsically large fluctuations are present, and therefore fractons cannot be well described by smooth functions, and that a proper definition of a localization length is not trivial.

2. Fractal Exponents

2.1. The Fractal Dimension

Let us suppose we want to measure the volume of a regular d-dimesional object. A way to do it is sketched on the left side of Fig.1, for the example of a two dimensional surface, and consists of covering the surface with N boxes of area v, and of defining the volume (i.e. the total area) as

$$V = N(l) \cdot v(l),$$

where l is the linear size of the boxes. In this way N and v depend on l, and so also V does. But if l is small enough this dependence becomes negligible and we approach the limit value of the volume

$$V = \lim_{l \to 0} N(l) \cdot v(l).$$

This is schematically shown by the curve (a) of Fig.2. Of course a length scale redefinition does not change V. In fact for small l we get

$$\begin{aligned} v(l) &\propto l^d \\ N(l) &\propto l^{-d}, \end{aligned}$$

so that the l dependence cancels out, and $V = const.$

This is no longer true for a fractal object, which possesses irregularities at all the length scales. If we try to cover it with cells of volume v, we shall find that the number of needed cells is not proportional to the inverse of the cell volume as in the case of a homogeneous object. A fractal possesses, so to say, a lot of "holes",

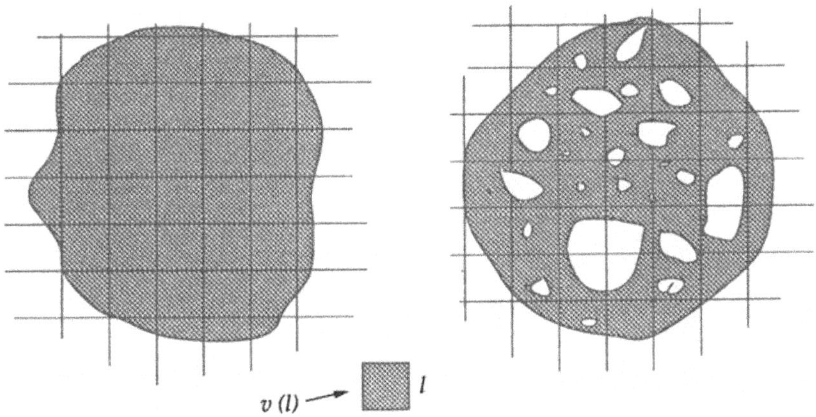

Figure 1: *Box covering of a homogeneous (left) and a fractal (right) two-dimensional plane surface.*

like the right side picture of Fig.1, and a smaller number of cells than in the usual homogeneous case suffices:

$$N_{fract}(l) \leq N_{hom}(l).$$

At the same time the holes are present at all the length scales, and this produces a regularity in the behaviour of *N(l)*. Actually for a fractal we have

$$N_{fract}(l) \propto l^{-D},$$

with $D \leq d$ as $N_{fract} \leq N_{hom}$, like in the curve (b) of Fig.2. This implies that for small l's:

$$V(l) = N_{fract}(l) \cdot v(l) \simeq l^{d-D}.$$

Thus if a structure is fractal up to infinitely small l's, its volume never approaches a constant value, but still scales as

$$V(l) \simeq l^{\alpha}; \qquad \alpha = d - D \geq 0.$$

This implies that a scale transformation, i.e. a different choice of the box-size l, does not leave the volume unchanged.

Different definitions of the fractal dimension can be given (see e.g. [1]), but this is not relevant for the discussion that will follow.

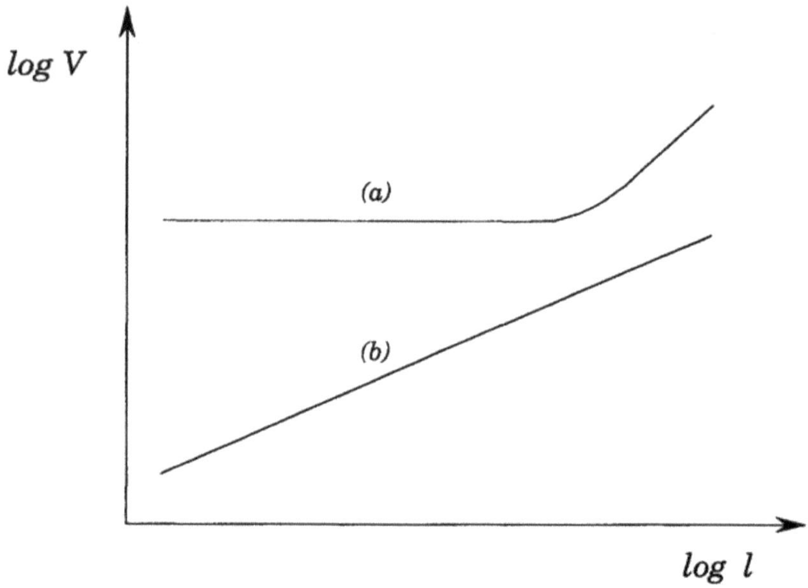

Figure 2: *Sketch of the volume behaviour as a function of the covering box size for a regular (a) and a fractal object (b).*

2.2. Scaling

As a consequence of self-similarity it is assumed that all physical quantities behave as powers of the relevant length scale for the considered quantity [3]. This is because in principle in a strictly self-similar structure there are no characteristic lengths, and therefore a change in the length scale cannot modify the functional behaviour of a physical quantity Q. If $Q = Q(l)$ relates the value of Q to the length scale l, then under a dilation like

$$l \to b \cdot l,$$

Q will transform without changing its functional form (up to a proportionality factor depending on b);

$$Q(l) \to Q(bl) = A(b) \cdot Q(l).$$

This implies

$$Q(l) \simeq l^{\alpha},$$

being now α the characteristic exponent of Q.

2.3. Finite Size Scaling

Real systems are fractal only within a certain range of lengths. The bounds of this range are therefore the lengths at which the physical properties of the system cross over to a homogeneous regime. If, for example, the system is fractal only up to an upper crossover length ξ, this will represent the only characteristic length for the system, appearing therefore in the expression of Q only through the ratio l/ξ:

$$Q(l) = l^{\alpha} \cdot f(l/\xi). \tag{1}$$

We can then distinguish two regimes:

- $l \lesssim \xi$:
  the system is fractal ($Q(l) \simeq l^{\alpha}$), and therefore

$$f(l/\xi) \simeq const.$$

- $l \gtrsim \xi$:
  the system is homogeneous. Its properties are now independent of the length scale l ($Q(l) \simeq const.$), and this implies to have in Eq.(1)

$$f(l/\xi) \simeq (l/\xi)^{-\alpha},$$

  and therefore

$$Q(l) \propto \xi^{\alpha}. \tag{2}$$

So it is seen that the properties of the system in the homogeneous regime are determined also by the crossover length and by the fractal exponent α. This implies that α can also be determined by measuring Q above the crossover in systems possessing similar structure but different crossover lengths ξ. Nice examples can be found for instance in the works by Courtens and coworkers on the radiation scattering from silica aerogels [4]. In real materials this crossover can actually occur in two opposite ways. For example, in silica aerogels the system is built by elementary grains that are locally compact. These units form, above some scale, a fractal structure that becomes again homogeneous at a larger scale.

One of the most investigated random fractal model is the percolation model, owing to its universality and wide applicability. In the following section we shall briefly recall the main fractal features of this model, which is very often used not only in the study of vibrational properties.

2.4. Percolation

Let us consider a lattice each site of which is occupied with probability p and empty with probability 1-p (Fig.3). For low values of p most of the sites are empty, so that the nearest neighbour occupied sites form several small separate clusters, as shown in Fig.3(a) for the case of a square lattice. By raising p the average size of the clusters increases and, at a certain critical value p_c, becomes infinite. By defining

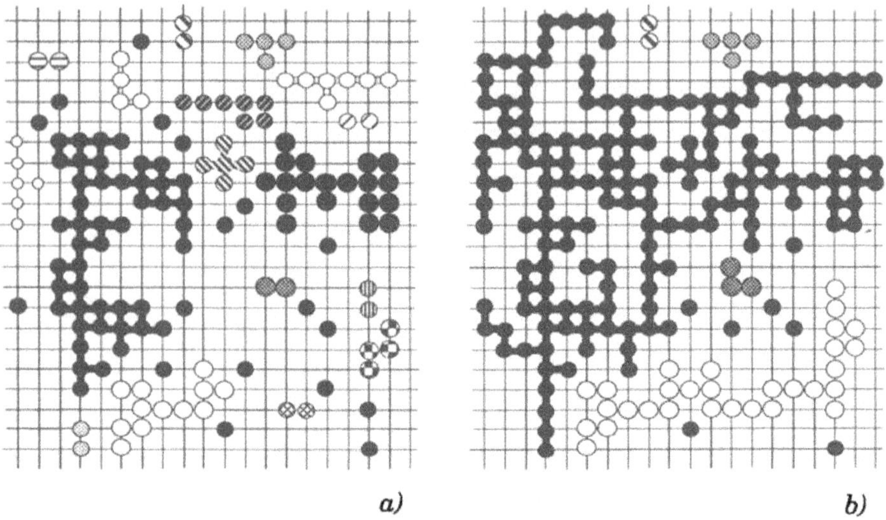

Figure 3: *Percolation on a square lattice. Particles belonging to different clusters are characterized by different symbols: a) for* $(p < p_c)$ *many separate clusters of different finite size are present; b) for* $(p > p_c)$ *a cluster spanning the whole lattice exists.*

the correlation length ξ as the average distance between two sites belonging to the same cluster, this turns out to be a well defined function of p, that for p approaching p_c from below behaves as:

$$\xi \approx | p - p_c |^{-\nu} \qquad \nu > 0, \tag{3}$$

i.e. ξ diverges when $p \to p_c$. Now the lattice is percolating, in the sense that there exists an infinitely wide cluster (called the percolating cluster at percolation threshold, or infinite incipient cluster) which includes sites at arbitrary large distance (Fig.3(b)). This cluster is actually fractal. In fact by counting the average number N of sites belonging to the infinite cluster and falling within a sphere of radius r, and averaging over all the possible origins (i.e. over all the occupied sites), one finds:

$$N(r) \simeq r^D \qquad \text{with } D < d, \tag{4}$$

instead of the usual $N(r) \simeq r^d$.

When $p > p_c$ Eq.(3) still holds partially, in the sense that the percolating cluster is fractal only up to length scales of the order of ξ:

$$N(r) \simeq r^D \qquad \text{for } r < \xi;$$
$$N(r) \simeq r^d \qquad \text{for } r > \xi.$$

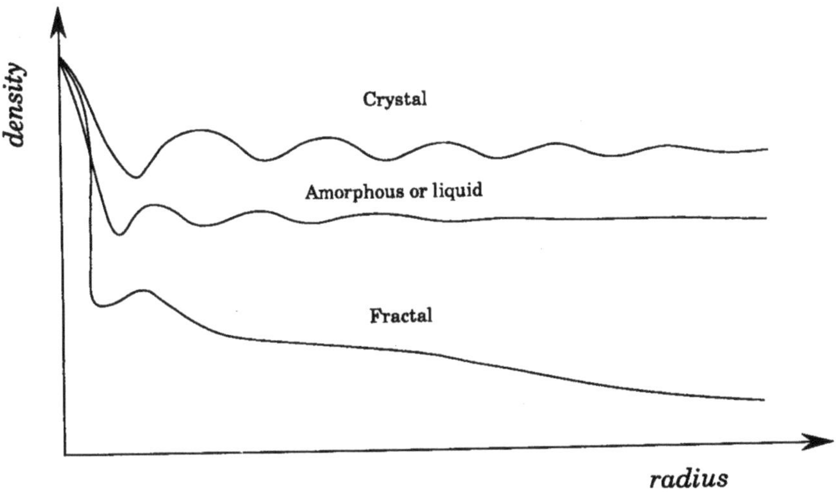

Figure 4: *Different behaviours of density, as defined by Eq.(5), for different condensed systems.*

By the way we note that, if we compute the density of a fractal object from Eq. (4), we find

$$\rho(r) \simeq r^{\alpha}; \qquad \alpha = D - d < 0, \tag{5}$$

implying a vanishing density when $r \to \infty$ (Fig.4). Therefore, owing to solely stability requirements, a finite range for the fractal regime in real structures follows.

The site percolation model can obviously be extended to other kinds of geometries (triangular lattices, honeycomb lattices, etc.). It is moreover possible to introduce different mechanism of percolation. In particular in the *bond* percolation model the lattice sites are all occupied, while the bonds between nearest neighbours are assigned with probability p, and different clusters result as sets of connected sites. Of course this last model seems to be more suitable for studying vibrational problems, and we shall often refer to it in the following; however, despite their differences, most of the scaling properties are the same in all models and, indeed, their universality lies in the universality of the fractal exponents, like D, ν, and others not discussed here. Neither do these exponents depend on the geometry of the model while on the contrary the values of p_c do depend on this and on the kind of model, as can be seen in Tab.1.

d	D	ν	p_c^{bond}	p_c^{site}
2	91/48	4/3	0.5	0.593
3	≈ 2.5	≈ 0.88	0.249	0.312

Tab. 1: *Values of some exponents and critical probabilities for square and cubic percolating lattices [3,5].*

3. VIBRATIONS OF FRACTALS

Vibrational properties of fractal structures have been widely investigated during the past few years, both theoretically and experimentally (see [4,6] and [20] for general reviews), and some of their spectral and spatial properties have been clarified, showing anomalous behaviours if compared with regular structures. In the following we shall discuss those properties of solutions to the scalar discrete elasticity equations

$$\omega^2 u_i = \sum_{(i,j)\in\{n.n.\}} K(u_i - u_j), \tag{6}$$

that can be considered to have a quite general validity. We refer to the bibliography for more complex models, with vector displacements or bending forces [7].

3.1. *Spectrum of Vibrations*

According to the concept of scaling it is assumed that the frequencies ω of *fractons* [8] behave as

$$\omega \approx l_\omega^{-d_\omega}, \tag{7}$$

being l_ω a characteristic length of the excitations, and d_ω the corresponding scaling exponent (the minus sign has been written explicitely for later convenience). The finite size scaling, Eq.(1), implies that if the system becomes homogeneous at a length scale ξ, then modes turn to the normal regime as soon as ω is larger than the crossover frequency given by Eq.(2)

$$\omega_{c.o.} \propto \xi^{-d_\omega}, \tag{8}$$

and the characteristic length assumes the usual meaning of a wavelength.

In the case of the percolating cluster above p_c, one can observe fractal properties from the minimal scale (lattice unit) up to the crossover length to a statistically homogeneous structure. This means that standard propagating wave solutions are possible up to the minimal wavelength corresponding to the correlation length ξ, and their velocity will be given by

$$v_\xi = \frac{\omega \cdot \lambda}{2\pi} = \frac{\omega_{c.o.} \cdot \xi}{2\pi} \propto \xi^{1-d_\omega}.$$

Alexander and Orbach [8] and Rammal and Toulouse [9] have shown that in many cases the relation

$$d_\omega = d_W/2 \tag{9}$$

holds, d_W being the random walk dimension, that is the exponent expressing the mean squared distance $< R^2 >$ covered in a time t by a particle diffusing on a fractal [5]

$$< R^2 > \approx t^{2/d_W}.$$

In regular cases $d_W = 2$, so that $d_\omega = 1$ and Eq.(7) simply expresses the relation between frequency and wavelength, while the sound velocity does not depend on the length scale.

As a direct consequence of (7) and of the self-similarity of the structure, the density of states $g(\omega)$, that is the number of modes with frequency between ω and $\omega + d\omega$, at low frequencies does not follow the usual power law

$$g(\omega) \approx \omega^{d-1},$$

but the similar law

$$g(\omega) \approx \omega^{\tilde{d}-1} \tag{10}$$

with $\tilde{d} \leq d$. $\tilde{d}$ is called the spectral dimension and can be related to the fractal dimension D and to the diffusion exponent d_W by [8,9]:

$$\tilde{d} = \frac{2D}{d_W}. \tag{11}$$

This exponent describes the mean behaviour of the density of states (which is generally a highly singular function of frequency) and it is therefore useful as an approximation for computing integral quantities in the low frequency limit such as specific heat and thermal conductance at not too high temperatures (Fig.5).

d	D	d_ω	d_W	$\tilde{d}$
2	1.9	1.4	2.8	1.36
3	2.5	1.78	3.55	1.41

Tab. 2: *Approximate values of characteristic exponents for the frequency (d_ω), the diffussion process (d_W), and the density of states ($\tilde{d}$) on the percolating lattice [8].*

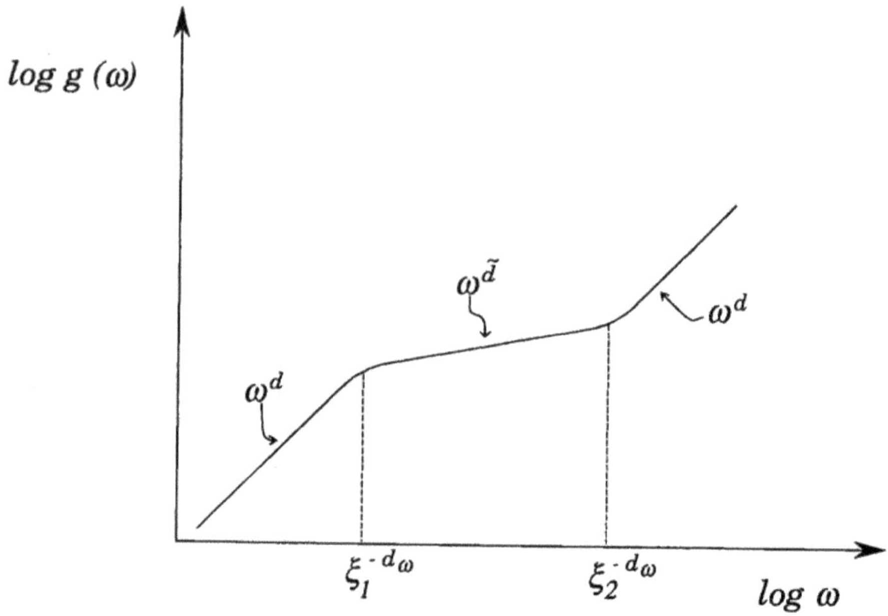

Figure 5: *Sketch of the different regimes of the density of states. Phonon-like densities of states with spectral exponent $\tilde{d} = d$ are observed at high and low frequencies, when the wavelength is respectively smaller or larger than the crossover lengths of the structure ξ_1 and ξ_2.*

3.2. Spatial Behaviour

We have just seen how the characteristic length l_ω can be identified with the wavelength of the extended modes in the homogeneous regime, and that it determines in some way the velocity of sound beyond the crossovers. Of course this is not true in the fractal regime, as in this situation eigenmodes do not possess a well defined wavevector, owing to the lacking of exact translational symmetry, and sound propagation cannot take place in the harmonic approximation because of the localized nature of fractons. In this regime l_ω assumes different meanings.

In particular it is believed that it can represent all the three lengths relevant for the description of waves in disordered media, that is the dominant wavelength, the scattering length and the localization length [10]. Limited space prevent us from entering here a detailed discussion, and we shall limit ourselves to some considerations concerning Eq.(7).

1. Fourier components.

   Each eigensolution of Eq.(6) can be Fourier transformed and expressed as a sum of plane waves. Then each single mode is represented by a packet in the transformed k space (Fig.6). Montagna et al. [11] studied the distribution of the k_M's with frequency, and found that they satatistically follow the scaling given by Eq.(7), with:

   $$k_M = \frac{2\pi}{l_\omega}.$$

   Thus l_ω still represents a wavelength, each mode exhibiting spatial oscillations mostly around it, and according to Eq.(6) it happens to shorten with frequency just as the usual wavelength does.

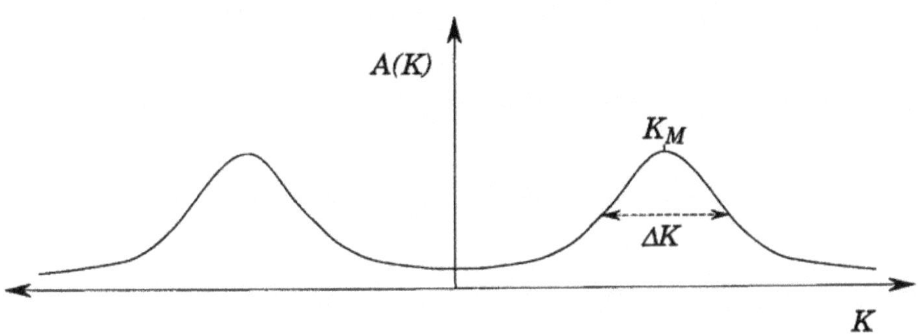

Figure 6: *Schematic representation of the Fourier transform of an eigenmode along one direction. A wavepacket in the Fourier space, with a maximum at* k_M *and width* Δk*, results.*

2. Scattering length.

   If a wave with well defined wavenumber is excited, it will not survive because it is not a proper mode of the structure, and it will be scattered into the eigenmodes into which it can be decomposed. It is believed that this occurs within a scattering length of the order of l_ω [10], although no rigorous proof of this conjecture seems to be available up to now. If, for instance, one computes the Fourier transform of the two-points correlation function along one direction x

   $$C(x) = \sum_r u(r)u(r+x),$$

   one obtains as a result that the average width Δk of its Fourier transform follows the above scaling at low and intermediate frequencies [12] with $\Delta k \propto 1/l_\omega$, as Δk is actually related to the inverse of the scattering length [13]. A similar result concerns the average width Δk of the Fourier transforms of single fractons [11], and the averarge sizes of fractons in the real space, [14] which are also related to Δk.

3. Localization length.

   Localization of electronic wave functions in disordered systems when $d \leq 2$ leads to believe that modes are essentially localized on fractals too [8,9,10]. Actually some studies have shown that states are *super*-localized on fractals, in the sense that their decay is faster then exponential. The main reason for thinking about superlocalization is that the physical distance l between two points on a fractal does not follow a straight line, but a rather winded path, as a fractal is a hole-rich structure (Fig.7). At the same time the self-similar nature of the structure imposes a scaling law for this distance, called *chemical length* or *minimum path*, along the structure:

   $$l \simeq r^\gamma,$$

   r being the euclidean distance between two points; e.g. $\gamma \simeq 1.68$ from numerical estimates for percolation in $d = 2$ [5]. This implies that, if any state decays exponentially along allowed paths with decay length η, then we can write:

   $$\psi \simeq e^{-(\frac{l}{\eta})} \simeq e^{-\frac{(r^\gamma)}{\eta}} \simeq e^{-(\frac{r}{\lambda(\omega)})^\gamma},$$

   with $\lambda(\omega)^\gamma = \eta$. The above argument, though far from rigorous, has been used in support of superlocalization for a few of years. The localization length $\lambda(\omega)$ is supposed to coincide with the characteristic length l_ω which controls the scaling of frequency given by equation (7). Numerical results for a percolating cluster give [15]

   $$\lambda(\omega) \approx \omega^{-0.705},$$

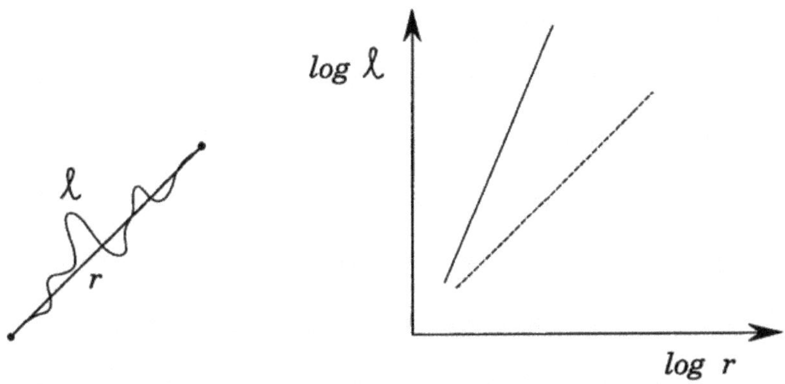

Figure 7: *On a fractal, the path length l between two points is distributed around a power of their euclidean distance r, thus increasing faster (continuous line) than on a homogeneous strucure, where they are simply proportional (dashed line).*

in very good agreement with the expected value derived by identifying $\lambda(\omega)$ with l_ω given by Eqs.(7), (9) and (11):

$$d_\omega \approx \tilde{d}/D \approx 0.71.$$

It must be stressed however that such an agreement is obtained by averaging the decaying patterns of more than hundred fractons.

Among others, existence of superlocalization has been theoretically supported also by Lévy and Souillard [16], who show that impurity states on percolating lattices are always superlocalized, with exponent $d_\phi \simeq 1.43$ (different from γ) for d=2; this conclusion is then extended by the authors to the case of general localized states on fractals, with no other kind of disorder. They nevertheless state that most of these states have a support much larger than the superlocalization length, so that the superlocalized behaviour actually occurs in this case up to some large cutoff length, and then it should be followed by some weaker decay. Harris and Aharony [16] give $1 \lesssim d_\phi \lesssim \gamma$ as a rigorous bound.

On the contrary, some recent works show absence of superlocalization both numerically and analitically, in favour of an average usual exponential pattern [17], [18], while different numerical approaches still seem to show evidence of superlocalization with $d_\phi = 1.13$ for $d = 2$ and $d_\phi = 1.39$ for $d = 3$ [19].

It must be stressed that different results can often have their origin in different ways of averaging, or of deriving the localization length (spatial decay, participation ratio, etc.), while shape of fractons and their fluctuations can have implications on heat transport, sound propagation by anharmonic couplings, local stress and strain distributions, etc. It is therefore important to decide correctly which kind of average, if one, is appropriate to the treatment of a given physical problem.

4. Multifractality of Modes

We have tried to sketch here a short overview on the properties of the normal modes of a fractal structure. With particular regard to the localization of modes, the wide range of cited results suggests that such properties cannot be uniquely asserted in a simple manner, and that they may eventually depend on the way of approaching the problem, also because of the presence of very strong fluctuations. We are therefore induced to think of multifractality as to a possible feature of these modes [20]. This property can, in principle, give rise to anomalous behaviours of average quantities, due to the presence of strong scale invariant fluctuations, and it can account for some features mentioned in the above picture, like evidences for a sensible extension of modes far from their characteristic lengths, and different possible definitions of localization lengths [21].

4.1. Multifractal analysis

Multifractality represents the situation in which the scaling properties of a measure (in the mathematical meaning) μ are different from point to point [21]. It is then possible to define a pointwise (local) dimension or singularity exponent α such that

$$\mu(x) \approx l^{\alpha(x)}, \tag{12}$$

in the limit of small values of l. This property implies high inhomogeneities and strong fluctuations in the distribution of the measure on the support, giving raise to a non gaussian scaling of moments $\langle \mu(l)^q \rangle$, so that:

$$\langle \mu(l)^q \rangle \neq const \cdot \langle \mu(l) \rangle^q .$$

As shown below, this fact allows to express the scaling properties of the moments of the μ distribution through a set of generalized dimensions $D(q)$. Such a property has been found to characterize some features of many critical dynamical systems, like disorder fluctuations of electronic wave functions in the Anderson model, and spatial fluctuations in incommensurate potentials or in Anderson model near the mobility edge, and in hierarchical structures. It also characterizes the voltage distribution in a random resistor network and the distribution of probabilities for finding a particle diffusing on a fractal (see [20] and [21] for general references).

Standard analysis of multifractality passes through determination of scaling properties of the partition function

$$Z(l,q) = \sum_b [\mu_b(l)]^q \approx l^{\tau(q)}, \tag{13}$$

where $\mu_b(l)$ is the contribution to $Z(l,q)$ of the measure contained in the b-th box into which the considered support (here the lattice) is divided, and in the present case consists of the sum of the squared amplitudes at sites i within the given box

$$\mu_b = \sum_{i \in b} u_i^2.$$

It is then possible to determine the scaling exponents of moments as

$$\tau(q) = \lim_{l \to 0} \frac{\ln Z(l,q)}{\ln l}. \tag{14}$$

One is usually interested in the scaling properties of the statistical distribution $N(\alpha)$ of the α's, the scaling exponents for μ in equation (12), and therefore assumes

$$N(\alpha) \approx l^{-f(\alpha)}.$$

Iit is seen that $f(\alpha)$ actually plays the role of the fractal dimension of the set on which the measure is characterized by a given singularity exponent α. [1] This quantity can be retrieved from $\tau(q)$ by means of a Legendre transformation [21]:

$$f(\alpha) = q\alpha - \tau(q); \qquad \left(\frac{df(\alpha)}{d\alpha} = q\right) \tag{15}$$

$$\alpha = \left(\frac{d\tau}{dq}\right). \tag{16}$$

For fractal sets on wich amplitudes are statistically homogeneously distributed we have

$$\tau(q) = D \cdot (q-1), \tag{17}$$

that is, τ is proportional to the fractal dimension D of the support. For non fractal sets the euclidean dimension d substitutes D in the above expression:

$$\tau(q) = d \cdot (q-1). \tag{18}$$

[1] It is pheraps useful to remark that this name derives from having a singular density $d\mu/d^d l$ when α is smaller than d.

Otherwise a whole family of exponents $D(q)$ appears through generalization of the previous relation,

$$\tau(q) = D(q)(q-1) \tag{19}$$

describing the different scaling of moments:

$$\langle \mu(l)^q \rangle \approx \langle \mu(l) \rangle^{\tau(q+1)}; \qquad (\tau(q) \neq D \cdot (q-1)).$$

In fact, $D(q)$ is the fractal dimension of the set that mainly contributes to Z for a given q. We have therefore the following particular cases:

- $q = 0$:
  all the lattice sites contribute to Z indipendently of their amplitude, and $D(q=0) = D$, the fractal dimension of the support;

- $q = 1$:
  all the modes are normalized to $\sum u_i^2 = \sum \mu^1 = 1$, and this yields $\tau(q = 1) = 0$, owing to the normalization of our measure; however $D(q = 1)$ can be computed through the expression

  $$D(1) = \lim_{l \to 0} \frac{\sum_i \mu_i(l) \, \ln \mu_i(l)}{\ln l}.$$

  Only the set of points on which amplitudes are different from 0 contributes to the sum; $D(1)$ can therefore be thought as the fractal dimension of this set, and for this reason is often called the *information* dimension;

- the others positive values of q "select" boxes of increasing density with ever increasingly strong singularities, while the negative values select those singularities present in the low density boxes.

It is straightforward to derive the different behaviours exhibited from $f(\alpha)$ by means of Eqs.(15) and (16), and by the form of $\tau(q)$ in the different cases :

homogeneous systems: $\tau(q)$ is linear in q, Eq.(18), and $f(\alpha)$ and α collapse to the single point:

$$f(\alpha) = \alpha = d.$$

This means that both the whole measure and the support scale in the ordinary euclidean way, with exponent equal to the euclidean dimension;

homogeneous fractals: this is the case of Eq.(17); $f(\alpha)$ and α still collapse to a single point,

$$f(\alpha) = \alpha = D,$$

and the measure scales with the same index of the support.

non homogeneous fractals: $\tau(q)$ is not linear, Eq.(19)

$$f_{min} \leq f(\alpha) \leq f_{max}; \qquad \alpha_{min} \leq \alpha \leq \alpha_{max}.$$

$f(\alpha)$ is in general a convex function of α, with a maximum $f_{max} = D(0)$ for $q = 0$, and

$$f_{min} \in (-\infty, D(0)]; \quad 0 \leq \alpha_{min} \leq \alpha_{max} \leq \infty$$

4.2. Results

It has been shown recently [18] that fractons are multifractal in the following sense. If one looks at sites that are at a given distance r from the site of maximum displacement, assumed to be at $\mathbf{r}_0$, and performs an average over all the fractons of (about) same frequency but with different $\mathbf{r}_0$, one get an amplitude distribution exhibiting anomalous moments:

$$\sum_{\mathbf{r}_0, r} u(\mathbf{r}_0 + \mathbf{r}, \omega)^q \propto \langle u(r,\omega)^q \rangle \neq \langle u(r,\omega) \rangle^q.$$

By assuming some kind of ergodicity, this procedure can be considered to be a sort of average over the possible realizations of disorder. This means that at a given spatial point the amplitude fluctuations due to possible different realizations of the lattice are multifractal. This property, however, does not imply self-similarity in the amplitude fluctuations of single realizations of a fractons [22], that must therefore be analyzed separately.

In order to characterize the scaling properties of single fractons, we have performed some numerical simulations on many two-dimensional lattices at the bond percolation threshold, with periodic boundary conditions [20]. Normal modes were then obtained by diagonalizing the dynamical matrix of the percolating cluster on several lattices of different sizes, up to more than 4000 sites wide; then multifractal analysis have been carried out on a large number of these modes. As a test against errors due to the limited size of our lattices, we also computed and analyzed the normal modes of homogeneous square lattices.

The main results of our analysis are exemplified in Figs. 7 and 8, reporting the computed $\tau(q)$ and $f(\alpha)$ curves for three typical cases.

At the top the curves inherent to one mode of the homogeneous lattice are shown. It is seen that within numerical errors they obey the behaviour discussed above, with a linear dependence of τ on q, and with $f(\alpha) = \alpha = d = 2$.

At the center the curves relative to a very low frequency mode of the percolating cluster are shown. Since its characteristic length l_ω is comparable to the lattice size, which in this case represents the upper correlation length, it looks like a phonon-like mode, with linear $\tau(q)$, and $f(\alpha) = \alpha = D \simeq 1.9$. Finally results for a typical fracton-like mode at intermediate frequency are shown. $\tau(q)$ bends very

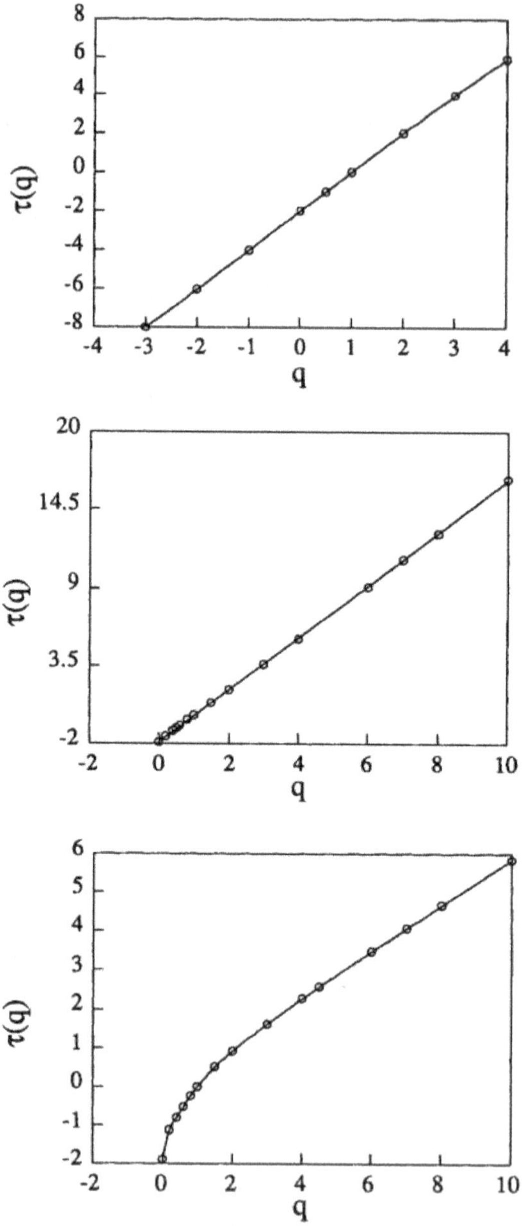

Figure 8: *The $\tau(q)$ curve of three representative modes (see text). A phonon mode (top); a phonon-like mode (center), and a fracton mode (bottom).*

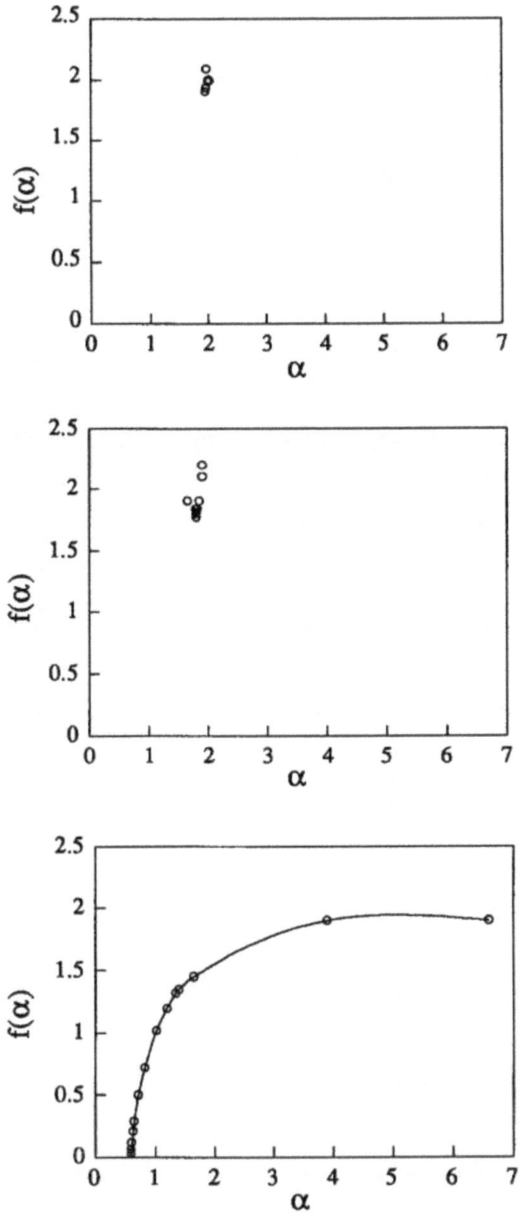

Figure 9: *The $f(\alpha)$ curve as derived from the $\tau(q)$ curves of Fig.8.*

clearly, and consequently $f(\alpha)$ and α widen their range. In particular we see that α assumes also values smaller than the euclidean dimension d, implying the presence of singular densities in the mesaure, and therefore of very strong fluctuations. Note that negative values of q have been considered only in the case of the homogeneous lattice, because estimates of negative moments did not seem reliable in the fractal case. Finally it must be said that the multifractal features of modes have also been found to depend on the frequency [20]. They exhibit a statistically stable behaviour, with the largest $f(\alpha)$ spectrum at intermediate frequencies.

4.3. Conclusions

We have summarized some properties of fractons, the localized vibrations of fractals, and examined the possibility that their amplitude fluctuations on the percolating cluster in two dimensions manifest multifractal features. We find clear evidence for this property that may help in clarifying the physical properties of these modes. In particular this is important in relation to the possibility of describing them in terms of smoothed convolutions and with respect to the proper definition of a localization length.

The results imply that a strongly fluctuating amplitude is an intrinsic characteristic of fractons, with consequent strong fluctuations in the local stress and strain distributions. The use of smoothed convolutions or averages over several modes can therefore lead to results that no longer correspond to the original properties, and that depend on how the averages are computed. It is hoped that a better understanding of these effects may finally lead to a clearer description of the physical properties of fractons. It is also hoped that more direct and experimental observations may verify the relevance of such model for the description of strongly disordered systems, and confirm or reject its predictions.

1. B. B. Mandelbrot: *The Fractal Geometry of Nature*, Freeman and Co., (New York 1983);
   L. Pietronero and E. Tosatti eds.: *Fractals in Physics*, North-Holland, (Amsterdam 1985);
   J. Feder: *Fractals*, Plenum Press, (New York 1988);
   A. Bunde and S. Havlin eds.: *Fractals and Disordered Systems*, Springer-Verlag, (Berlin 1991).

2. F. Craciun, A. Alippi, A. Bettucci, in these Proceedings;
   W. Lauterborn, ibid.

3. D. Stauffer: *Introduction to Percolation Theory*, Taylor & Francis, (London 1985).

4. E. Courtens, R. Vacher and E. Stoll, Physica D **38**, 41 (1989) and references therein;

5. S. Havlin in H. E. Stanley and N. Ostrowsky eds: *Random Fluctuations and Pattern Growth: Experiments and Models*, Martinus Nijhoff (Dordrecht, 1988);
   S. Havlin, D. Ben-Havraham, Adv. in Phys. **36**, 695 (1987).

6. S. Alexander, Phys. Rev. B. **40**, 7953 (1989) and references therein.

7. See e.g. I. Webman and G. S. Grest, Phys. Rev. B **31**, 1689 (1985).

8. S. Alexander and R. Orbach, J. Phys. (Paris) Lett. **43**, L625 (1982).

9. R. Rammal and G. Toulouse, J. Phys. (Paris) Lett. **44**, L13 (1983).

10. S. Alexander, Physica **140A**, 397 (1986).

11. M. Montagna, O. Pilla, G. Viliani, V. Mazzacurati, G. Ruocco and G. Signorelli, Phys. Rev. Lett. **65**, 1136 (1990);
    V. Mazzacurati, M. Montagna, O. Pilla, G. Viliani, G. Ruocco and G. Signorelli, Phys. Rev. B, to be published.

12. A. Petri, at LXXVII Conference of the Italian Physical Society, L'Aquila (Italy), Oct. 1991, unpublished;
    A. Petri and G. Ruocco, in progress.

13. S. Alexander, E. Courtens and R. Vacher, submitted to Phys. Rev. B (1991).

14. E. Stoll and E. Courtens, Z. Phys. B - Condensed Matter **81**, 1 (1990).

15. K. Yakubo and T. Nakayama, Phys. Rev. B. **40**, 517 (1989).

16. Y. E. Lévy and B. Souillard, Europhys. Lett. **4**, 233 (1987);
A. Brooks Harris and A. Aharony, Europhys. Lett. **4**, 1355 (1987).

17. P. de Vries, H. De Raedt, and A. Lagendijk, Phys. Rev. Lett. **62**, 2515 (1989).

18. H. E. Roman and A. Bunde, Phyl. Mag. B **63** (1991).

19. C. J. Lambert and G. D. Hughes, Phys. Rev. Lett. **66**, 1074 (1991).

20. A. Petri and L. Pietronero, to be published on Phys. Rev. B.

21. See e.g. G. Paladin and A. Vulpiani, Phys. Rep. **156** 147 (1987), and references therein.

22. L. Pietronero, A. P. Siebesma, Phys. Rev. Lett. **57**, 1098 (1986).

II. CONTRIBUTED PAPERS

B. Characterization through ultrasonic waves

REMOTE ACOUSTIC SENSING OF THE SEA BED

R C Chivers
Physics Department, University of Surrey, Guildford GU2 5XH, Surrey

D Burns
Marine Micro Systems, Kinsale Road, Cork, Ireland

ABSTRACT

The principles of sea bed classification and their relation to acoustic parameters are discussed at a fundamental level. The mechanisms of formation of the first and second echoes received by an echo sounder are analysed. It is shown that careful signal processing can provide two parameters for classification of the sea bed. An instrument which performs this task is described, as is its successful use in hydrographic surveying.

1 Introduction

Knowledge of the nature of the sea bed at a given location is of ecological and economic importance to most of humanity. It represents one of the boundaries of our existence whose conditions have to be satisfied as we seek a satisfactory solution in terms of a modus vivendi that will be valid for all time. A knowledge of the status quo and its dynamics is the basis of understanding, and a prerequisite to sensitive strategies. In addition there are a number of more direct reasons for knowing the nature of the sea bed at a given time and place.

Marine engineers need information to identify suitable sites for construction. Suitability is defined both in terms of meeting the demands of the project and in terms of causing least interference to the ecology of the area. Cable-laying is usually achieved by blasting a channel in sediments with a high pressure jet. The presence of regions of rock or cohesive clay precludes this approach and a knowledge of the sea bed in this situation permits an optimal route to be planned.

In the fishing industry it is of vital importance to many types of fishermen to be aware of the presence of rock or even of stones, to avoid wear or damage to their expensive fishing gear. The sea bed appears to be one of several environmental factors[1] that affect the habits of fish, some species showing a preference for certain types of ground. Further, in the case of shell fish, they may be present in sufficient abundance as to change the effective characteristics of the sea bed[2].

Finally there are 'defence' requirements of a knowledge of the sea bed which are generally ill-defined in public terms.

The navigational charts on which sailors depend have clear limitations as far as their accuracy is concerned, as is demonstrated by the occasional unexpected grounding. They have generally been dependent for their production on direct sampling of the sea bed by grab sampling or coring. Both of these techniques are expensive and time-consuming and give only information at the point of the sample. It must then be assumed that the point is characteristic of the local area. In more recent years, technological progress has provided sophisticated advances in side-scan sonar systems and in video-cameras which can be mounted on remotely operated vehicles. These have the capability of surveying large areas of the sea bed relatively quickly. The major limitation is that the classification which is achieved with either of these techniques is subjective. In the case of side-scan sonar, even with carefully adjusted equipment and a skilled analyst, the classification, based on texture changes, is inferential.

There is a clear need for a procedure which will permit rapid, objective sea-bed classification based on information from a large enough area for the characteristics to be typical. The basis of such an approach is described in the present summary. Before discussing the acoustic principles and instrumentation, however, it is appropriate to consider the primary basis of sea bed classification.

2 Elements of ground classification

It is conventional and convenient to try to ascribe a simple label, such as rock, gravel, fine sand, silt, mud, etc., to the sea bed at a particular location. Apart from rock, for which a geological classification may be appropriate, classification is objectively achieved by ascribing a label to a range of particle sizes, as is shown schematically in figure 1. Unfortunately there is not universal agreement on the ranges of particle sizes to be given a particular label, although there is a general similarity between the different schemes.

We may now relate these schemes to acoustical parameters such as the wave velocity, the attenuation, the backscattering at a particular angle of incidence, etc. At a specific value of the independent parameter (eg frequency or angle of incidence) the acoustical parameter will, in principle, itself exhibit a range of values, corresponding to a particular ground type, as is shown schematically in figure 2.

The question now arises as to how many acoustical parameters are needed for classification. The argument that the more that are measured, the better the classification will be is almost certainly a false one. Firstly because of the inevitable relations between the different acoustical parameters, and secondly because of the intrinsic uncertainties of the problem mentioned shortly below. It is certainly logical to extend the concept of figure 2 to two variables (figure 3) where the locus of combinations of parameter values corresponding to a particular type of ground can be seen to be a rectangular box. Although more parameters can be used,their display and interpretation is not so readily achieved - even if the 'layer cake' approach which is currently popular in sea bed roughness analyses is adopted.

There are two fundamental practical problems associated with the discussion of the last two paragraphs. The first is that even if only one type of material is present, it may be present at different concentrations. Thus the neat mutually complementary and distinct ranges of values of the acoustic parameter for the different ground types in figure 2 become a series of overlapping bands. This does not necessarily cause a major problem since in transferring to the display of figure 3, the rectangular boxes will have some degree of overlap, representing an intrinsic ambiguity. The second problem is less resolvable. It is simply that at any given location there are usually a number of different ground materials present, in different proportions. Thus the ascription of a single label, such as 'fine sand' can be, at best, only an identification of the principle component present.

3 Mechanisms of formation of echo-sounder echoes

3.1 The first echo

Almost all sea-going vessels are equipped with an echo sounder. The main use of this device is to indicate the depth of the sea bed below the vessel. To this end, the first returning echo, which has been reflected once from the sea bed is often amplified into saturation to provide a clear boundary indication. In the process potentially valuable information is lost. The impulse response of the sea bed which comprises the first echo can be written, on the basis of a rather simple model, as a time function of the acoustic pressure returning to the echo-sounder transducer.

$$p_1(t) = \frac{P^2(0)\ T_{01}T_{10}\Re[\alpha(z),c(z),\rho(z)]\ \exp - [2\alpha_w d + 2\int_0^z \alpha(z)dz]}{4(d+z)^2} + \frac{P^2(\theta)\ S(\theta)\ \exp - (\alpha_w\ c_w t)}{c_w^2 t^2} \tag{1}$$

where

$$ct = 2d\sec\theta\ . \tag{2}$$

The model assumes two contributions to the returning echo - the first is due to sub-bottom penetration and reverberation of the normally incident sound, which re-emerges and returns to the transducer. A similar process may occur for sound obliquely incident on the sea bed but it is explicitly ignored here. The second term describes the oblique back reflection (backscattering) of rays from the sea bed. The following notation is used:

d is the depth of the sea to the sea bed, z is the vertical distance into the sea bed, θ is the angle between a ray emerging from the transducer and the vertical, $P(\theta)$ is the transducer directivity (assumed to be the same in transmission as in reception), c_w and α_w are respectively the speed and attenuation of sound in sea water, T_{01} and T_{10} are the normal amplitude transmission coefficients at the sea bed surface for waves travelling down and up respectively, and $S(\theta)$ is the oblique backscattering function.

$\Re$ is the sub-bottom reverberation function which depends on the way in which the density ($\rho(z)$), the sound velocity ($c(z)$) and the attenuation ($\alpha(z)$) vary with depth into the bottom. This has no explicit time dependence. It is implicitly present in that the variable z represents the distance into the sea bed from which an echo returns to the transducer at time t. Formally we have:

$$z(t) = \frac{1}{2}\int_{t_0}^{t} c(z)dt = \frac{1}{2}\int_{t_0}^{t} c(z(t))dt \qquad (3)$$

where $t_0 = 2d/c_w$. It is clear that this implicit dependence of the variable z is a major complication. Even if $\rho(z)$, $c(z)$ and $\alpha(z)$ are known, an analytical solution to the reflection problem is only available in a perturbation form[3], if these acoustic parameters vary continuously with z. The layered problem is also still under discussion[4].

As far as ground classification is concerned, we may observe that while a full analysis of the reverberant signal might be a rich source of information, it is impossible with most common instrumentation to distinguish between the reverberant echoes and those returning by oblique surface backscattering at small angles, because both return to the transducer over the same period of time. Noticing, however, that the function $S(\theta)$ has been found empirically to be strongly dependent on ground type (eg McKinney and Anderson[5]), a sensible strategy would be to gate out the initial part of the first echo until such a time as the reverberant echo may be considered to have decayed to insignificance. (This can be determined only by order of magnitude calculations and experiment.)

3.2 The second echo

Although it is usually ignored in conventional echo-sounders, the second echo is sometimes used subjectively by fishermen to identify hard ground[6]. The second echo comprises those rays which have travelled to the sea bed, returned to the sea surface, been back-reflected there, and again at the sea bed, to return to the transducer. The possibilities for multiple paths are legion, and often unassessable because of a general shortage of relevant backscattering data. Assuming only specular reflection at the sea bed and a negligible reverberation contribution, the pulse returning to the transducer for the second echo can be written as:

$$p_2 = \frac{P^2(\theta)\, R^2(\theta) e^{-\alpha_w c_w t} S_w(\theta)}{c_w^2 t^2} \qquad (4)$$

where $c_w t = 4d \sec\theta$, $R(\theta)$ is the specular amplitude reflection coefficient at the sea bed, $S_W(\theta)$ is the oblique backscattering function for the sea surface, and other notation is as given above. Clearly the dependence on the square of R - which, by being related to changes in characteristic acoustic impedance (in the simplest model) is directly related to the hardness of the sea bed - offers an excellent opportunity for characterisation.

4 Instrumentation

Two parameters, E1 and E2 were derived respectively from the first and second echoes, by gating and integrating, as shown in figure 4. To avoid the reflections changing with depth, E1 was obtained by integrating between a fixed pair of angles, θ_1, and θ_2. θ_1 is determined using equation (2) from the time after which sub-bottom reverberation is considered negligible. Absorption in the water was considered to be negligible. The instrument was designed to utilise as much as possible of the existing instrumentation on a vessel but it soon became apparent that the commercial swept gain circuits used to compensate for spherical spreading are of very variable quality. The approach adopted was therefore to build a "parallel receiver" which is connected across the transducer terminals, and incorporates the swept gain. An advantage of this procedure is that the dynamic range of the receiving and processing electronics can be optimised[7].

A block diagram of the system is shown in figure 6, where it can be seen that the parallel receiver is linked to a personal computer for data collection and storage. The computer is also linked to the navigational system of the ship. This permits the 2-parameter classification shown in figure 5 to be linked to a display on the PC-screen of the actual course of the vessel. Each box with a ground type classification in figure 5 is assigned a colour. At a particular location, the parallel receiver and processing electronics derive a value of E1 and a value of E2. This identifies a point on the E1:E2 diagram of figure 5 and determines the colour to be given to the track point on the plot of the vessel's course. A typical PC display is shown in figure 7, which contains not only the course plot (top left), but also the E1:E2 plot (bottom right), a depth plot (bottom left) - also colour coded - and other relevant numerical data. More details of the instrumentation are given elsewhere[8].

5 Results

Thirteen major cruises have been used to collect over ¼ million data points, many of which have been classified by one of the independent methods mentioned in section 1. The quality of the classification is clearly high although there are a number of instrumental refinements that can be identified. An example of the quality of the data obtained is shown in figure 8 where the boundaries between different types of material determined using the present (Roxann) system and a sophisticated side-scan sonar system are compared. The present system gives excellent agreement, with perhaps even more detail.

Apart from representing one of the major advances in hydrographic surveying of the last decade, the instrument developed has shown the importance of being careful not to discard potentially useful information. More basically it has identified a number of areas of ocean acoustics where our empirical knowledge is inadequate and others where the data should be viewed more critically in the light of basic concepts of ground classification.

6 Acknowledgements

The authors gratefully acknowledge support from the Directorate General of Energy of the CEC, and permission from The Hydrographic Journal to reproduce some of the figures.

7 References

1 A Orlowski, *Proc I.O.A.* **11 (3)** (1979) 179-185

2 D Burns, (1990) *private communication*

3 R C Chivers, *Proc I.O.A.* **11 (5)** (1989) 470-476

4 P Cobo Parra and A J Berkhout, *Acustica* **75** (1991) 40-50

5 C M McKinney and C D Anderson, *J Acoust Soc Am* **36** (1964) 158-163

6 D Burns, C B Queen and R C Chivers, *Ultrasonics* **23** (1985) 189-191

7 R C Chivers and W J Mullarkey, *Acoust Lett* **12** (1989) 151-156

8 R C Chivers, N Emerson and D Burns, *The Hydrographic Journal* **56** (1990) 8-17

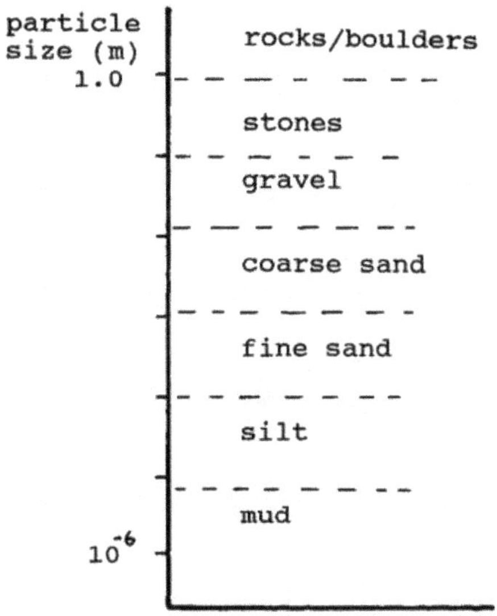

Figure 1 Conventional ground classification by particle size range

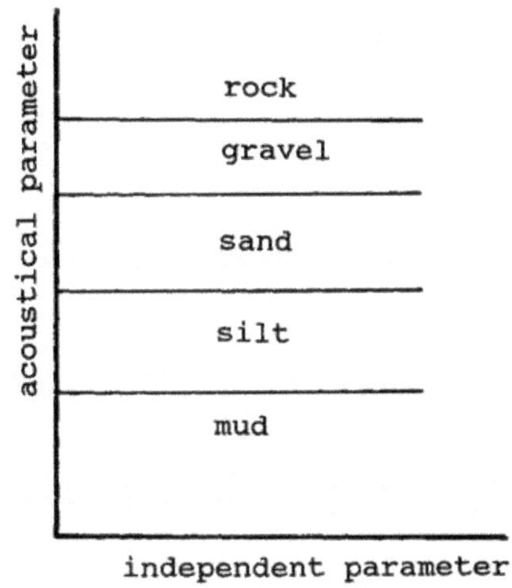

Figure 2 Relation of acoustical parameter values to ground classification

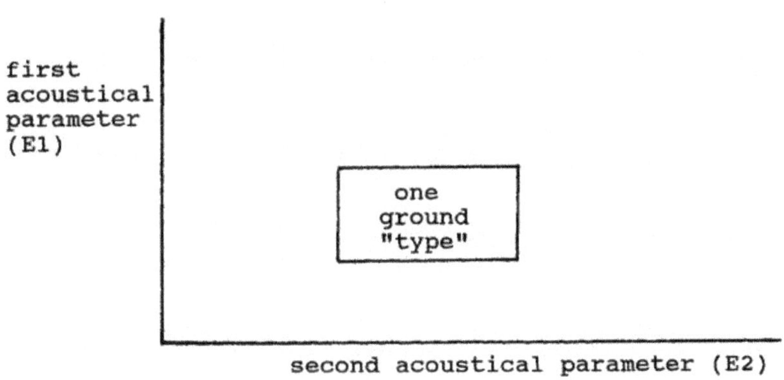

Figure 3 A two-parameter classification scheme

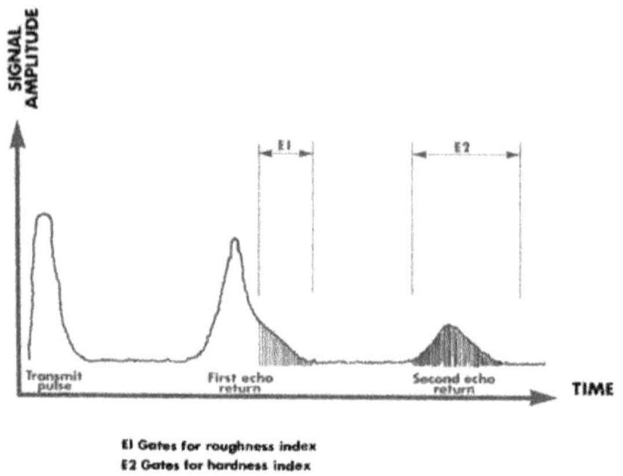

Figure 4 Portions of returning echoes gated and integrated to obtain E1 and E2

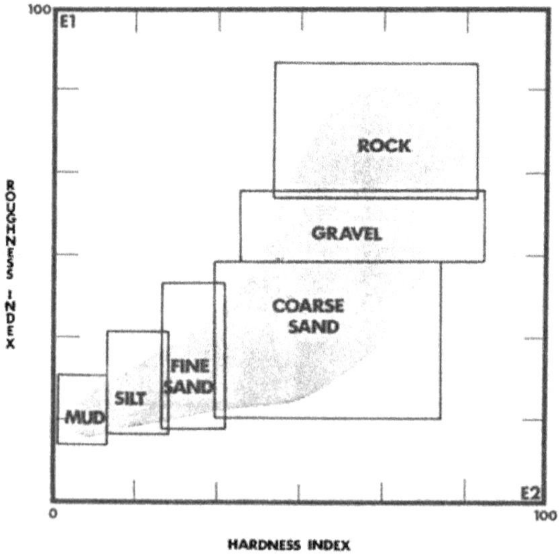

Figure 5 A two-parameter classification scheme based on E1 and E2. The scales of the axes depend upon the transducer and gain settings used

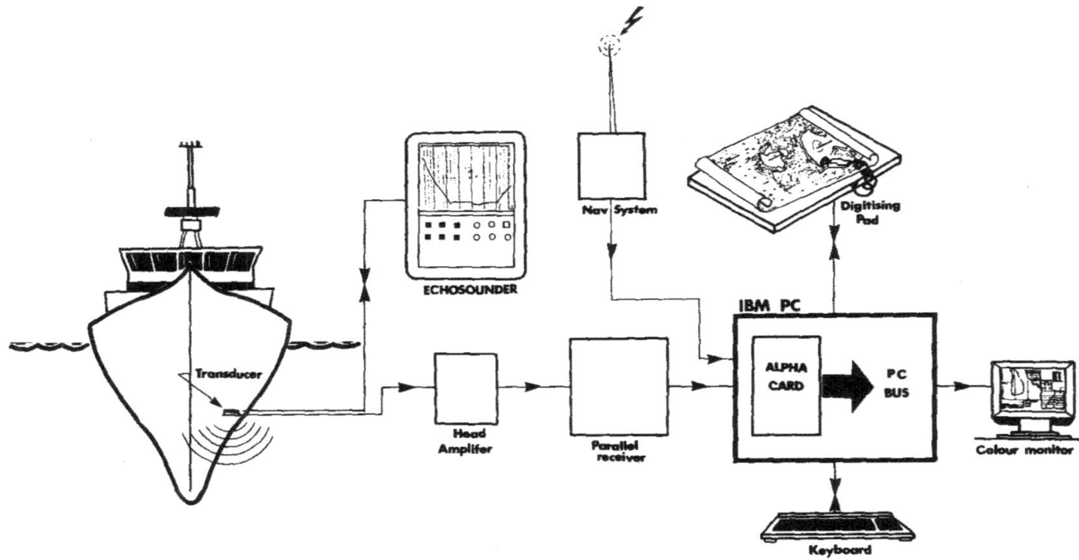

Figure 6 Schematic form of the sea-bed discrimination system

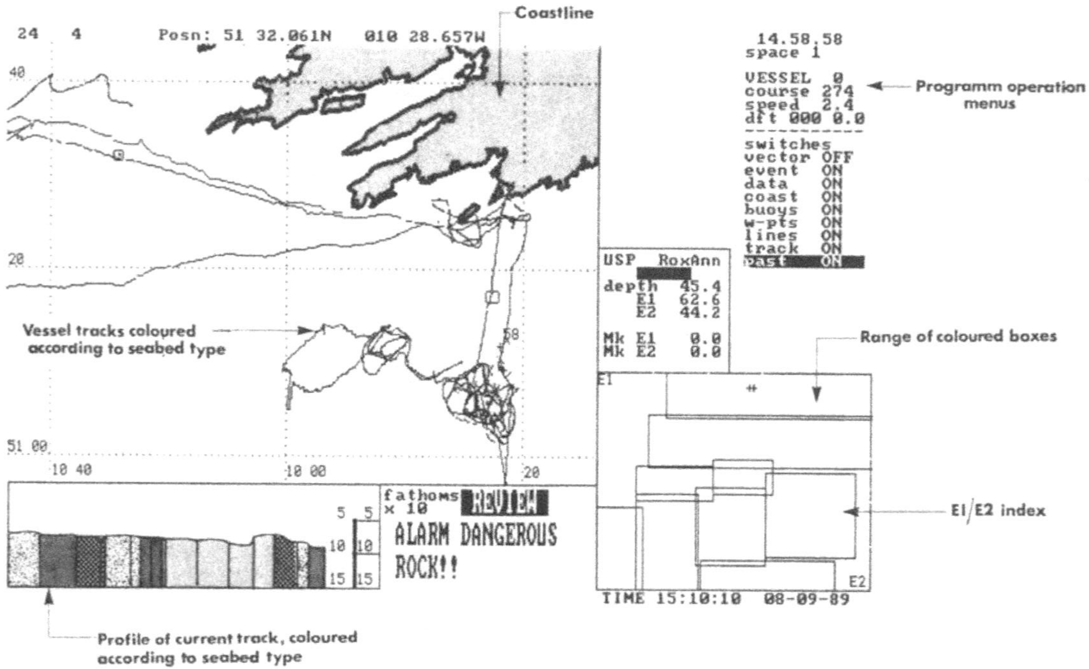

Figure 7 Typical PC colour display seen on board the vessel

Figure 8 Comparison of different techniques on a 7km x 1km test ground

SOME THOUGHTS ON SUPERRESOLVING IMAGING OF EXTENDED OBJECTS

H.W. Jones
Dept.of Medicine, Medical School, Dalhousie University,
Halifax, Nova Scotia, Canada.

ABSTRACT

Since it has been demonstrated that superresolving acousto-optical systems can be realized the question of their applicability to the imaging of other than point sources arises. This paper reviews the experimental and theoretical evidence for the possibility of the application of superresolution to ultrasonic imaging and explains why it is of particular interest to ultrasonic systems. The problems relating to the limits imposed by information theory to the imaging of distributed objects is examined in some detail by reference to particular cases such as double slits and diffraction gratings. The applicability of holography is discussed by reference to their storage capacity and it is shown that significant gains in resolution do not appear to be available by this route. A method of imaging, which appears rather cumbrous is suggested to indicate that there may be, in principle, some possibility of devising schemes which trade signal for noise ratio for increased resolution. The use of multiple frequency ultrasound is given some limited consideration.

Introduction

In an earlier publication[1] we described experimental evidence demonstrating the realisation of superresolved images of a small ultrasound source. The image was produced by a lens which had very small aberration and was self-shaded by the interfacial effects at the refracting surfaces. This demonstration indicates that it is possible to think in terms of superresolving imaging of point sources at least under some conditions. The practical usefulness of superresolution is, however, another matter altogether and the prospects for the application to extended objects is the subject of this paper.

The need for superresolution in imaging comes from the problems arising from absorption of ultrasound at high frequencies. Higher resolution at present is only obtainable, in many circumstances by increasing frequency, but this degrades the signal to noise ratio of the received signal, consequently there comes a point at which higher resolution becomes impossible because of the poor signal to noise ratio. In ultrasonics superresolution may provide for better resolution by allowing imaging at lower frequencies with signal to noise ratios which are better than those which would be obtained in matching conventional systems. The hope for this approach depends on the fact that absorption of ultrasound, generally, increases at least as the square of the frequency i.e. signal to noise degrades as the inverse of this ratio. The ratio of signal to noise in superresolving systems is not sufficiently explored for the signal to noise ratio to be expressed as a function of frequency but elementary arguments suggest that the relationship may be linear in some circumstances. This gain in superresolution will allow increased image resolution, provided other factors permit, which could not be obtained otherwise. This being said such imaging requires a process by which the simple Nyquist conditions which lead to aliasing are circumvented.

Historical background

Apart from the use of diffraction patterns to obtain data about a source (e.g. stellar interferometers) the first clear recognition of superresolution appears to be that of Luneburg[2] who in 1944 showed that the Airy diffraction integral:

$$F(x,y,z) = \frac{1}{2\pi}\int\int \phi(p,q)e^{ik(W+xp+yq+zr)}dpdq \qquad (1)$$

allowed a multiplicity of solutions and that the classical solution was simply that which gave the central lobe the maximum energy. Luneburg gave other solutions which reduced the halfwidth of the main lobe to satisfy particular conditions and which were readily applicable to a form of the diffraction integral which could be readily solved. Later this problem was addressed in a series of papers by Toraldo di Francia[3] particularly in relation to radar beamforming. He showed that there was an arithmetic technique which allowed the central lobe to be of any arbitrary width and simultaneously provided control of the sidelobe maxima. This work led to a bitter controversy in which it was claimed that the technique advanced was in conflict with the uncertainty principle, however it was finally demonstrated that zero beam width was coincident with zero energy in the beam and that the approach to zero beam width led to conditions which satisfied that principle. Since that time much work has been done on various aspects of apodisation. Resolution for ultrasonic imaging can use these techniques but recognition must be given to the fact that the work of the pioneers necessarily invoked energy whereas in ultrasound pressure or velocity can be used. The work relating to the control of the diffraction pattern which has just been described is unfortunately only one aspect of the problem and applies particularly to a point source. In general an extended array of sources constitutes the objects we need to image and we arrive at a situation in which it is necessary to consider the interaction of the various main and sidelobes associated with this array. This problem was considered by Abbe[4] and Rayliegh[5] who used a regular array to determine the resolving power of microscopes. Much later with the advent of Fourier optics the problem was revisited and considered in terms of the spatial frequencies of wave fronts and the bandpass of the aperture involved. This development was associated with the discoveries in information theory which allowed the Nyquist sampling theorem to be applied (and arrive at Rayliegh's conclusion by another method) The new treatment was extended to describe the use of filters in the Fourier domain in order that particular images could be recognised, such filters in effect being apodisation devices.

High resolution ultrasonic imaging

Some commentary on the application of the optical theory to ultrasonic imaging of small sources is now presented. Suppose there is a circular disc of radius ρ_0 which is shaded so that the transmission is symmetrical about the axis of the aperture or lens system so that the transmission factor (which may be complex) is given by:

$$T = A(\rho) \tag{2}$$

We know that for an aperture in which T=1 that the pressure (in the case of ultrasound) at the focus of the lens gives a diffraction pattern which is the Airy formula:

$$F(r) = 2J_1(k(\rho_0)r)/k(\rho_0)r \tag{3}$$

where $k=2\pi/\lambda$, ρ_0 is outer radius of the aperture and r is the distance from the centre of the diffraction pattern.

If the lens is coated then it can be shown that:

$$F(r) = \int_0^{\rho_0} \rho A(\rho) J_0(\rho r)\, d\rho \tag{4}$$

Many solutions to can be obtained for F(r) which depend on the shading function which can be chosen arbitrarily.

The effect of absorption in the medium between the object and the aperture implies a shading condition which is frequency dependent.
The discussion above indicates that as:

$$p = p_0 e^{(-\alpha x)} \quad \text{and usually}$$

$$\alpha = \text{Const.}\, f^2$$

Consequently we need to solve the equations for the diffraction pattern which is produced taking this into account. We need to compare two conditions; first, that in which a higher frequency f_1 is used without special shading; this frequency is the maximum which can be used in any particular circumstance and produces images at the signal to noise limiting condition and second, some lower frequency f_2 which will pass through a superresolving aperture or system. Clearly:

$$f_1 > f_2 .$$

We can now compare the point image resolution of the two systems. For example, it has been shown[1] that if a form of shading given by:

$$A(\rho) = \Lambda(\sigma_0 - \tau_0 J_0(\alpha\rho)) \tag{5}$$

where $\sigma_0 = J_0(\rho_0\alpha) + J_1(\alpha\rho)$,

and $\tau_0 = 2\,(J_1(\rho_0 . \alpha))/(\rho_0 . \alpha)$,

is a possible choice provided:

$$\alpha < 0.31, \quad (2\pi/\rho_0).$$

It remains now to determine the signal to noise ratio for the diffraction patterns (making suitable assumptions for the noise, of course) first, for the higher frequency and second, for the lower frequency at various values of α. It is noted that for both frequencies there is a shading effect due to absorption in the object medium and this produces a shading factor. The determination requires a numerical evaluation of many shading regimes and this is presently in process and results will be published at a later time. As initial results appear to indicate a better signal to noise ratio at lower frequencies for the same overall resolution.

Turning to the matter of extended sources and considering the effect of shading a circular aperture, it is noted that the imaging system always implies an aperture limitation at some place in the system. It is also noted that point sources can have superresolved diffraction patterns if this aperture is suitably shaded. Extended images cannot be superresolved unless some further factor is introduced. In introducing such factors it may be necessary to make them specific to ultrasound systems in that: real time processing can be envisaged, the pressure amplitude is usually measured by the receiving system and the receiving system often has a bandwidth which is comparable (i.e. a reasonable fraction of) to the frequency of the ultrasound being used. If these factors are accepted then we can illustrate a procedure which might be used to obtain a superresolved image by reference to Rayliegh's treatment of a microscope which uses coherent light. Given the space limitation that which follows uses the notation of Born and Wolf[6].

Figure 1 shows the geometry of a grating and figure 2 the geometry of an imaging system (it is noted however the latter can be generalised).

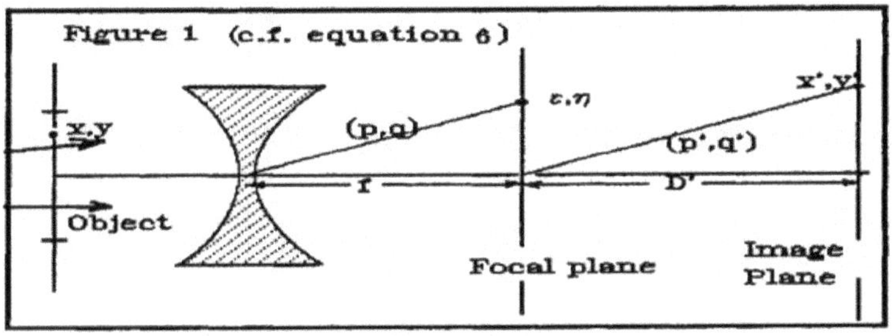

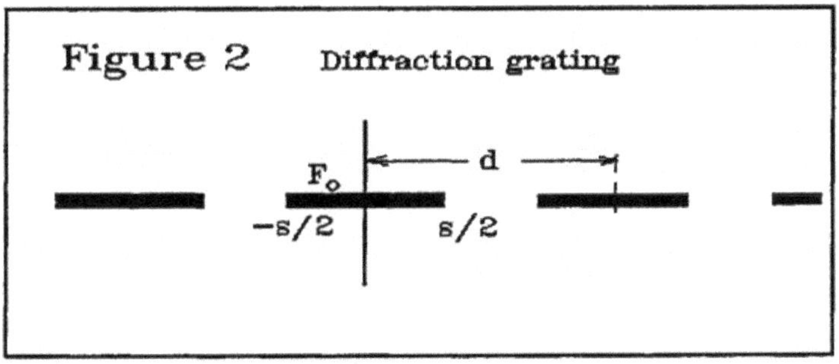

Supposing, initially, that we concern ourselves with coherent insonification then we see that:

$$U(\xi,\eta) = C_1 \iint_{\mathcal{A}} F(x,y)e^{-ik\left[\frac{\xi}{f}x+\frac{\eta}{f}y\right]} dxdy \qquad (6)$$

describes the pressure in the focal plane of the lens where $\mathcal{A}$ is the aperture of the lens. Using the diffraction integral again we find that in the image plane the image is described by:

$$V(x',y') = C_1C_2 \iint_{\mathcal{A}} \iint_{\mathcal{B}} F(x,y)e^{-i\frac{k}{f}\left[\left[x+\frac{f}{D'}x'\right]\xi+\left[y+\frac{f}{D'}y'\right]\eta\right]} dxdyd\xi d\eta \qquad (7)$$

We can suppose the original object was a simple grating with the dimensions shown in figure 1. It follows from grating theory that the pressure in the focal plane is given by:

$$U(\xi) = C_1 \left(\frac{\sin\frac{k\xi s}{2f}}{\frac{k\xi s}{2f}}\right) \frac{1-e^{-iNkd\xi/f}}{1-e^{-ikd\xi/f}} \qquad (8)$$

If the aperture has dimensions in the ξ direction of: $-a << \xi << a$, then in the image plane the diffraction pattern is described by:

$$V(x') = C_1C_2 \int_{-a}^{a} \frac{\sin\frac{ks\xi}{2f}}{\frac{ks\xi}{2f}} \frac{1-e^{-iNkd\xi/f}}{1-e^{-ikd\xi/f}} e^{-ikx'd\xi/D'} d\xi \qquad (9)$$

Given that N is large and that V_0 is defined by the term below:

$$V_0 = C_1C_2 \int_{\Omega_m}^{\Omega_{m+1}} \frac{1-e^{-iNkd\xi/f}}{1-e^{-ikd\xi/f}} d\xi$$

and $\overline{m}=ad/\lambda f$ then it follows that:

$$\frac{V(x')}{V_0} = 1 + 2 \sum_{1<m<m} \frac{\sin\frac{m\pi s}{d}}{\frac{m\pi s}{d}} \cos\frac{2\pi m x'}{d'} \qquad (10)$$

This series (10) is the series for the grating transmission factor F (which = 0 or 1) provided that sufficient terms are taken. If insufficient terms are taken then additional lines appear and thus "false" detail appears. This treatment links to directly to that of information theory

We now consider the factors which lead to the results discussed above which were natural to the constraints arising from the use of light, constraints which do not necessarily apply to ultrasound. First equations (7) and (8) relate to coherent monochromatic waves, it is noted that an alternative treatment for incoherent monochromatic illumination exists. There is no particular reason why, in ultrasound we need accept these constraints. we can use sources in which the frequency of the insonification can be more or less any function of time. If this is the case then the integrals referred to become time dependent and we acquire an additional degree of freedom. The effect of such a frequency swing is that the elements of the diffraction pattern in the image plane will move. There may, therefore, be a method, by recognising the movement, of resolving the original object. In fact by quite a simple argument it is possible to show, using Rayliegh theory, that recognition of the details of a diffraction grating is possible. For more complex images at would seem that a direct analytic approach is difficult and it may be that a series of studies using computer modelling would allow this aspect of the superresolving process to be tested (c.f. the work of Bates et al[7]). The principle can be extended to provide for if a series of extended or point ultrasound sources which effectively insonify the object with asymmetric time varying ultrasound of variable frequency. We have undertaken some limited study of such arrangements but space limitation prevents their being reported here.

An alternative approach suggests itself if we recall that Gabor[8] invented holography in an attempt to overcome resolution limitations in electron microscopy. It is interesting to see if there is anything to be gained by applying such ideas. The most direct approach appears to be to consider the "storage capacity" of a hologram. It can be shown[9] that only in lensless systems can the storage capacity appear to exceed normal resolving limits. However, the consideration of this process suggests that there is, in principle, a direct method of achieving superresolution of an extended object. The scheme of such a system is shown in figure 3. Several widely spaced points in the image are illuminated (or examined through a suitable set of moving pinholes) and the image of these points recorded in the image plane of the lens (figure 3a). The image is recorded and the sidelobes suppressed by subtracting a signal which effectively raises the baseline by an amount which is proportional to the amplitude of the central maximum of the diffraction pattern. By progressively examining all the image points and summing the processed results it seems that there may be a chance of obtaining a better resolved image. This is not to suggest that the process is without its practical difficulties or that some false detail may still remain after processing.

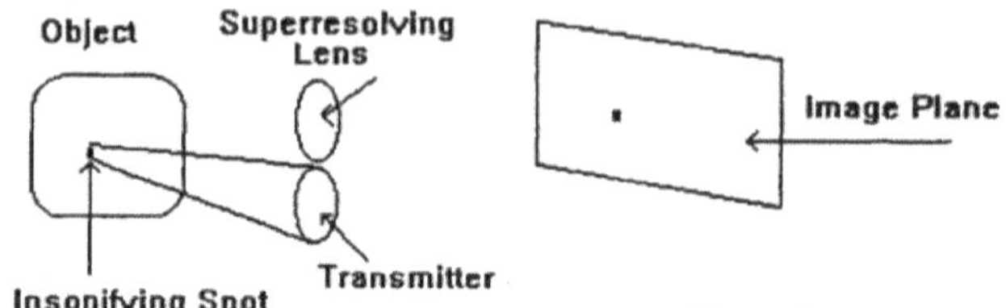

Figure 3a

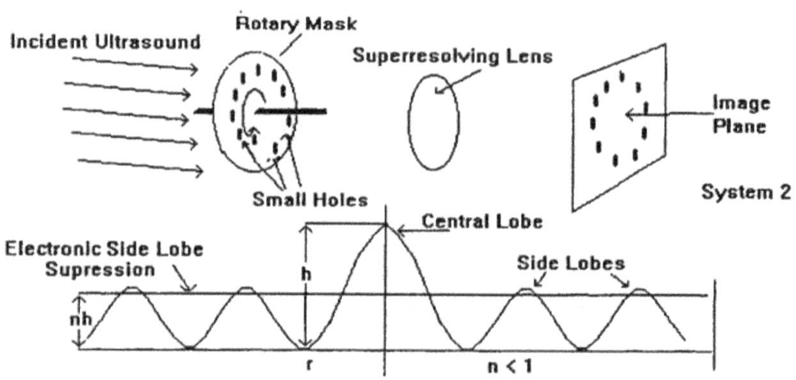

Figure 3b

Conclusion

A review of the prospect for the production of superresolved images by ultrasound systems of extended objects has been given. A reason has been advanced for such high resolution being especially applicable to ultrasonic systems. Suggestions are made of methods which might prove of interest for further study. Very clearly much work is required to find the applicable limits of superresolution in this context.

References

1. Jones & Kwan "*Ultrasonic Signal Processing*" Ed. A.Alippi 389 et seq, World Scientific, Singapore.
2. Luneburg "*Mathematical Theory of Optics*" Press U.Calif. 1964
3. G. Toraldo di Francia "*Del Nuovo Cimento*" pp426-438, **N3**,1952.
4. Abbe "*Archiv.f.Mickro.*" Anat,.**9**, 413, 1873.
5. Lord Rayliegh, "*Phil.Mag*".**(5)**,42, 189, 1676
6. Born and Wolf "*Principles of Optics*" Pergamon Press.
7. R.H.T.Bates et.al. "*Acoustical Imaging*", **12**, 185-191, Plenum,(1982)

8 D.Gabor."*Proc. Roy. Soc.*", **A 197**, 454 et.seq., (1949)

9.H.M.Smith, "*Principles of Holography*", Wiley, (1969)

ACOUSTOELECTRONIC READ-OUT OF THE OPTICAL IMAGES

I.B.YAKOVKIN, V.A.VYUN, V.V.PNEV and A.V.TSAREV

Institute of Semiconductior Physics
Siberian Branch of the USSR Academy of Sciences
Novosibirck 630090, USSR

ABSTRACT

The present work is the experimental study of the longitudinal acoustophotoelectric effect under the surface acoustic waves (SAW) propagating in the GaAs semi-insulated Cr-compensated substrate with resistivity of 10^6 Ω*cm.

1. Introduction

Acoustic charge transport is considered to be an alternative to electron switching while reading electric potentials. They both imply the transmission and accumulation of the generated charge with the aim of its further recording in the sensing elements for physical quantities such as temperature, pressure, radiation etc[1, 2]. In particular, in order to realize the acoustic transport it is supposed to make the use of the acoustoelectronic (AE) interaction with the charge induced by the action of a certain external factor on the sample surface. To take an example, for light flux registration of particular interest are those acoustoelectric effects which come into being only with the acoustic waves propagating in the sample exposed to light [3]. Such effects as these can be called acoustophotoelectric effects by analogy with the well-known acoustomagnetoelectric ones.

The present work is the experimental study of the longitudinal acoustophotoelectric effect under the surface acoustic waves (SAW) propagating in the GaAs semi-insulated Cr- compensated substrate with resistivity of 10^6 Ω*cm. This GaAs posseses the piezoeffect needed to produce the direct SAW excitation. The electric fields accompanying surface waves ensure the coupling between acoustic waves and free charge carriers in the subsurface layer. Another fact of interest is the absence of the acousto-emf without light exposure due to the infinitesimal concentration of dark-charge in the subsurface layer where the SAW electric fields exist. Unlike the bulk waves the SAW have the advantage of enabling the investigation of the SAW interaction processes with the surface photoinduced charges (in the presence of light absorption in the thin subsurface layer).

2. Results and discussion

In Fig.1 is shown the experimental set-up. An amplitude-modulated SAW of the wavelength λ_s=30 μm, the modulation frequency of 100 Hz and the modulation percentage of 0.4 is propagating in the [110] direction on the (001) GaAs plane [4]. A pair of aluminium electrodes is placed on the GaAs surface across the SAW pathway for the purpose of the acoustoelectronic potential detection. The AM-SAW operation together with the harmonic acousto-emf detection [5] by means of synchronous detector offer the significantly large reduction of the threshold sensitivity. The light beam of the wave length λ_l is directed into the interelectrode region. There in the layer with the thickness of the order of the light absorption length in GaAs there are generated the electron-hole pairs. More free electrons excitated by light are being carried along by the SAW towards the potential electrode on which an acoustoelectric voltage arises at the frequency of SAW modulation.

Fig.2 displays the acoustophoto-emf and the photosensitive conductivity as functions of the light flux intensity I_l at the SAW power P of about 0.5 mW at the edge of the light absorption band in GaAs at λ_l of 0.87 μm. In a logarithmic scale the acoustophoto-emf dependences versus light flux intensity show three clearly pronounced regions; of steady level at the small light intensity, then linear and sublinear parts (Fig.3). The steady part does not depend on the SAW presence and arises from the high frequency signal applied to the interdigital transducer generating the SAW both in and out of the transducer frequency bandwidth. The mechanism of the signal generation is apparently conditioned by the detection of the applied exciting HF signal on the Schottky barrier formed between the transducer electrodes and the semiconductor. After being detected the signal is transmitted further along the GaAs substrate and is mixed with the valid signal on the potential electrode. The acoustophoto-emf conduct in the linear region is submitted to the Weinreich relation . It will be recalled that the relation can be obtained from the quantum representation of the phonon- charge carries interaction together with the classical representation .

We assume the phonon flux intensity to be

$$I=n_p(h\omega/2\pi)V_s,$$

where n_p is the concentration of phonons with the angular frequency ω and the velocity V_s.

The momentum transferred to an electron is

$$p_y=h\omega/2\pi V_s.$$

An increase in the electron velocity is

$$\Delta V=h\omega/2\pi mV_s,$$

where m is the electron mass. The acoustoelectric current

$$J_{AE}=-en_e\Delta V_e=en_e h\omega/2\pi mV_s.$$

Expressing in a conventional way the charge carrier mobility as $\mu=(e/m)\tau$ and the electron attenuation as $\alpha=n_e/(n_p\tau V_s)$ we obtain the relation in the form:

$$I_{AE}=\alpha\mu I/N_s$$

and for the acoustophoto-emf $E_{AE}=\alpha\mu I/\sigma V_s$,

where σ is the sample conductivity. On the linear and sublinear parts of curves the acoustophoto-emf behaviour obeys this relation with the attenuation α in the case of the thin layer (far less than the SAW length λ_s) of semiconductor [6].

$$\alpha = \frac{1}{2} K^2 \frac{\omega}{V_s} \frac{\omega_\sigma/\omega}{1+(\omega_\sigma/\omega)^2(1+(\omega^2/\omega_\sigma\omega_D))^2} ,$$

where K - coefficient of electro-mechanical coupling, ω - SAW angular frequency, V_s- SAW velocity,

$$\omega_\sigma = \frac{\omega}{V} \frac{1}{(\varepsilon_p+1)\varepsilon_0} \frac{1}{R} -$$

frequency of Maxwell relaxation, ε_0- space permittivity around substrate, ε_p- fractional permittivity of substrate, R - film resistivity per square, $\omega_D = V_s^2/D_e$ - diffusion frequency, D_e- diffusion coefficient of electrons in film.

It should be noted that the region of maximum attenuation is in a good agreement with the sublinear part. The linear part slope, that is the photodetector sensitivity to the light flux , when the photodetector based on the acoustophoto-emf is used, is equal to $0.31*10^{-3}$ mV*cm^2/W. The linearity range is about 20 dB. The change- over from the steady to the linear region takes place at the light flux power $P_0 \sim 10^{-9}$ W which is the photodetector threshold sensitivity. The estimation shows that for GaAs with the life time of charge carriers about 10^{-8} s at the photodetector threshold sensitivity there exist as low as a few tens of electron-hole pairs in the AE interaction region only. Such a high efficiency of the AE interaction at the number of charge carriers being small can be achieved because of their high concentration in the thin surface layer since, as it was mentioned above, the generation process comes about at the length of light absorption which equals around 1 μm for GaAs. For large values of the absorption length the interaction efficiency is much lower. It should be emphasized that for some samples, first of all with different doping levels, one can observe an anomalous region on the representative curves for the acoustophoto-emf dependence versus the light flux intensity (Fig.3). It exhibits itself as a fall-through in going from the constant level to the linear part. Besides such samples have a peak maximum at $\lambda_1 \sim 1.2$ μm in their spectral response and the acoustophoto-emf of the opposite sign which evidences the presence of impurity extrinsic conduction of p-type in the surface layer. Fig.3 also shows the above relationship at the same SAW power and the light wavelength of 1.2 μm. In the case of impurity-type conduction there appears a nonlinear behaviour and significantly reduced level of sensitivity. It is likely that the light absorption length at $\lambda_l \sim 1.2$ μm is greater than λ_s and in this way the charge carriers concentration in the surface layer is lower. Also for such "anomalous" samples it is typical to have the lower level of spurious signals.

A few words should be added about the transformation of a photosensitive line into an electrical signal. As Fig.4 pictures, the mask with transparent sections, is positioned above the photosensitive layer (spatial dimension of the photolayer aperture is plotted on the upper abscissa axis). With the SAW frequency being changed there occurs the acoustophoto-emf below the illuminated region as a result of the SAW beam scanning.

As it is obvious, the acoustophoto-emf of the working region exposed to the varying illumination describes adequately the spatial distribution of illumination.

3. Conclusions

From the point of view of the application of the effect considered above of major interest are such quantities as sensitivity and the region of linearity. Changing the power of sound being pumped into the sample, it is possible to vary the sensitivity level. Potential sources of the threshold sensitivity reduction and, thus, the linear range extention lie in the suppression of spurious signals when the values of the light flux intensity are small.

4. References

1. G.S.Kino, Proc. IEEE **64** (1976) 724.
2. Yu.V.Gulyaev and A.V.Medved, Izv. VUZov. Radiofizika **26** (1983) 911.
3. I.B.Yakovkin, Proc. of Intern. Colloq. (Ilmenau. GDR. 1990) p. 83.
4. M.D.Levin, G.A.Lobanova, N.S.Paschin and I.B.Yakovkin, J. Acoust. **31** (1975) 68.
5. V.A.Vyun, V.V.Pnev and I.B.Yakovkin I.B., Proc. Intern. Symp. "Surface Waves in Solids and Layered Structures" (Novosibirsk, USSR, 1986) **2**,. p.354.
6. R.Adler, IEEE Trans. **SU-18** (1971) 115.

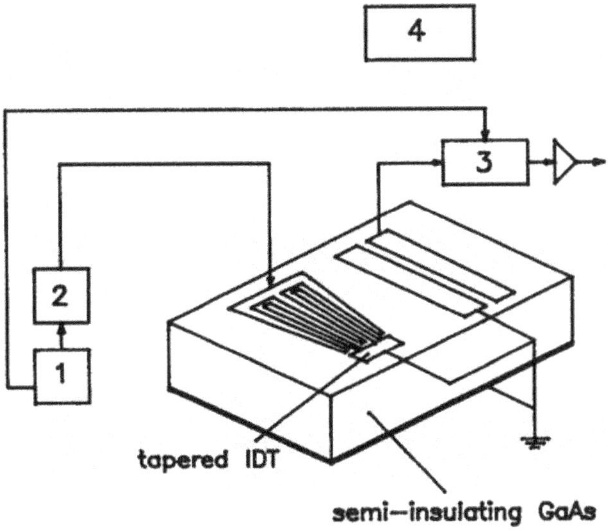

Fig. 1. Experimental set-up.
1 - generator of broadband signal, 2 - HF generator, 3 - matched filter, 4 - monochromator.

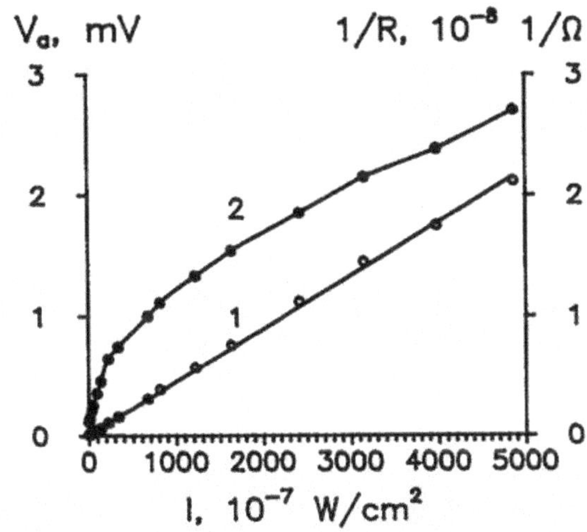

Fig. 2. The coustophoto-emf (curve 1) and the photosensitive conductivity (curve 2) as functions of the light flux intensity

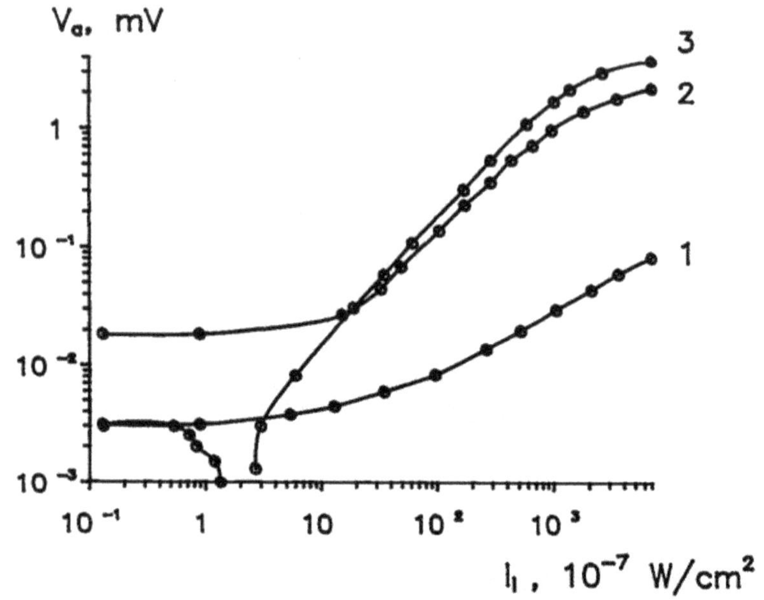

Fig. 3. The coustophoto-emf as functions of the light flux intensity for $\lambda=1.2$ μm (curve 1), $\lambda=0.87$ μm (curves 2, 3)

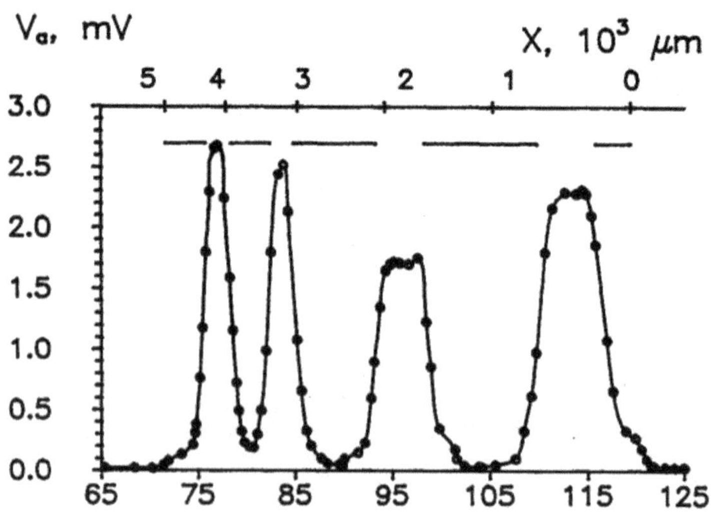

Fig. 4. Reading of the mask with transparent sections

MODELLING, CONSTRUCTION AND PERFORMANCE TEST OF MULTIFREQUENCY COMPOSITES

F.R. Montero de Espinosa, M.S. Sanchez and P.T. Sanz

UEI Ultrasonidos, INSTITUTO DE ACUSTICA
Serrano 144, 28006 Madrid, Spain

ABSTRACT

Piezoelectric composites are well suited materials to ultrasonic transducer applications. Their material properties varies in a wide range making then very useful for medical ultrasonic imaging. This paper describes a new composite geometry which, keeping the general advantages of the normal composite structures, permits to increase the composites acoustic frequency band.

1. Introduction

Piezoelectric composites have been a research topic since they were introduced in 1978 by the Materials Group of the Penn University. A considerable amount of scientific and technical contributions to this subject have been performed since then[1]. The most important practical feature of these new materials is their design flexibility. There are two big application areas in which the composites are now widely used: hydrophones and medical ultrasonic imaging. While for the first area several composite structures are used, composites with (1-3) ceramic polymer connectivity seems to be the most adequated for the second area. Summarizing the properties of the (1-3) composites, the thickness-mode electromechanical coupling coefficient exceeds that of the primary ceramic material, ranging then up to K_t=0.75. The specific acoustic impedance ranges from that of the ceramic down to 10 Mrayls or less, keeping high values of K_t. The dielectric constant presents a values range quite wide ($20 < \varepsilon^S < 1000$) with low dielectric losses -tg $\delta_E \cong 0.05$-. The Q_m can be lowered down to 10 or less, depending of the performances wanted in each case.

This paper describes a new composite geometry which, keeping the general advantages of the common (2-2) and (1-3) composites, permits to increase the composites acoustic frequency band up to a 7%.

2. Multifrequency composites concept

Bowen et al proposed the use of the stepped shape concept to design a multifrequency composite structure having non-parallel faces[2]. The composites designed following this geometry present a complementary bandwidth that can be made very large. This design, able for (2-2) composites, adds more difficulties in the transducers construction.

Another approach to design multifrequency composite structures is that of Bui et al[3]. The idea consist of using piezoceramic bars having lateral dimensions similar to their thickness. This concept leads to degraded impulse responses with a clear signal lengthering.

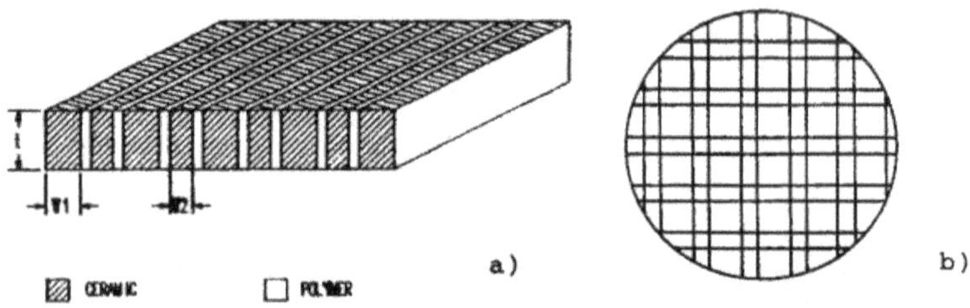

Fig. 1.- Multifrequency composites geometry (a) (2-2) composite with two different types of ceramic strips. (b) (1-3) composite with three different types of ceramic bars.

Figure 1 shows the multifrequency composite structure that we have introduced[4]. The idea consist of constructing composites having ceramic bars with different shapes. In the case of (2-2) composites, the ceramic elements are strips with the same thickness and length but with different width. The (1-3) connectivity multifrequncy composites have ceramic bars having in common only the thickness. The lateral dimensions are less than their thickness to work with single and clear vibration resonances. To choose the actual ceramic dimensions we uses the Onoe coupling modes theory, chosing width to thickness ratios between $0 < w/t < 0.8$.

3. 2-2 Composites

3.1 Design criteria and theoretical model

The Onoe coupling modes theory gives the resonance frequencies behaviour of ceramic strips in function of their width to thickness ratio[5]. Figure 2 shows the application of this theory to PZ-27 piezoceramic. If we observe the lowest coupled branch, we can see that a resonance frequency dispersión of about a 16% can be obtained when varying the ceramic strip w/t ratios between 0 and 0.8. In this size range, the strips present an isolated single resonance. This frequency dispersion will be different to that presented by the corresponding composites made with those strips.

To be able to predict this resonance frequency dispersion in the composites we have linked the Onoe theory with the (2-2) composites model of Hashimoto[6]. To make possible this linking some approximations must be made. The Hashimoto model assumes

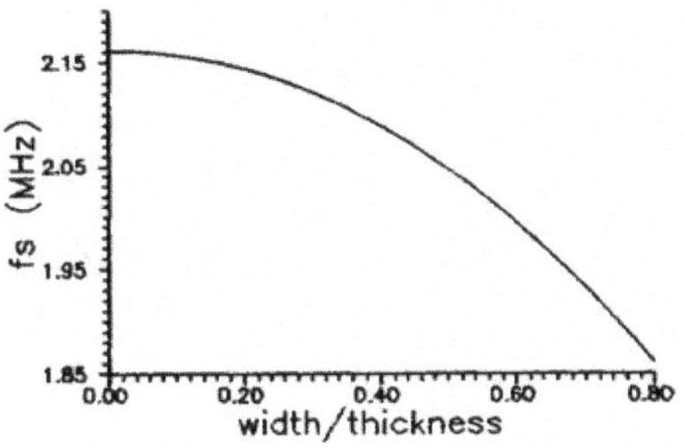

Fig. 2.- The lowest branch of Onoe approach for PZ-27 ceramic strips

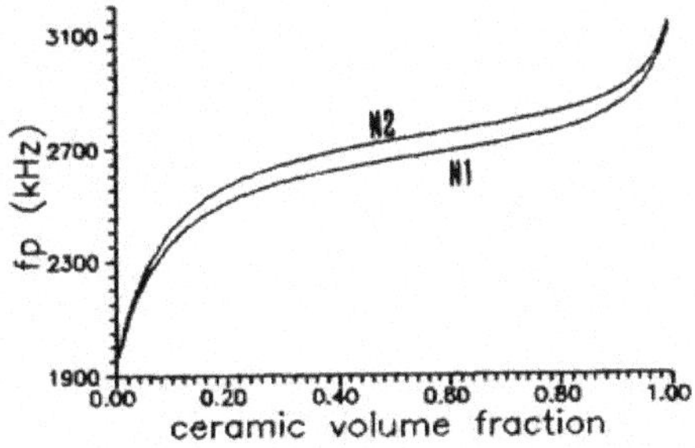

Fig. 3.- Variation of resonance frequency with the volume fraction for N1 and N2 composites.

that the composite ceramic strips vibrate in the pure array mode, whereas we use ceramic strips which present different resonance frequencies because they do not vibrate in a pure array mode. In fact, the resonance frequencies depend on the w/t ratios. We will assume that each of the ceramic strips that can be cutted with lateral sizes in the range commented before $-0 < w/t < 0.8-$ are, for the model purposes, strips with pure array mode geometry but belonging to different ceramic materials. The problem is to define these "differences". We know that the only appreciable change between the different strips is the resonance frequency. In consequence, we can hypothesize that the bars have some different elastic constants. After studying several possibilities, the simplest one consists on supposing that c_{13} is dependent on the dimensions following the next expression:

$$c_{13}^2 = c_{11}^E\ c_{33}^E \left(1 - \frac{(1 - K_t^*)^2 \rho}{c_{33}^E}\ f_d^2 \left(\frac{t\pi}{X_t^*}\right)^2\right)^2$$

where f_d corresponds to array resonance frequency k_t^*, ρ and t are the array-mode electromechanical coupling coefficient, density and thickness, and X_t^* is the first positive nonzero root of:

$$\tan X = X/(K_t^*)^2$$

The other elastic constants are assumed to be the same. Figure 3 shows, as an example, the variation of the resonance frequency with respect to the ceramic volume percent for two composites with different strips. Composite N1 has strips with w=0.45,l=15 and t=0.69 mm whereas composite N2 has w=0.25 mm and the same thickness and length. The piezoelectric material is PZ-27 (Ferroperm) and the polymer is Araldit-H (Ciba-Geigy). The other material parameters needed to simulate the impedance of these composites with the frequency can also be obtained using the modellization approach presented here. So, K_t, ρ, V^D and C_o are calculated for each single and, after measuring their respective mechanical Quality Factors Q_m, their impedance behaviour with the frequency is simulated by using the well-known thickness resonance models. Fig. 4 shows the conductance in the resonance zones for the composites N1 and N2. In the same figure, the experimental results are also presented. Summarizing up to this point, we are able to predict the transducer characteristics of (2-2) composites with ceramic bars vibrating in not pure array mode, by linking with some approximations the Onoe coupled modes theory, the Hashimoto composites approach and the monodimensional thickness mode equivalent circuits.

3.2. Multifrequency (2-2) composites

The easiest way to start the study of multifrequency composites is to use the geometry of fig 1-a, formed by only two different ceramic strips. This structure could be considered as an electrical and mechanical connection of two single composites, having each of them one of the two different ceramic strips. The electrical connection is in parallel throught the common electrodes, being not clear the type of the mechanical connection. So, we start assuming that this mechanical influence is the same that the one existing in the single composites. If we do that, the equivalent circuit able to describe the electric impedance of the double composite is the parallel connection of the two equivalent circuits which correspond to each of the single composites, balanced for their volume contributions. Figure 5 shows the predicted and measured conductance frequency bands for a double composite formed with N1 and N2 single composites -ND1- with the same volume percent. The disagreement is evident, showing that the mechanical constraints of the double composite are not the assumed ones. Looking into the double composite resonance mechanism, we can arrive to a physical explanation of this disagreement. Let us do an imaginary frequency scan of the composite with an ideal voltage source. When we arrive to the series resonance of the N1 type strips, they vibrate with a maximum of displacement. The mechanical coupling throught the polymer will originate an over-vibration of the N2 type strips which will then bring back an extravoltage to the N1 type strips. If we continue the frequency scan, at the series resonance of the N2 strips, the situation will be phenomenologically the same with only one difference. The N2 type strips bring back the extravoltage opposite in phase with the one originated by the N1 type. The N2 strips generate the voltage at a frequency before their series resonance whereas N1 strips generate the voltage at a frequency later to their series resonance. The consequence is that the extravoltage will be constructive in one case, increasing the conductance of the overall composite, and destructive in the other case, lowering the conductance.

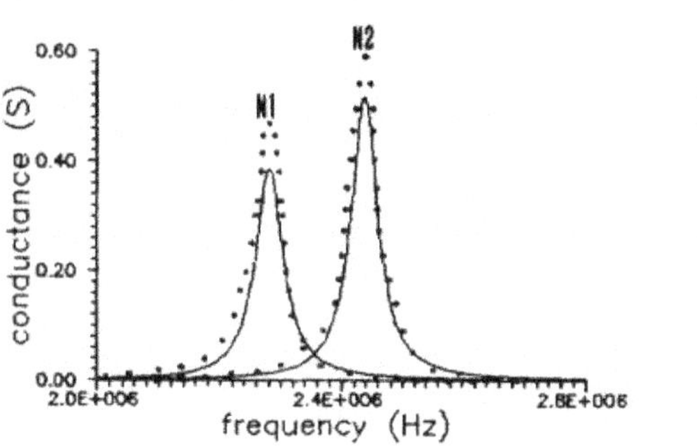

Fig. 4.- Experimental and calculated conductance of N1 and N2 composites.

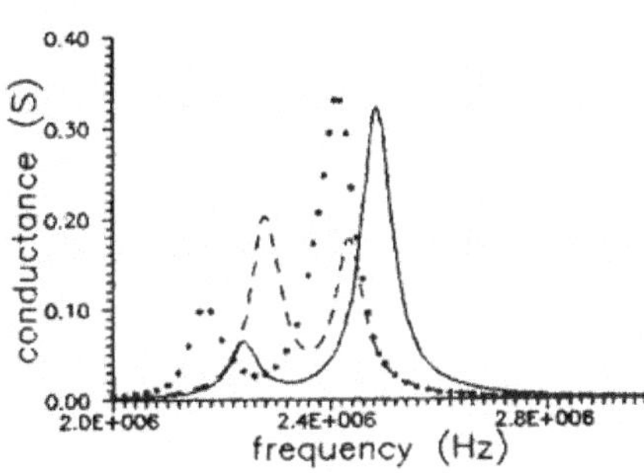

Fig. 5.- Input conductance of ND1 single. ——— calculated with coupling.
--- without coupling.
*** experimental.

To take into account these physical explanation, a simple circuital way can be used, coupling the two inductances of the series branches by means of a mutual induction coefficient. To know the value of the induction coefficient, we have followed an empirical fitting process. In the case of the double composite ND1, we have found a value of K=0.08. If we compare now, the predicted and measured conductance curves, we can see how we have lowered the desagreement -Fig. 5-. The shift to lower frequencies in the measurements indicates for an overall softening of the single.

This result can be also physically explained having in mind that not all the strips vibrate in resonance at the same frequency, being each of the ceramic strips rounded by two strips that resonates at different frequeny. Then when the first one resonates, the two neighbours are only passive elements which do not comunicate active stiffness to the system. We have also constructed (2-2) composites having more than two different sized strips. Fig. 6 shows the conductance frequency bands of three singles having three, four and five different sized strips -singles NT, NF NV-. The frequency bands take smoother shapes when increasing the number of different strips, without broadening more the bandwith obtained with only two different strips.

4. (1-3) composites

4.1. Theoretical model and design criteria

The design criteria followed for this geometry is similar to the proposed one for the (2-2) case. The Onoe coupled modes theory gives the resonance frequency of ceramic bars with different lateral size. Then, using the approximation described previously to convert different sized bars of the same ceramic material on bars of different ceramic materials, with the same bar geometry, the Smith (1-3) composites model[7] gives us the material constants needed to modellize, with the help of the thickness circuital models, the impedance behaviour of those composites. We have constructed and tested three (1-3) simple composites with the next dimensions B1: w=0.27, l=0.27, t=0.68mm, B2: w=0.4, l=0.27, t=0.68 mm and B3: w=0.4, l=0.4, t=0.69 mm. All singles have, the same ceramic volume percent (62%).

A tripple composite formed with three different ceramic bars of the B1, B2 and B3 types was studied. This composite have the same ceramic volume percent than the simple composites.

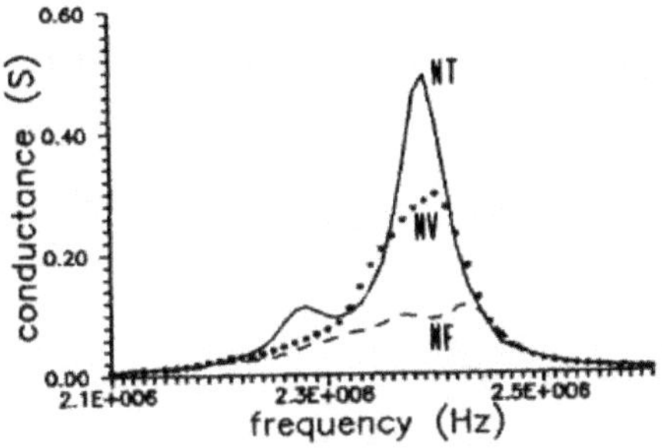

Fig. 6.- Measured conductance of NT, NF, NV composites with three, four and five different sized strips respectively.

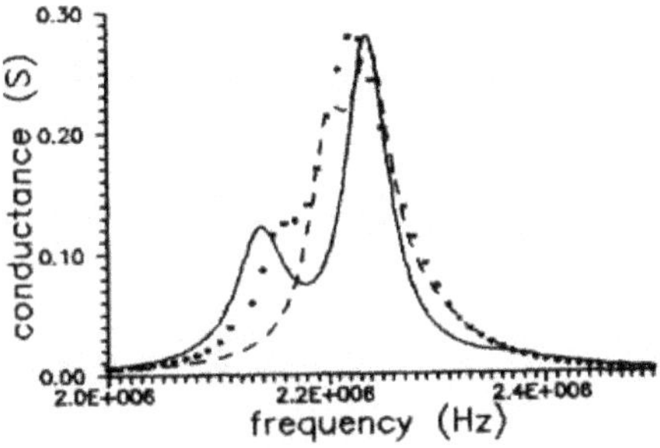

Fig. 7.- Conductance band of T1 single ——— simulated with coupling, - - - without coupling, *** experimental.

Now, connecting in parallel the three different equivalent circuits corresponding to each of the single composites, balanced with their volumen content in the tripple structure, using the stiffness and mechanical coupling corrections as in the (2-2) composites case, we are able to predict its impedance behaviour. Fig. 7 gives a comparation between the measured and calculated conductance bands. The agreement is good enough to support the approach presented.

Finally, some experimental acoustic impulse responses are shown to confirm the influence of this geometry design on the acoustic frequency band broadening. Fig. 8 shows the impulse response of singles B3 and T1 showing how the frequency band is increased without lowering the sensitivity. Fig. 9 shows the conductance frequency band and the pulse-echo response of another single.

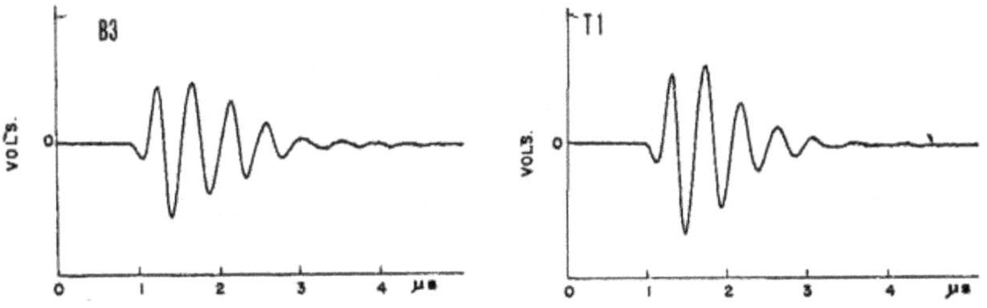

Fig. 8.- Measured acoustic impulse responses of B3 and T1 composites.

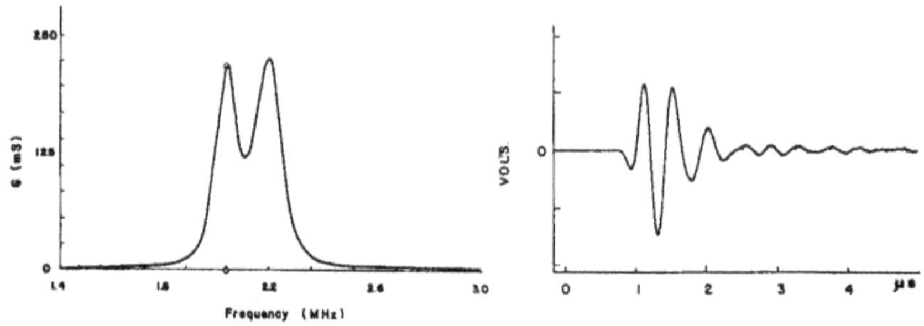

Fig. 9 Conductance band, and pulso-echo of T2 composite.

-T2- with the same design that T1 but with different sized bars. In that last case the pulse-echo is narrower, with the same amplitude than those of single T1. All the pulse-echo measurements were made with the same excitation and experimental set-up conditions.

5. Conclusions

A new concept of composites has been shown. The link between well-known models has been used to modellize their transducer behaviour. This multifrequency concept is useful to enlarge the acoustic bandwidth of regular (2-2) and (1-3) composites without lossing sensitivity. The (2-2) multifrequency composites results can be applied to the design of linear ultrasonic arrays.

6. Acknowledgements

This work has been supported by a PRONTIC-CICYT program number 999/070. The assistance of M. Antonia Garcia Olias has been, as usual, essential.

7. References

1. W.A. Smith, IEEE *Ultrasonics Symposium* (1989)
2. L.J. Bowen and T.R. Gururaja *J. Appl. Phys.* **51** (1980) 5661
3. T. Bui, H.L.W. Chan and J. Unsworth *IEEE Proc. Ultrasonic Symposium* (1988) 627
4. M.S. Sanchez and F.R. Montero de Espinosa, *Ultrasonic International Conference Proceedings* (1989) 641
5. M. Onoe and H.F. Tiersten *IEEE Trans. on Ultras.(1963) 32*
6. K.Y. Hashimoto and M. Yamaguchi *IEEE Ultrasonic Symposium* (1986)
7. W.A. Smith, A. Shaulov and B.A. Alud *Ultrasonics Symposium* (1985) 642.

ULTRASONIC SENSING AND PROBING WOOD MATERIAL

V. BUCUR
Centre de Recherches Forestières de NANCY
Champenoux, 54280 Seichamps, F R A N C E

ABSTRACT

The understanding of the mechanisms of ultrasonic waves propagating in solid wood and wood based composites permitted the development of peculiar technics for quality assessment of trees, of solid wood small clear specimens and of flakeboards.

On trees stress was put on the nondestructive measurement of the slope of the grain, with ultrasonic velocity method (longitudinal waves). On small clear specimens the development of direct transmission ultrasonic method enabled the measurement of all elastic constants. Comments are given on the preparation of specimens, on coupling media used during the measurements and on signal processing.

The anisotropic behaviour of wood composites plates (structural flakeboards) was deduced from measurements on full size boards and on specimens.

1. Introduction

The problems related to sensing and probing wood material with ultrasound are most convieniently divided in three main groups that could be related to the tress, to small clear specimens of solid wood and to wood based composites. It has seemed appropriate to review briefly some of specific requirements for these materials.

2. Ultrasound on trees

The methods employed to evaluate the quality of trees are based on the assumption that some physical properties can be used to give a reasonable indication of the characteristics which determined it.

The potential of ultrasonic velocity method was proved for the nondestructive measurement of the slope of grain on living trees[1]. This parameter is of pertinent interest for forestry genetics and is related to big tree populations.

The elliptical path of ultrasonic waves propagating in wood, deduced from Christoffel equations for orthotropic solids, and the parameters of acoustic anisotropy of this material were used for the measurement of the slope of grain. Three velocity vectors can determine the grain angle, having in mind that along the fibers the velocity value is maximum and along the tangential direction to the annual rings, the value is minimum.

3. Ultrasound on small clear specimens

On small clear specimens two ultrasonic techniques can be used: immersion technique and direct contact transmission technique. The principal advantage of these techniques is the flexibility in measuring all terms of stiffness matrix, using longitudinal and transverse waves.

It is appropriate to observe that the measurements of ultrasonic velocity and attenuation are conditioned by requirements related to sample preparation, to the bonding of the transducer to the sample and to signal processing.

3.1. Preparation of specimens

The range of ultrasonic velocities measured in wood at 1MHz is between 6000m/s for longitudinal waves in fibers direction and 400m/s for shear waves in radial tangential plane versus the annual rings. The values of attenuation coefficients are roughly 2dB/cm for longitudinal waves in fibers direction and 15dB/cm for shear waves in transversal anisotropic plane[2].

The conditions for satisfactory specimen preparation depend essentially on the attenuation magnitude of ultrasonic waves into the wood species under test. In general, it can be stated that the lower attenuation and the higher the wave frequency, the greater care is required with respect to flatness and parallelness of the surfaces of specimen. In addition the sample must be accurately perpendicular to the principal direction or to the required direction out of main symmetry axes along which the ultrasonic waves are propagating.

Typical specimens used for laboratory measurements with ultrasound are rectangular parallelepipeds, cubes, rods, disks or plates. The choise of one of the geometrical shape is determined by the research purposes. Generally speaking, for immersion technique plates are needed. For direct contact technique all other types of specimens could be utilised.

If all stiffnesses of wood are required, the constrains for the sample size and shape are more severe. We need to manipulate rather small specimens, to limit or to avoid the spatial inhomogeneity of wood material induced by its anatomical structure and to neglect the ring curvatures related to tangential anisotropic direction.

The influence of the natural variability of specimens, due to the biological nature of wood material, on velocities and attenuations, may be studied by choosing the frequency of the source so that the wave length in the material lie in a range roughly between the maximum dimension of main anatomical elements (fibers) and the minimum of the specimen size.

For the experimental efficiency, for the economy of sample material and of time for producing specimens, the multifaced disks seem to be ideal samples for ultrasonic measurements with direct contact technique. Only 3 disks are needed for the complete characterization (18 constants) of a wood species.

The accuracy of the velocity and attenuation measurements depends on several factors. The parameters as the probes diameter, the maximum pulse width, the attenuation of ultrasound in all symmetry planes, or the time required for the separation of various modes need to be considered in sampling strategy.

The deviation of flux energy vector should not be ignored when determinig the size of the specimen tested. A lower limit to the size of the specimen is also imposed by the requirement that waves should have the character of plane wave in an infinite medium. For disk type specimens the appropriate diameter seems to be 35mm with corresponding thickness 20...25mm. For cubes satisfactory results were obtained with specimens on 20x20x20mm.

The physical properties of wood that can influence ultrasonic measurements at constant temperature and moisture content are: the density, the annual ring width, the corresponding proportion of latewood and earlywood, the main anatomical elements - fibers, proportion of rays, or microfibrils angle in cellular wall.

The influence of the curvature of rings is a pertinent question in relation with correct measurements of velocity, when compressional waves propagates in direction perpendicular to the annual ring. For this particular case WOODHOUSE underline the sample size contraint and point out that the continum theory employed in ultrasonic characterization using orthotropic model ignores this curvature[3].

Experiments on disks with different ring curvatures (360°, 90°,etc.) show that for compression waves there is no influence of the geometry on the measurement of radial velocity. However, for shear waves, the modulation of wave by the structure is strongly related to the propagation and polarisation directions. It was proved [4] that the shear velocity V_{RT}, where R is propagation direction and T is the polarisation direction, is not affected by ring curvature. The opposite effect is observed on V_{TR}, namely on species with pronounced difference between latewood and earlywood. This effect could be related to the guide wave phenomenon induced by the layered medium at annual ring scale, on which the latewood has thick walls and high density (900kg/m3). For softwoods the difference between shear velocities is of 10...15% and for ring porous oak of 17%. For diffuse porous hardwoods the discrepancy is only of 5%.

The influence of the main structural elements of wood at macroscopic and microscopic anatomical scale on velocities and attenuation coefficients could be observed with ultrasonic spectroscopy [5]. The annual ring spacing, the width of latewood layer or the fiber length can be determined in non destructive way.

3.2 Coupling media

Several types of coupling media from transducer to sample are possible, as well as several arrangement of the transducer with respect to specimen. In the peculiar case of small clear wood samples, the suitable testing disposition seems to be with two transducers.

Coupling media are necessary to ensure the bonding of the transducer to the wood specimen. Depending on temperature and moisture content of sample, silicone resins, wax, mineral greases, glycerine or alcohol were employed. On hardwoods, when the specimen is big enough, the transducer may be wring into the sample and no coupling media is needed.

In any case, the losses due to the transducer bond and to the system can be fairly well measured. The effects of the thickness of the coupling medium could be neglected if a very thin coating couplant was used. For precise measurements, a well defined pressure applied to the specimen and to the transducer is required for reproducible measurements of attenation.

In practice, the changes induced by the penetration of the coupling medium into wood can produce unexpected errors in experimennts (3...14% differences in velocity measurements). To avoid such situations, the thin layer of coupling medium on the transducer surface could be covered with a cellophane sheet, that has an acoustic impedance that match well that of wood. Recent progress on ultrasonic transducers were achieved with dry coupling layer.

3.3. Signal processing

Elaborate signal processing could help for the study of dispersive ultrasonic propagation phenomena in wood, due to the interaction of ultrasonic waves with microstructure. In this case the wave length or the cut-off frequency must be considered. Up to this point it was assumed that the wave length is much larger than the material structural dimensions. However, as soon as the wave length matches the dimensions of earlywood, of latewood or of cells, the velocity and attenuation are dependent on pulse frequency.

In FIGURE 1 typical time domaine and frequency domaine signals are presented for poplar wood on pulse excitation with longitudinal and shear waves. More elaborated signal

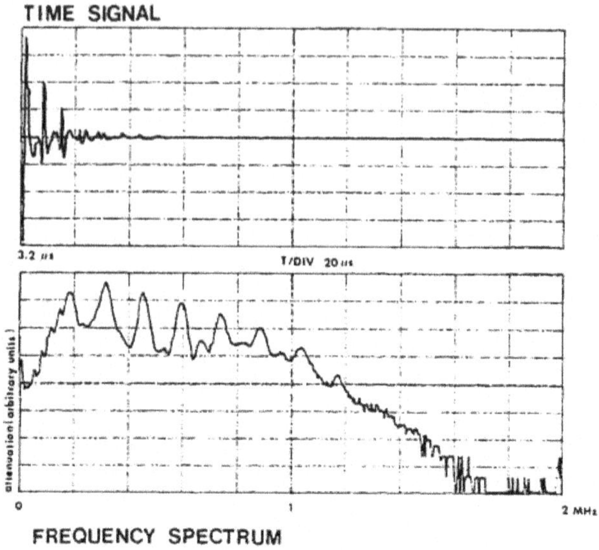

a) longitudinal waves in longitudinal direction.

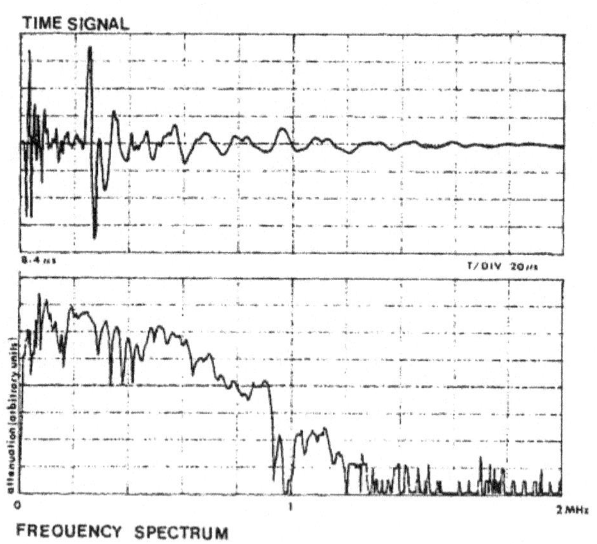

b) shear waves, propagation R polarisation T

FIGURE 1: Pulse excitation with longitudinal and shear waves on poplar wood. Time and frequency domaine[5].

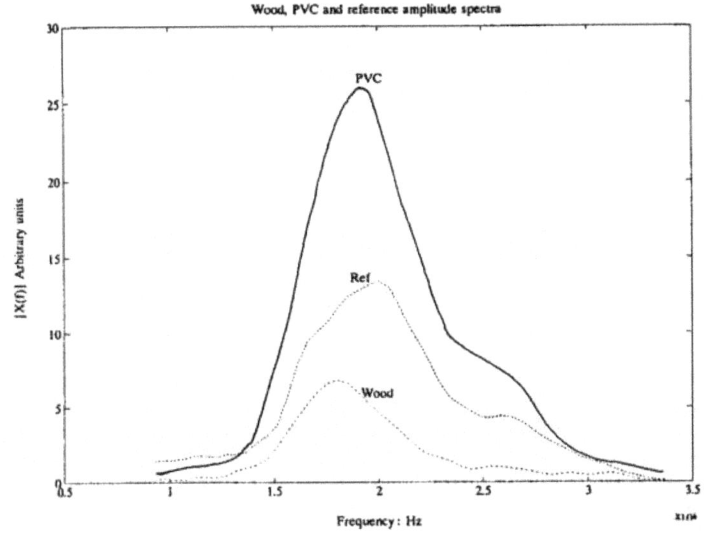

FIGURE 2: Amplitude spectra on spruce compared with pvc[6].

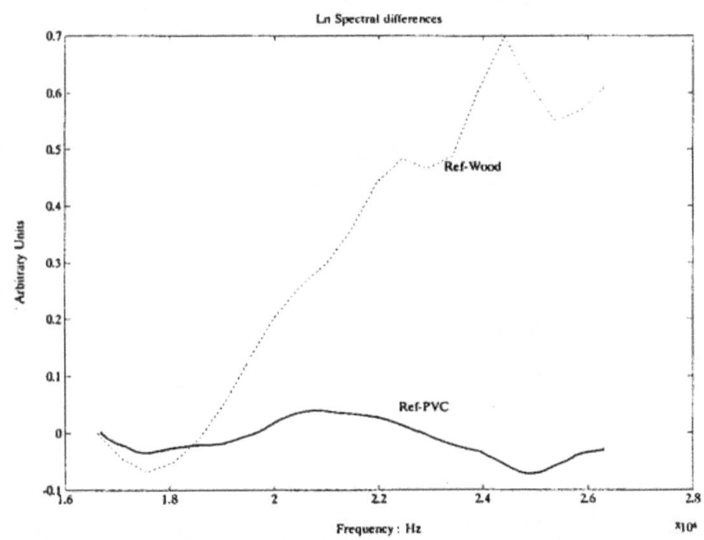

FIGURE 3: Power spectra on spruce compared with pvc[6] .

processing [6] of pulse excitation with longitudinal waves, along the longitudinal anisotropic direction of spruce is shown in FIGURE 2. The amplitude spectra and eaven more the power spectra FIGURE 3 show the complexity of propagation phenomena in wood. In frequency range exceeding 2MHz the vibration of the fine anatomical structure of wood can be observed.

4. Ultrasound on wood based composites

Probing boards of wood based composites is a less difficult task than probing solid wood. From boards of commercial size (approximatively of 3000x1800 mm. and 12...45 mm thickness) specimens (400 mm length x75 mm width x the thickness of the board) are cut, parallel or perpendicular to the direction of the flakes or particles alignment.

Ultrasonic measurements can be performed on full size boards or on specimens. Bulk waves (longitudinal or shear waves) or surface waves can be launched in samples, in a frequency range inferior to 5MHz.

Moreover on wood based composites acoustic emission activity can be stimulated very easily by breaking a 0.5mm pencil lead on the surface of the specimen. Parameters of acoustic emission as: duration, count number, peak amplitude or rise time can be determined. Finally the anisotropy of wood composites can be expressed by ultrasonic velocities and acoustic emission parameters.

The manufacturing characteristics of boards can be related to the anisotropy of samples exprsssed bu acoustic parameters as is shown in TABLE 1[7]. Surface velocity measured on boards is related to the properties of the skin. The internal bond and the shear behaviour of layers, related to the thickness interlaminar heterogeneity, could be tied together with the ratio of shear velocities or to the acoustic invariants. The global board anisotropy is expressed by the ratio of acoustic invariants.

5. Concluding remarks

The main advantages of ultrasonic techniques proposed for a very wide variety of samples (trees, small clear specimens, boards)are the real time nature of testing and the possibility of continous monitoring of all specimens in nondestructive way.

6. References

1.Brevet Francais No 90 06589: Procédé ultrasonore pour l'estimation de l'angle du fil du bois.I.N.R.A Paris 28 Mai 1990 **2.Bucur,V.; Fenney,B.**:Attenuation of ultrasound in solid wood. ULTRASONICS 1991(in press) **3.Woodhouse,J.**, Proc.Institut of Acoustics 8 (1986) Part 1,99-106, Edinburgh, UK **4.Bucur,V.; Perrin, J.R.**:, Proc. Institut of Acoustics 10(1988) part 2, 199-206, Edinborough, UK **5.Bohnke, I.**: Ph.D. Thesis (in preparation) Ecole de Mines de St. Etienne, France. **6.Leeman, S.**: Signal processing on wood specimens. Personal communication, 1990 **7.Bucur V.**: Anisotropy characterization of structural flakeboards with ultrasonic methods. submitted to WOOD AND FIBER SCI.1991

TABLE 1: Manufacturing characteristics of flakboards (19mm) and acoustic parameters [7]

Characteristics	ratio of	symbols	values minim	values maxim
Flakes alignment				
	-longitudinal velocities	V_{11}/V_{22}	1.11	1.13
	-surface velocities	(on 1/on 2)	1.03	1.10
	-acoustic invariants	I_{12}/I_{13}	1.32	1.46
Skin damage (axis 1/axis 2)	acoustic emission at 5cm.			
	-rise time		1.07	1.10
	-peak amplitude		1.04	1.08
	-duration		1.18	0.80
	-count		0.15	0.11
	-energy		1.72	1.68
Interlaminar heterogeneity				
	-in planes 12 and 13	V_{66}/V_{55}	1.86	1.91
	-in planes 12 and 23	V_{66}/V_{44}	1.87	1.99
Global board anisotropy				
	-acoustic invariants	Iratio	0.72	0.82

ACOUSTIC CHARACTERIZATION OF THIN FILM MATERIALS

V.I. ANISIMKIN
I.M. KOTELYANSKII
Institute of Radioengineering and Electronics
Academy of Sciences of the USSR

ABSTRACT

An experimental technique for measuring mass density, elastic and piezoelectric constants as well as velocity, attenuation coefficient and temperature coefficients of delay of acoustic waves, propagating in thin film materials, are developed. The technique is applied to:
1) acoustic characterization of ZnO and Ta_2O_5 piezoelectric films;
2) comparison of ZnO film with ZnO single crystal, and
3) Surface acoustic wave gas sensor characterization. The possibilities of improving the sensor sensitivity is discussed.

1. Introduction

Thin layers of metals, dielectrics, semiconductors, magnetics and piezoelectrics have been investigated extensively over the past years because they play an important role in microelectronics, in surface acoustic wave (SAW) devices. So, it is important to know their elastic and acoustic properties as they are related to device design parameters.

In this paper the elastic constants are measured and SAW propagation characteristics for the film materials are defined, which are in mechanical contact with the surface of substrate. The properties of substrate are assumed to be known.

2. Experimental methods

The strategy of the measurement is shown in Fig.1. As a first step, the phase velocity V_{ls} of Rayleigh mode as a function of film thickness to wavelength ratio h/λ are measured. Then, using the experimentally determined dispersion curve $V_{ls} = V_{ls}(h/\lambda)$ of the layered structure, the elastic parameters of substrate (s), the mass density $\hat{\rho}$, the Rayleigh wave velocity $\hat{V}_R$ and the transverse wave velocity $\hat{V}_T$ of the film are evaluated by means of a least squares fit of the experimental data to the theoretical dispersion curve. After that, the Poisson's ration $\hat{\sigma}$, the longitudinal wave velocity $\hat{V}_L$ and the isotropic elastic constants $\hat{C}_{11}, \hat{C}_{44}$ of the film are obtained:

$$\hat{\sigma} \approx \frac{\hat{V}_R/\hat{V}_T - 0,87}{1,12 - \hat{V}_R/\hat{V}_T} \tag{1}$$

$$\hat{V}_L = \hat{V}_T \sqrt{\frac{2(1-\hat{\sigma})}{1-2\hat{\sigma}}} \tag{2}$$

$$\hat{C}_{44} = \hat{V}_T^2 \hat{\rho} \tag{3}$$

$$\hat{C}_{11} = \hat{V}_T^2 \ \hat{\rho} \ \frac{2(1-\hat{\sigma})}{1-2\hat{\sigma}} \tag{4}$$

Next step is the calculation of the relative fraction r of the energy of Rayleigh mode localized in the film (power ratio). This parameter was shown [1] to be responsible for the relationship of the temperature coefficient of delay (TCD) and the coefficient of Rayleigh wave attenuation (α) in the film (Λ), substrate (s), and layeres structure (ls):

$$\widehat{TCD}_R = \frac{1}{r}\left[TCD_{ls} - (1-r)TCD_s\right] \tag{5}$$

$$\hat{\alpha}_R = \frac{1}{r}\left[\alpha_{ls} - (1-r)\alpha_s\right] \tag{6}$$

The calculation of r is performed by the procedure given in [1,2] using the material constants of the film, which have already been determined, and the material constants of substrates, which are know a priori. After measuring TDC_{ls} and α_{ls} the values of $\widehat{TDC}_r$ and $\hat{\alpha}_R$ are recalculated form (5) e (6).

In addition, the absorbtion coefficients of longitudinal ($\hat{\alpha}_L$) and transverse ($\hat{\alpha}_T$) bulk waves as well as the piezolelectric constant $\hat{d}_{33}$ of the film are independently measured.

The procedure outlined above is used for acoustically characterizing the films vs. thickness (0,5 - 5) μm, frequencies (100-300) MHz, and ambient temperatures $(0-100)^\circ C$.

The phase velocity V_{ls} is measured in a traditional way - by means of an acoustic microscope (local value) and from the central frequency and period of interdigital transducers (mean value). In both cases the uncertainties of V_{ls} are of the order of 1% because of the uncertainties of the film thickness ($\approx$ 3%) and of the SAW dispersion.

The temperature coefficient TCD_{l_s} is measured by the experimental setup shown in Fig.2. Here, the acoustic pulse overlaps the pulse of RT-leakage and interferes with it. The value of TCD_{l_s} is determined as a temperature gradient of the frequency of constructive or destructive interference of these two pulses in the overlap area. The precision of the method is about 5 %.

The piezoelectric constant $\hat{d}_{33}$ is measured by means of the method presented in [3], where the test film is used as a sound recever for probing ultrasonic pulse.

The experimental setup shown in Fig.3 permits to measure both the local value, and the destribution of the piezoelectric properties everywhere on the film surface (Fig.4). The precision of the measurements of the absolute values is about ±15%, while the relative measurements are much more precise.

The coefficients of attenuation of the longitudinal $\hat{\alpha}_L$ and transverse $\hat{\alpha}_T$ bulk waves are estimated from the ratio of the reflection coefficients of the waves from the surface of the sample covered with the test film and without it [4]. In turn, the reflection coefficients are measured by the comparison of the amplitudes of the echo-pulses for the free and the covered surfaces. For the experimental setup used in our experiment (Fig.5) typical number of the echo-pulses were about 25 and the precision was about 10 %.

At last, the absorption coefficient of Rayleigh wave in layered structure α_{l_s} is evaluated from insertion losses of four interdigital transducers located on the film surface along the SAW propagating direction [5]. Because of the uncertainties of the film thickness ($\approx$ 5%) the uncertainties of α_{l_s} measured along 20 mm path are about ±20%.

3. The localization of elastic energy in the film and the Rayleigh wave dispersion in layered structures

Because of the importance for acoustic characterization of thin film materials, the power ratio r and the initial slope S of the dispersion curve are analysed in more details. To cover the total system of film/substrate combinations the analysis were performed by the scheme proposed by Adler and Farnell in 1972 [2] (Fig.6).

The power ratio r is calculated as a function of the film thickness h/λ: it presents some general properties which can be summarized as follows:

1) For "slow" film on "fast" substrate (structures 1,2,3 in Fig.6) the value of the power ratio r is a monotonical function of h/λ , the function being about the same for all structures of this type.

2) For "fast" film on "slow" substrate (structures 4,5,6,7 in Fig.6) the power ratio r is a nonmonotonical function of h/λ : at small film thickness the value of r first increases, then it reaches a maximum value, and after that is falls drown to zero when a surface mode transforms into a leaky one.

To characterize the dispersion of Rayleigh mode in the layered structures the linear slope parameter S is calculated for the structures shown by dashed lines C, D, E in Fig.6. The calculations are performed as a function of h/λ for the mass densities and the elastic constants of the film which are changed in the range ± 10 % of the mean values. The calculated results are summarized as follows:

a) The relative changes of S separately produced by decrease (increase) of the elastic constants $\hat{C}_{11}, \hat{C}_{44}$ and mass density $\hat{\rho}$ are opposite for any structure. For example, 10 % decrease of $\hat{C}_{11}, \hat{C}_{44}$ and $\hat{\rho}$ in Pd/SiO_2 structure, results in relative changes of the slope S = +5 %, 12 %, and -26 %, respectively.

b) For the "slow" film on "fast" structures ($\hat{V}_T < V_T^S$) the relative change of the density produce the dominant change of the slope S, while for the "fast" on "slow" structures ($\hat{V}_T > V_T^S$) the dominant contribution is due to the elastic constants changes.

c) For "intermediate" structures ($\hat{V}_T \approx V_T^S$) the change of the film parameters can change both value and the sign of S. In this case the layered structure can "transform" from "fast" on "slow" to "slow" on "fast", or viceversa.

More clearly, all these properties of SAW dispersion can be deduced from the formula for SAW velocity perturbation due to thin, lossless isotropic overlay [6]:

$$-\frac{\Delta V_R}{V_R} = \frac{V_R h}{4P_R}\left[\hat{\rho}|v_{Ry}|^2 + \left(\hat{\rho} - \frac{4\hat{\mu}}{v_R^2}\frac{\hat{\lambda}+\hat{\mu}}{\hat{\lambda}+2\hat{\mu}}\right)|v_{Rz}|^2\right]_{y=0} \tag{7}$$

where $\hat{\rho}, \hat{\lambda}, \hat{\mu}$ are the mass density and Lamé constants of the film, $|V_{Ry}|^2/\sqrt{P_R}$ and $|v_{Rz}|^2/\sqrt{P_R}$ are the normalized particle velocities (time derivatives of the instantaneous displacements). Using the analytic expression for the normalized particle velocity components at the surface [6] and the relationships between the Lamé constants and the elastic constants ($\hat{\mu} = \hat{C}_{44}, \hat{\lambda} + 2\hat{\mu} = \hat{C}_{11}, \hat{\lambda} + \hat{\mu} = \hat{C}_{11} - \hat{C}_{44}$) Eq.(7) can be transformed into

$$-\frac{\Delta v_R}{v_R} = \left[\frac{\pi}{2}\left(\frac{v_R^2}{v_T}\right)^2 (f_y + f_z)\right]\frac{\hat{\rho}}{\rho}\frac{h}{\lambda} - [2\pi f_z]\left[1 - \frac{\hat{C}_{44}}{\hat{C}_{11}}\right]\frac{\hat{C}_{44}}{C_{44}}\frac{h}{\lambda} \tag{8}$$

$$-\frac{\Delta v_R}{v_R} = F_1(\hat{\rho}/\rho)(h/\lambda) - F_2F_3(\hat{C}_{44}/C_{44})(h/\lambda) \tag{9}$$

where F_1, F_2 and F_3 are some slowly changing functions of the Poisson's ratioes of the film $\hat{\sigma}$ and substrate σ materials. If the mass density $\hat{\rho}$ and the elastic constants $\hat{C}_{11}, \hat{C}_{44}$ of the film are changed, then the initial slope $S = (-\Delta V_R/V_R)(\lambda/h)$ is also changed:

$$\Delta S \approx F_1(\hat{\rho}_0/\rho)(\Delta\hat{\rho}/\hat{\rho}) - F_2F_3(\hat{C}^{\circ}_{44}/C_{44})(\Delta\hat{C}_{44}/\hat{C}^{\circ}_{44}), \tag{10}$$

here $\hat{\rho} = \hat{\rho}_0 + \Delta\hat{\rho}, \hat{C}_{44} = \hat{C}^{\circ}_{44} + \Delta\hat{C}_{44}$, and all above mentioned properties of the S parameter are clearly visible.

4. Experimental results

First, the experimental technique, described in section 11, has been applied for characterization of the elastic, piezoelectric and acoustic properties of textured piezoelectric films made of ZnO and Ta_2O_5. The experimental results are shown in Table 1:

Table 1.

Material	$\hat{\rho}$ $10^3 kg/m^3$	$\hat{C}_{44}$ $10^{10} N/m^2$	*velocity*, $10^3 m/s$ $\hat{V}_T$	$\hat{V}_R$	$\hat{V}_L$	$\hat{\alpha}_R dB/cm$ $(100MHz)$	$\widehat{TCD}$ *ppm*
Ta_2O_5 film	8,6	5,8	2,6	2,4	4,8	35	+25
ZnO film	5,0	3,3	2,6	2,5	4,9	2	+34
ZnO film to ZnO single crystal	−12%	−22%	−5%	− 5%	− 20%	+800%	+5%

It is shown that elastic and acoustic properties of Ta_2O_5 film are very close to those of ZnO film. The only difference is the acoustic attenuation that can be decreased by improving sputtering techniques. Both films have a polycrystalline structure with texture axis aligned perpendicular to the surface of substrate (the misorientation was less than 3°). The crystallite size was 300°. The thickness of the films was $h = 0,5 - 5\mu m$.

Second, we compare the properties of ZnO film and ZnO single crystal. The mechanical anisotropy of textured film is significantly increased due to the presence of boundaries between the crystallites. In this case, the film is similar to the multilayer single crystal with layers whose thickness is much smaller than the wavelength and equal to the thickness of the crystallites, and interlayer gaps, equal to the boundary region. So, the elastic properties, as well as the propagation characteristics of acoustic waves in textured film, are assumed to differ from the ones of the single crystal. Our experiments confirm this assumption, i.e. the elastic constants and the mass density of ZnO film are decreased of about 5 - 20 %, while the acoustic attenuation in the film is dramatically increased about 9 times compared with the single crystal (see Table 1).

On the contrary, if the acoustic wave propagates along the crystallites (no influence of the weak coupling between the crystallytes), the acoustic attenuation in the film is about the same as in a single crystal (Fig.8). The piezoelectric constant $\hat{d}_{33}$ of textured ZnO film was also decreased. This can be attributed to the crystalline misorientation. On the other hand the temperature coefficient of delay for textured films is increased of about 5 % (Table 1), which is not understandable now.

The acoustic characterization of thin films was also applied to SAW gas sensors. As an example, we compare the hydrogen sensors fabricated on Pd/SiO_2, $Pd/LiNbO_3$ and Pd/Si structures. The effective isotropic elastic constants of substrate materials are taken from [2]. The effective isotropic second order elastic constants of the polycrystalline Pd film are calculated from the constants of single crystal using simple arithmetic average of Reuss's and Voigt's approximation (Hill's average). To estimate the mass density change and the elastic constant change due to the hydrogen exposure, the elastic properties of palladium hydride are also calculated (Table 2). It worths to be pointed out that all material constants of polycrystalline Pd are smaller than those for single crystal Pd. Also, all material constants of palladium hydride are smaller than these of polycrystalline Pd, i.e. $\Delta\hat{\rho}/\rho \approx -10\%, \Delta\hat{C}_{11}/\hat{C}_{11} \approx -6,7\%, \Delta\hat{C}_{44}/\hat{C}_{44} \approx -3,2\%$.

Table 2.

Material		$\hat{\rho}$	$\hat{C}_{11}$	$\hat{C}_{12}$	$\hat{C}_{44}$
Pd	single crystal	12,0	224,5	173,5	71,1
	polycrystal	—	253,4	159,0	47,2
PdH	single crystal	10,84	212,2	157,0	64,0
	0.66 polycrystal	—	236,3	145,0	45,7

($\hat{\rho}$in 10^3 kg/m^3, $\hat{C}_{ii}$ in 10^9 N/m^2).

The comparison of SAW hydrogen sensors frabricated on different structures are show in Table 3, where the magnitude of the sensors response is shown to increase with increase of the relative fraction r of the energy of the mode, localized in Pd film and with increase of the change of the initial slope ΔS of the dispersion curve due to the change of the material constants of Pd film.

Table 3.

Structure	$-\Delta\hat{V}_R/\hat{V}_R, ppm$	r	$\Delta S, 10^{-2}$
Pd/ZnO/Si	600	0,07	−320
Pd/SiO_2	200	0,06	−40,5
Pd/$LiNbO_3$	25	0,04	−28

This result as well as Eq.(9) enable one to choose properly the material of the substrate if the material of the film of SAW gas sensor is determined by the gas which has to be detected. The preferable combinations of the materials are the ones which provide the sensitive gas detection mainly due to only one of two factors: mass density change (first term in Eq.(9)), or the mechanical property change (second term in Eq.(9)):
1) the heavy film with small elastic constant $\hat{C}_{44}$ on the light substrate with large elastic constant C_{44}, or
2) the light film with large $\hat{C}_{44}$ on the heavy substrate with small C_{44}.

5. Conclusion

Crystallographic defects of thin film materials can be estimated using the mass density, the elastic and piezoelectric constants of the film together with its acoustic parameters. The power ratio r, the initial slope S of dispersion curve, and acoustic attenuation α play an important role in such estimations. Using the same parameters, the sensitivity of SAW gas sensors can be improved by means of a proper choice of the film and substrate materials combination.

6. References

1. V.I. Anisimkin, I.M. Kotelianskii, Sov. Phys., J. Techn. Phys., 1985, v.55, n.12, p.2420; V.I. Anisimkin, I.M. Kotelyanskii, M.A. Magomadov, V.I. Fedosov, Proc. Int. School on Phys. Acoust., Italy, Erice, 1988, p.347.
2. G.W. Farnell, E.L. Adler, Physical Acoustics, ed. by W.W. Mason, R.N.Turston, New York, London, 1972, v.9, p.35.
3. V.A. Vyun, V.N. Umashev, I.B. Yakovkin, Sov. Phys., PTE, 1986, n.6, p.192.
4. K.I. Baranskii, V.A. Magomedov, S.V. Pavlov, S.V. Pustovoyt, Sov. Phys., Microelectronics, 1982, v.11, n.5, p.418.
5. F.S. Hickernell, Proc. IEEE Ultrason. Symp., 1982, v.1, p.325.
6. B.A. Auld, Acoustic waves and fields, 1972, v.2.

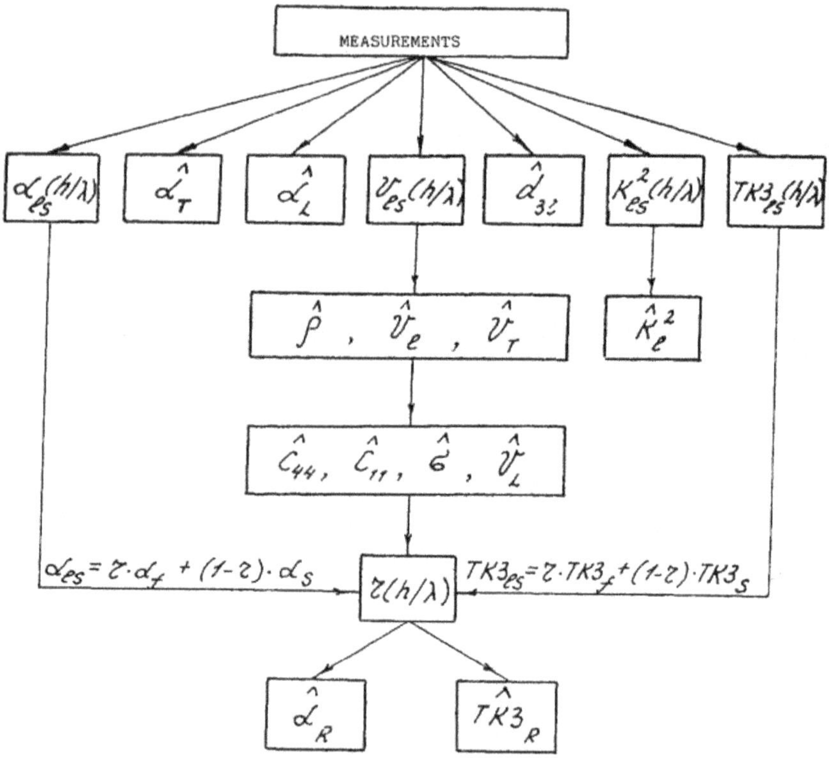

FIG.1. The strategy of acoustic characterization of thin film materials.

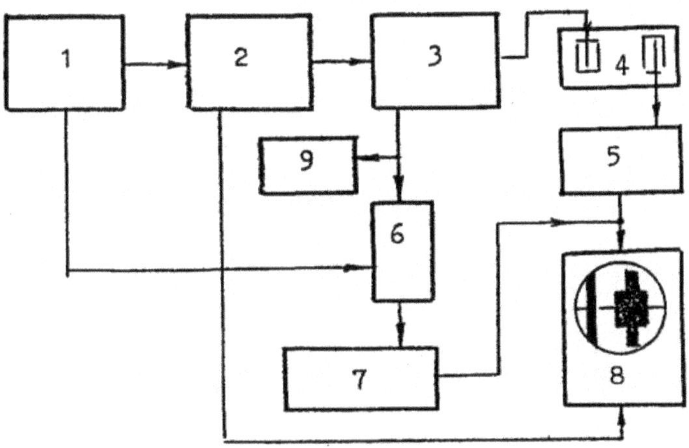

FIG.2. The experimental setup for measurements of the temterature coefficient of delay of surface acoustic waves in layered structures $(TCD)_{ls}$.

1,2 - pulse generators, 3 - RF generator, 4 - test sample, 5 - RF amplifier, 6 - electronic key, 7 - attenuator, 8 - oscilloscope, 9 - frequency meter.

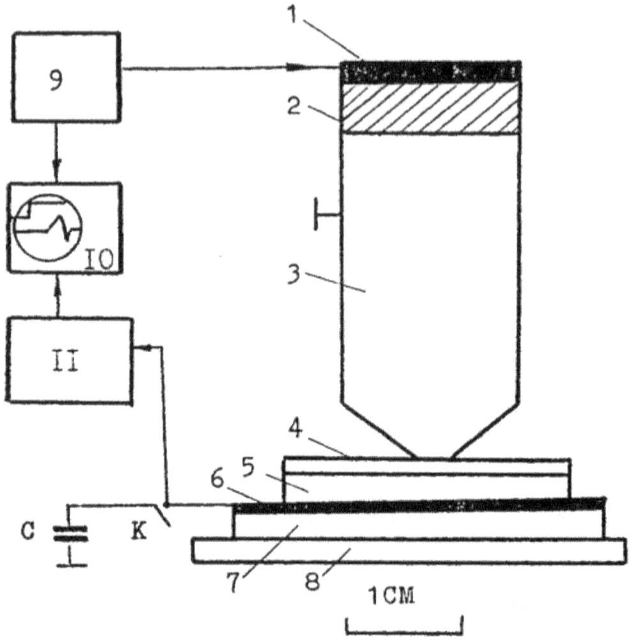

FIG.3. The experimental setup for measurements of the piezoelectric constant d_{33} of the films/3/.

1 - electrode, 2 - piezoelectric plate (acoustic transducer), 3 - probe, 4 - test film, 5 - substrate, 6 - metal foil, 7 - resin plate, 8 - two-coordinate table, 9 - pulse generator, IO - oscilloscope, 11 - video pulse amplifier.

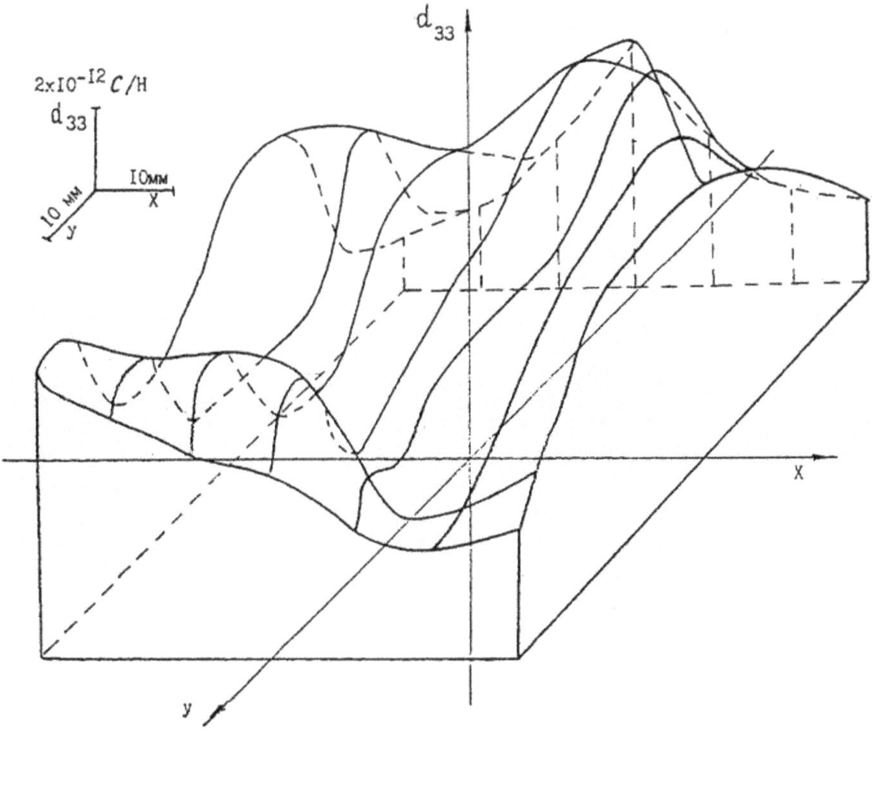

FIG.4. The destribution of the piezoelectric constant $\hat{d}_{33}$ on the surface of ZnO film on Si substraye (h = 3 μm).

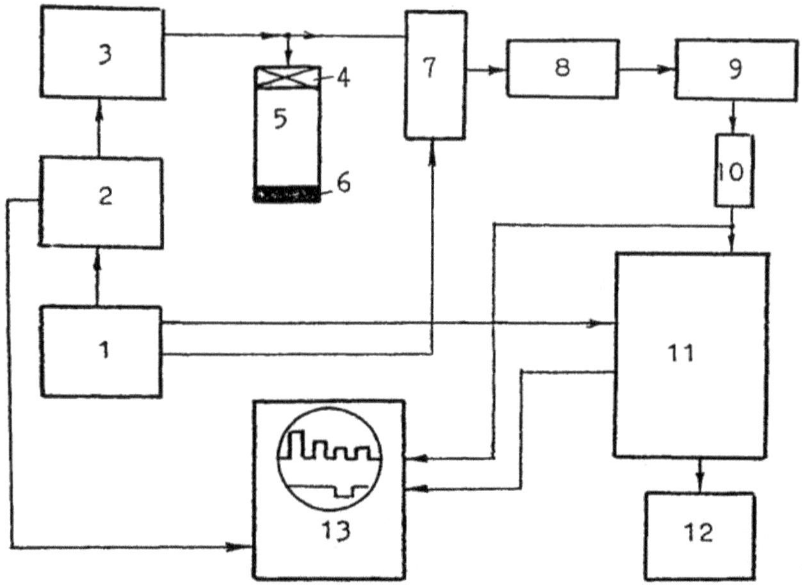

FIG.5. The experimental setup for measurements of the attenuation coefficients of bulk acoustic waves.

1,2 - pulse generators, RF generator, 4 - acoustic wave transducer, 5 - sample, 6 - test film, 7 - electronic key, 8 - amplifier, 9 - attenuator, IO - detector, 11 - Box-Car integrator, 12 - digital voltmeter, 13 - oscilloscope.

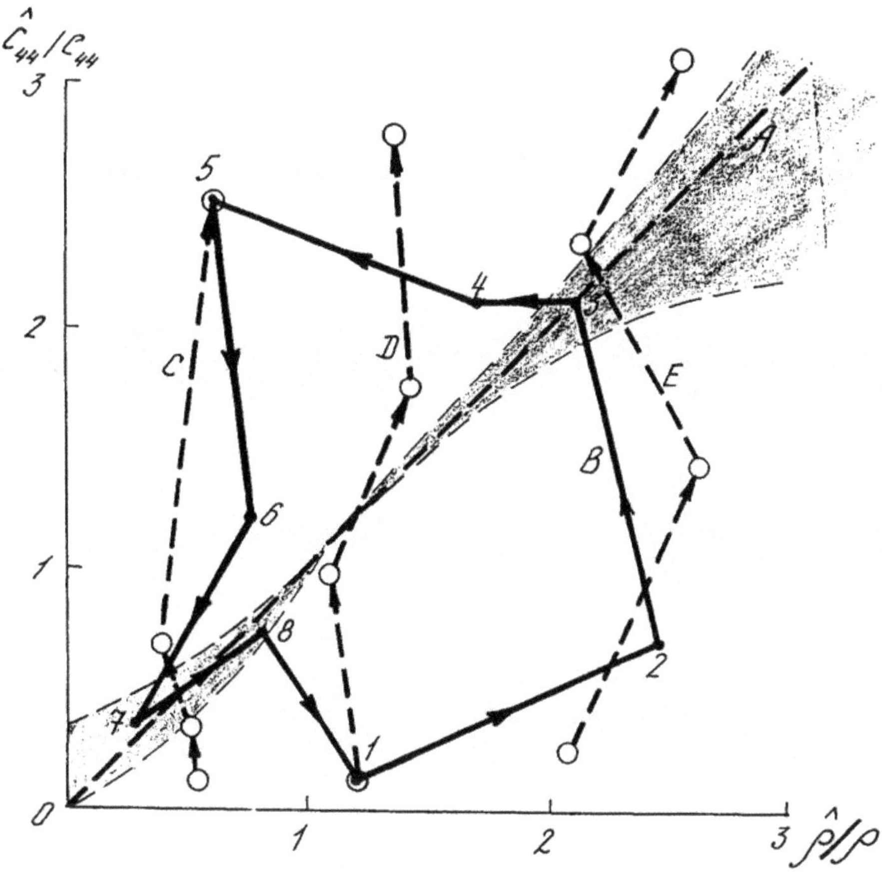

FIG.6. A useful type of coordinate system within which to classify different material combinations in layered structures /2/.

Line A - the positions of layered structures with equal transverse wave velocities in the film ($\hat{V}_T$) and substrate (V_T).

Lines B,C,D,E - the positions of layered structures used for analysis of the power ratio r and the dispertion slope S.

Shaded region - the positions of layered structures supporting Stoneley wave.

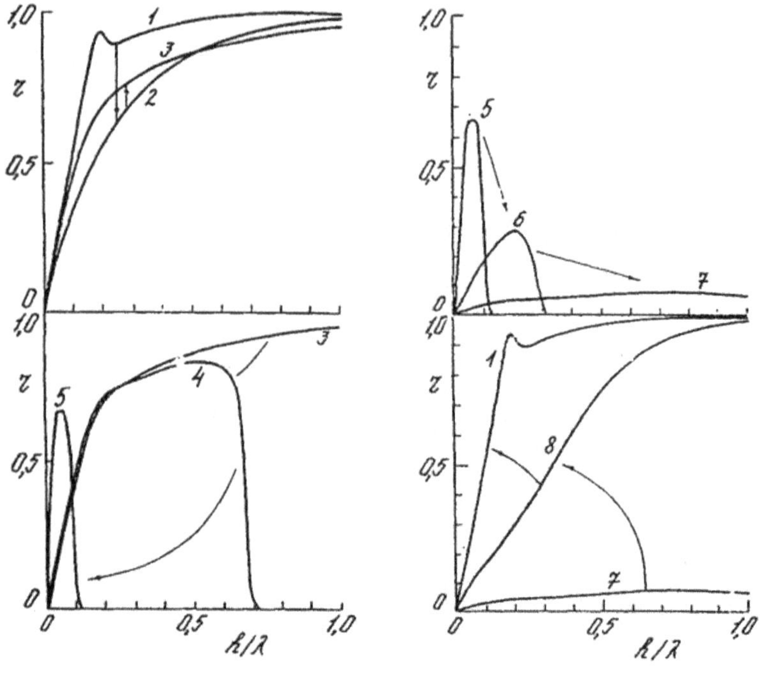

FIG.7. The relative fraction r of the energy of Rayleigh mode localized in the film as a function of film thickness/wavelength ratio h/λ.

1,2 - "slow" film on "fast" substrate ($\hat{V}_T < V_T$),
4-6 - "fast" film on "slow" substrate ($\hat{V}_T > V_T$),
3,7,8 - "intermediate" structures.

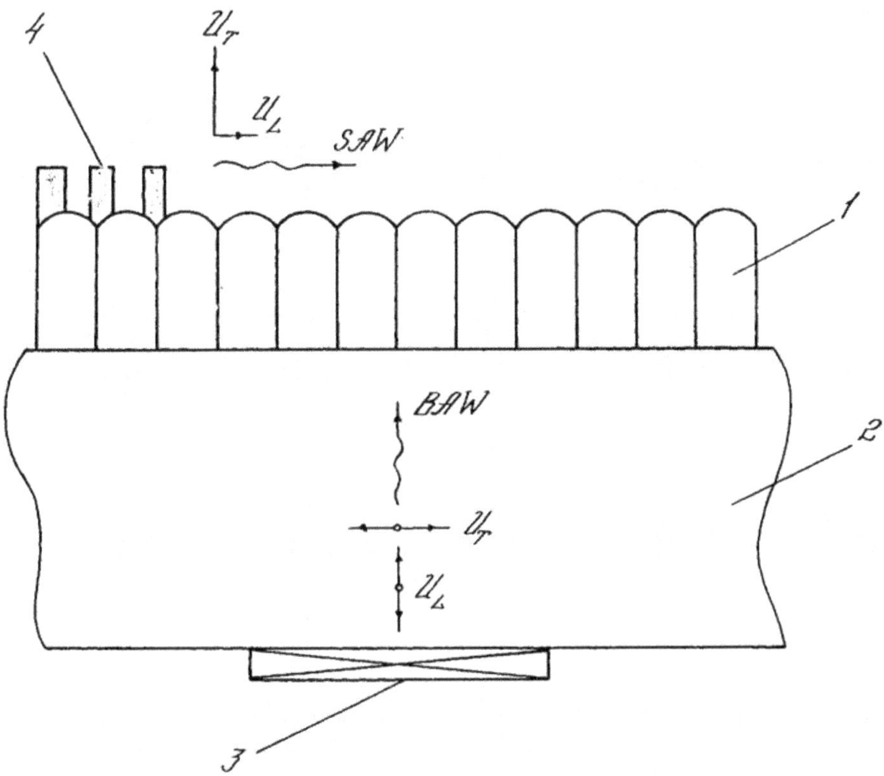

FIG.9. The experimental setup for measurements of attenuation coefficients of bulk acoustic waves (BAW), propagating along, and surface acoustic waves (SAW), propagating across the crystallites of the textured films.

1 - textured film, 2 - substrate, 3 - BAW transducer, 4 - SAW transducer.

SPECTRAL MAPS BY ARMA MODELLING FOR ULTRASONIC TISSUE CHARACTERIZATION: APPLICATION IN OPHTHALMOLOGY

ADA FORT, CLAUDIA MANFREDI, LEONARDO MASOTTI, SANTINA ROCCHI
Electronic Engineering Department
University of Florence, Via S.Marta 3
50139 Florence, Italy

ABSTRACT

The aim of this study is to produce maps based on ultrasonic spectral parameters that can be used to diagnose ophthalmic pathologies. Low-order autoregressive models have been used to estimate the power spectral density of ultrasonic signals backscattered by pathological tissues. This technique yields smoothed estimates of the power spectra of short signal segments, by emphasizing only higher energy frequencies and by eliminating random spectral peaks and side-lobe effects. It, therefore, allows local characterization of the investigated inhomogeneous medium.
This approach has been tested with different sets of simulated backscattered signals. Results showed that this method is able to detect scatterers of different sizes, providing easily legible topological maps.

1. Introduction

Ultrasonic broadband pulse-echo techniques are widely used in ophthalmology for intraocular tumors diagnosis.
In many cases the pathological region behaves as a random suspension of small particles in a homogeneous fluid, so, when a disease is present, the investigating ultrasonic pulse interacts with a scattering medium characterized by specific acoustic properties. For this reason the ultrasonic backscattered signal is a diagnostic tool as it contains thorough information on the structure of a particular pathology.

Spectral characterization supports the conventional ultrasonic imaging techniques (A-mode and B-mode displays), since it exploits the frequency dependence of the radiofrequency (RF) backscattered signal, providing further information about tissue microstructure. In fact, it is well known [1] [2] that the shape of the backscattered power spectrum depends on the mean size, shape and elastic properties of the scattering particles, while the spectrum magnitude is related to concentration, acoustic impedance and size of the scattering particles.

A common approach to the spectral characterization of the pathology, [3][4], consists of obtaining some parameters that describe the frequency dependence of the RF backscattered power spectrum. The most commonly used parameters are the spectrum centroid shift and the spectral magnitude.

The power spectrum of the backscattered signal is customarily estimated using an averaging process based on the periodogram algorithm. When the region of interest has a small volume, as for local characterization, this approach can not eliminate all the random spectral features related to the spatial distribution of the scatterers. Therefore, the estimated spectral parameters can be affected by the residual randomness.

In order to avoid these problems, the autoregressive (AR) modelling technique [5] was proposed to estimate the power spectral density and to evaluate the spectral centroid shift. Low-order AR models are able to smooth the random peaks and pits of the backscattered signal spectrum while tracking the spectral centroid shift. Moreover, this technique is intrinsically insensitive to the windowing side-lobe effects. Thus with low-order (2 or 3) AR models the centroid frequency estimate is consistent also when the number of samples of the RF signal is drastically reduced, and the canonical algorithm becomes inapplicable. The use of low-order AR models has the additional advantage of requiring low processing-time. These aspects make the AR technique particularly suitable for the real time production of pathology maps based on spectral parameters.

Spectral maps are a useful tool for tumoral diagnostic because they emphasize local inhomogeneities affecting frequency behaviour of echo signals and complete the information concerning the echo signal magnitude provided by A-B-mode displays.

2. Tissue Model

The ultrasonic signal coming from a random particle distribution can be described by the superimposition of each single scatterer contribution, provided that concentration is weak enough to ignore multiple scattering effects.

Moreover a mono-dimensional model is adopted for isotropic media when the ultrasonic beam is narrow and a plane-wave propagation is assumed [6]. This is often the case in medical analysis when a weakly focused ultrasonic transducer is used and the region of interest lies in the focal zone.

Applying the Born approximation [1] backscattered signal is expressed in the frequency domain by:

$$E(f)=\sum_i E_i(f)$$

$$E_i(f)=T^2(f,z_i)A^2(f,z_i)B_i(f) \qquad (1)$$

In Eq. 1, f is the frequency; z_i and $B_i(f)$ are the axial position and the frequency response of the i-th scatterer; T(f,z) is the transducer response and A(f,z) is the attenuation of the biological medium. T(z,f) is squared because the transducer is used as transmitter and receiver. When the absorption phenomenon is only considered, A(f,z) can be expressed by the following formula:

$$A(f,z)=e^{-\alpha fz} \qquad (2)$$

α is the absorption coefficient. The square of A(f,z) in Eq.1 represents the attenuation along the direct and backscattered path.
Assuming the validity of the Rayleigh-Born approximation, $B_i(f)$ can be obtained as follows. The velocity potential of the acoustic field at point $\underline{r}$ (Φ_{scatt}) scattered by an object placed at point $\underline{r}_i$ is:

$$\Phi_{scatt}(\underline{r},\underline{r}_i)=\Phi_{inc}(\underline{r}_i)F_i(\underline{K},\underline{K}')\frac{e^{jK|\underline{r}-\underline{r}_i|}}{|\underline{r}-\underline{r}_i|} \qquad (3)$$

Φ_{inc} is the velocity potential of the incident field; $\underline{K}$ is the propagation vector of the impinging field and $\underline{K}'$ is the propagation vector of the scattered field. The backscattered field can be found by setting $\underline{K}'=-\underline{K}$:

$$B(f)=F(\underline{K},\underline{K}')\big|_{\underline{K}'=-\underline{K}} \qquad (4)$$

In the present model spherical fluid particles have been considered. Therefore B(f) is a function of frequency as follows:

$$B(f)=F(\underline{K},-\underline{K})=\mu\left[\frac{\sin(2Ka)}{8K}-a\frac{\cos(2Ka)}{4}\right] \qquad (5)$$

μ is a parameter related to the particle acoustic impedance and a is the particle radius.

In ophthalmology, intraocular tumors such as certain melanomas and retinoblastomas, [2], are adequately described by the spherical particle model. In fact, some types of melanoma are characterized by a fine homogeneous microstructure containing spherical particles with a diameter of few tens of micrometers. On the other hand, retinoblastomas are structurally coarse, because small particles sometimes form aggregates which can be approximately considered as larger spherical particles (diameters up to 100 μm).

In the adopted model, a monodimensional spatial distribution of particle is considered and a uniform or Gaussian random distribution is used for particle distance. The scatterer radius a and strength μ can either be deterministic values or random variables with uniform or Gaussian distribution.

Figure 1 shows a time domain simulated signal. Here, scatterers are assumed to be identical spheres with a uniform random spatial distribution. The attenuation A(f) was not taken into account.

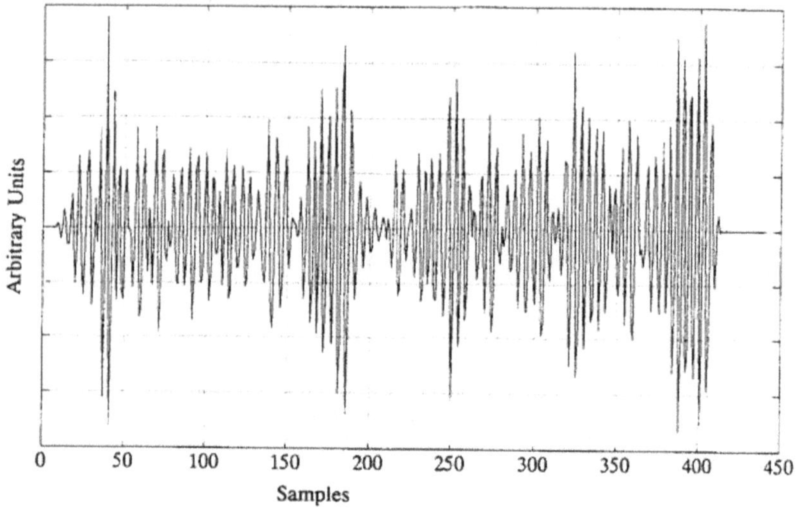

Figure 1. Time simulated backscattered signal. Identical spheres with 20 μm dia, and uniform spatial distribution are considered.

3. Spectrum Estimation by ARMA Modelling

The estimation of the power spectral density (PSD) of a finite discrete data sequence is commonly performed via the Fast Fourier Transform (FFT) based periodogram technique. Despite the wide range of its possible applications, the periodogram approach has two main limitations: an increase of the estimation quality decreases the frequency resolution and the windowing of data, implicit in the method, causes the leakage effect in the spectral domain, which may obscure and distort weak signal spectral responses.

The above limitations are particularly troublesome when analyzing short data records. Therefore, alternative spectral estimation procedures, based on the fitting of the measured data to an assumed model, [8][9], have been proposed over the last two decades.

In this section, after a brief review of the main characteristics of the periodogram method, modelling approach will be outlined. Finally, the characteristics of the algorithm used in this work will be described, showing its suitability for the diagnosis of ophthalmological tumors.

It is well known that the PSD:

$$P(f)=\int_{-\infty}^{\infty} R_{xx}(\tau)e^{-j2\pi f\tau}d\tau \tag{6}$$

can be estimated using the periodogram method, when the measured process can be considered stationary and ergodic.
In Eq. 6 $R_{xx}(t)$ is the autocorrelation function.

When applying the periodogram method, the PSD is evaluated by splitting the data sequence x(n) into M subsegments of length L, weighted with the sequence w(n) (window) and computing the average of the squared DFT (Digital Fourier Transform) of each segment. The segments can be adjacent or partially overlapping. Hence:

$$\hat{P}_{PER}(k)=\frac{1}{M}\sum_{r=1}^{M} P_r(k)$$

$$P_r(k)=\frac{1}{L\Delta T}\left|\Delta T\sum_{n=0}^{L-1} x_r(n)w(n)e^{-j\frac{2\pi}{L}kn}\right|^2 \tag{7}$$

As mentioned before many of the limitations of the periodogram approach are caused by the implicit hypothesis that data outside the window w(n) are equal to zero. Such hypothesis is usually not verified. Therefore, the overall transform is the convolution of the desired signal transform and the window transform. When the signal is concentrated in a narrow bandwidth, this operation spreads the power into adjacent frequency regions (side-lobe leakage) and limits the frequency resolution of the periodogram. Therefore, smeared spectral estimates are a consequence of windowing.

On the other hand, the use of a priori information or assumptions based on some knowledge about the process under consideration, may permit the selection of a model that is a good approximation of the actual underlying process: it is then possible to obtain better spectral estimates. Thus spectrum analysis, in the context of modelling becomes a three step procedure:

1) Selection of a time series model;
2) Estimation of the parameters of the assumed model;
3) Substitution of the above parameters into the theoretical PSD implied by the model.

The first step is fundamentally connected to estimation and identification techniques applied in linear system theory, with the difference that only the model's output process is available, while the input is assumed unknown.

With this approach, more realistic assumptions are made on the measured process outside the measured interval and the windowing is intrinsically eliminated; thus the improvement over the conventional periodogram approach can be quite noticeable, especially for short data records.

The most basic ARMA (AutoRegressive Moving Average) model with input {u(n)} and output {x(n)} is:

$$x(n)=\sum_{l=0}^{q} b_l u(n-l)-\sum_{k=1}^{p} a_k x(n-k) \tag{8}$$

where the a_k are the parameters of the AR part (of order p) and the b_k are those of the MA part (of order q).

If the driving process, u(n), is assumed to be a white noise sequence with zero mean and variance σ_u^2, the PSD of the ARMA output process is:

$$P_{ARMA}(f)=\sigma_u^2 \Delta T \left|\frac{B(e^{j2\pi f\Delta T})}{A(e^{j2\pi f\Delta T})}\right|^2$$

where:

$$A(z)=1-\sum_{k=1}^{p} a_k z^{-k}$$

$$B(z)=\sum_{l=0}^{q} b_l z^{-l} \tag{9}$$

Thus the spectrum of the process $\{x_n\}$ is entirely specified by the coefficients $\{a_k, b_k, \sigma_u^2\}$.

If all the $\{b_k\}=0$, except $b_0=1$, the sequence {x(n)} is a linear regression on itself with u(n) representing the error. In this case, the process is simply called AR (all-poles).

In this research an AR process was selected to describe the backscattered ultrasonic signal. This choice is particularly convenient since the only linear equations must be solved to estimate the AR parameters. If an ARMA model had been selected, solution of non-linear equations would have been required. Moreover, a close relationship exists between a linear prediction filter and an AR process; the forward linear prediction coefficients are the solution of normal equations whose coefficient matrix is structurally identical to that of the Youle-Walker equations [8]. Thus the theory of one is applicable to the other. Furthermore ARMA models are only needed when strong observation noise is present.

As mentioned above, the backscattered echo is described by an AR process of order p:

$$x(n) = -\sum_{k=1}^{p} a_k x(n-k) + u(n) \tag{10}$$

where x(n) represents the measured data and u(n) is the unknown input sequence. Hence, the power spectral density (AR PSD) is:

$$P_{AR}(f) = \frac{T\sigma_u^2}{\left|1 + \sum_{k=1}^{p} a_k e^{-j2\pi fkT}\right|^2} \tag{11}$$

T is the signal sampling period and σ_u^2 is the variance of u(n) assumed to be a zero mean, white noise process.

It should be noted that AR PSD estimators perform an implicit extrapolation of the autocorrelation sequence by recursively extending it beyond the interval of known samples. In this way, AR estimators provide high frequency resolution PSD estimates.

Since the filter order p is usually unknown, it is often necessary to postulate several model orders. An exaggerated model order produces spurious peaks in the AR PSD estimate, while a low order does not reproduce spectral details (number of spectral peaks $\leq$ p-1). Thus, the selection of the AR filter order represents a trade-off between increased resolution and decreased variance of the prediction error.

In the present work, a low p value was chosen, the main interest being in emphasizing the frequency range of higher energy, regardless of spectrum details.

In order to evaluate the AR parameters a variety of batch estimation methods, based on the least squares technique, have been developed. These techniques minimize different functionals related to the energy of the error.

The modified covariance method (MCM) was used in this research. This method, regardless of the selected AR order, utilizes the information contained in the whole selected data sequence. Basically, the MCM minimizes the average of the forward and backward linear prediction squared errors ($e_{fp}(n)$ and $e_{bp}(n)$) over the N available data, with respect to all the AR parameters:

$$r_p^{fb} = \frac{1}{2}\left[\sum_{n=p+1}^{N} |e_{f\,p}(n)|^2 + |e_{b\,p}(n)|^2\right] \tag{12}$$

where:

$$e_{f\,p}(n) = x(n) - \hat{x}^f(n) = x(n) + \sum_{k=1}^{p} a_p^f(k)x(n-k)$$

$$e_{b\,p}(n) = x(n-p) - \hat{x}^b(n-p) = x(n-p) + \sum_{k=1}^{p} a_p^{f*}(k)x(n-p+k) \tag{13}$$

$x^f(n)$ and $x^b(n\text{-}p)$ are the standard forward and backward linear predictors for samples x(n) and x(n-p), and $a^f{}_p(k)$ are the forward linear prediction coefficients for p order models (the * indicates the complex conjugate). This leads to a p+1 order linear system with a coefficient matrix whose entries are a combination of all the available data. A fast computational solution of the MCM is obtained by expressing the coefficient matrix as sums and products of Toeplitz and Hankel matrices. This solution only requires $Np+6p^2$ operations and N+4p data storage, making the algorithm especially suitable for real-time applications in the biomedical field.

In Figs. 2a and 2b a dotted line represents the squared Fourier transform of a simulated RF backscattered signal. The presence of noise, represented by random peaks and pits (fluctuations) , is due to the scatterers' spatial distribution. In Fig. 2a the theoretical power spectrum is compared to the AR estimate, while in Fig. 2b it is compared to the periodogram estimate.

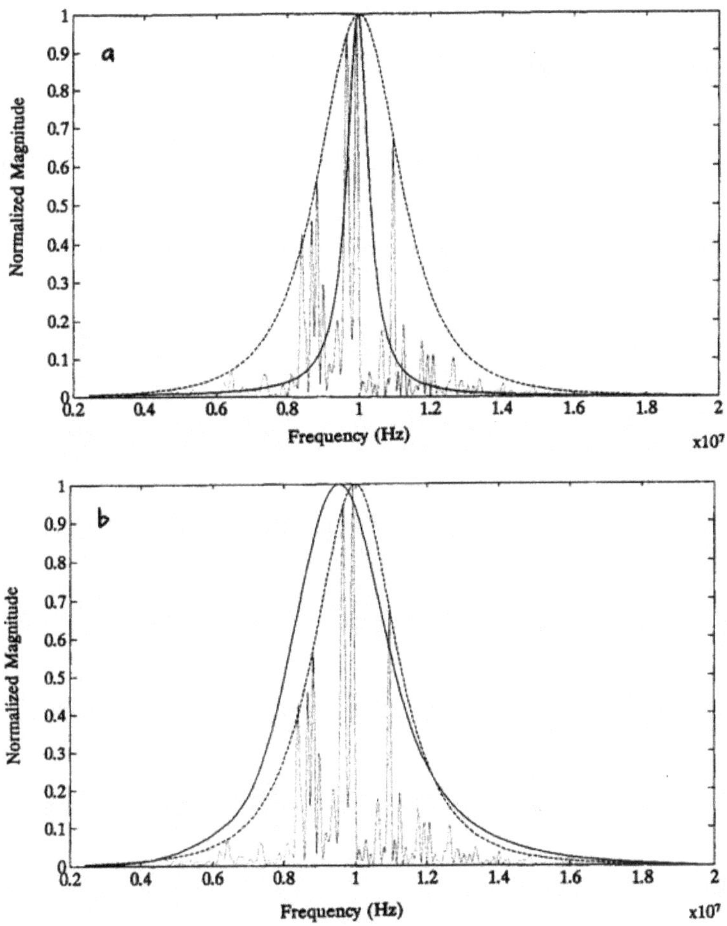

Figure 2. a) Dotted line; squared FFT of the backscattered signal. Dashed line; theoretical power spectrum. Solid line; power spectrum estimation via AR modelling. **b)** Dotted line; squared FFT of the backscattered signal. Dashed line; theoretical power spectrum. Solid line; power spectrum estimate via periodogram technique.

3. Simulation Results

Maps are produced considering simulated RF echo signals. For each simulation the random parameters are independently generated, but their statistical properties are maintained, as well as the characteristics of the pulse-echo system and an echographic analysis of a macroscopically homogeneous bidimensional tissue zone is simulated.

Simulations were obtained using a broadband transducer frequency response, T(f), with a 10 MHz center frequency and a 7 MHz bandwidth at -6 dB. The transducer response was previously determined experimentally by measuring the echo coming from a planar reflector in water.

Each echo line was divided into 1÷4-pulse-long segments (analysis of zones of 250 μm÷ 1 mm. An AR modelling technique was used to extract the spectral parameters. A bidimensional spectral image was produced by associating the above parameters to a color coded or grey level legend. Maps of the centroid frequency are presented below.

A second order AR filter was selected when the echographic system's response $T^2(f)$) exhibited a quasi-symmetrical shape. A third order filter was used to represent highly asymmetrical responses.

The centroid frequency was approximated by the maximum energy frequency, evaluated using the second and third order AR model parameters, according to the following expressions [7]:

$$f_m^{(2)} = \frac{1}{2\pi T} arcos\left(\frac{a_1 a_2 + a_1}{-4a_2}\right) \tag{14}$$

$$f_m^{(3)} = \frac{1}{2\pi T} arcos\left\{\frac{1}{2a_3}\left[-\frac{1}{3}(a_1 a_3 + a_2) + \sqrt{\frac{1}{9}(a_1 a_3 + a_2)^2 - \frac{1}{3}(a_1 a_2 a_3 + a_2 a_3^2 + a_1 a_3 - 3a_3^3)}\right]\right\} \tag{15}$$

Maps obtained by AR spectral estimates were then compared with maps produced by the parameters evaluated using the periodogram algorithm.

A similar comparison was carried out on the results obtained by short pass filtering the Cepstrum transform. The performance of these two methods proved to be similar, but AR modelling was less computationally complex than Cepstrum.

In Fig. 3 spectral grey-level maps obtained with AR (a), periodogram (b) and Cepstrum (c) techniques are shown. 32 and 64 sample long segments were used (twice and four the pulse length). The simulated tissue consisted of three zones with different scatterer sizes (20 μm, 50 μm, 80 μm), having random and uniform spatial distribution.

Figure 4 represents the maps of two simulated intraocular pathologies. The first one is a homogeneous melanoma, modelled as a quasi-regular array of spheres.

The spheres' radii were uniformly distributed between 20 μm and 40 μm. The second pathology is a retinoblastoma characterized by a disarrayed microstructure with cellular agglomerates and evident necrosis. In this simulation both the spatial position of the sphere centers and radii had uniform distribution. Radii varied from 20μm to 80 μm. These spectral maps emphasize the differences between the simulated pathologies by pointing out both the average scatterers' size value and the degree of local disarray.

5. Conclusion

A parametric method for characterizing pathologic biological tissues by forming ultrasonic spectral maps was presented. Echoes backscattered by the tissue convey information on the average pathology features, but this information is embedded in random components due to the stochastic nature of the tissue microstructure. Thus, the power spectral density (PSD) centroid frequency is used to estimate the mean scatterer size. In particular, the low order autoregressive (AR) modelling technique [5] is proposed for PSD centroid evaluation. This technique, being intrinsically insensitive to the windowing side-lobe effects and requiring low processing time, is suitable for local characterization of the investigated inhomogeneous medium and for real time production of spectral maps.

The presented maps obtained by processing simulated signals show that this technique provides information both on the mean scatterer size and on the tissue local properties and can support the conventional A- and B- mode displays for intraocular tumors diagnosis.

6. Acknowledgments

The authors aknowledge the contribution of Dr. R.Frezzotti and his coworkers for their support in clinical field and Dr. Vignoli for the useful discussions.

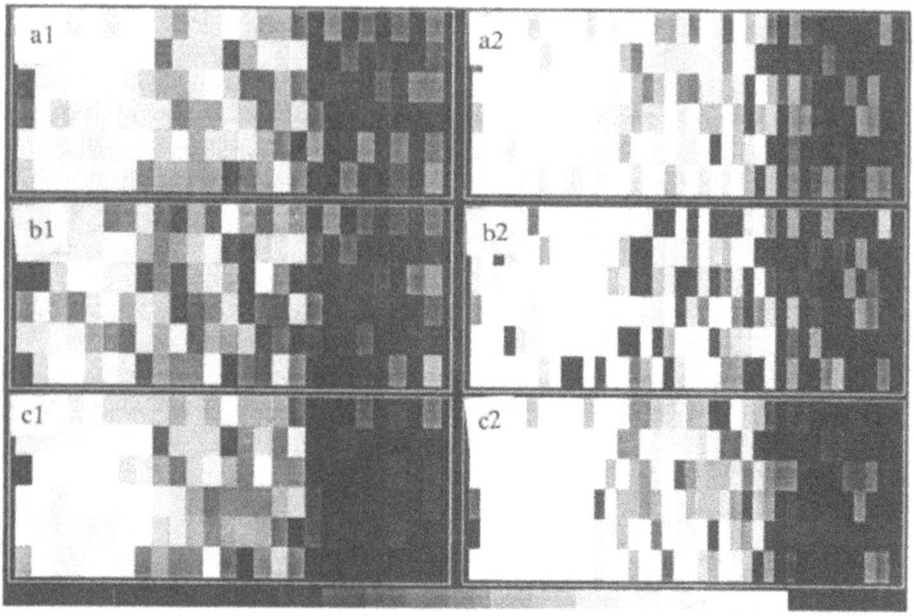

Figure 3. Spectral maps based on evaluation of the centroid's frequency. The simulated tissue consisted of three zones with scatterers of 20, 50 and 80 μm [dia]. Maps a1) and a2) AR spectral estimation. Maps b1) and b2) periodogram technique. Maps c1) and c2) Cepstrum transform. Maps a1) b1) c1) are obtained processing 64 sample data segments. Maps a2) b2) c2) are obtained processing 32 sample data segments.

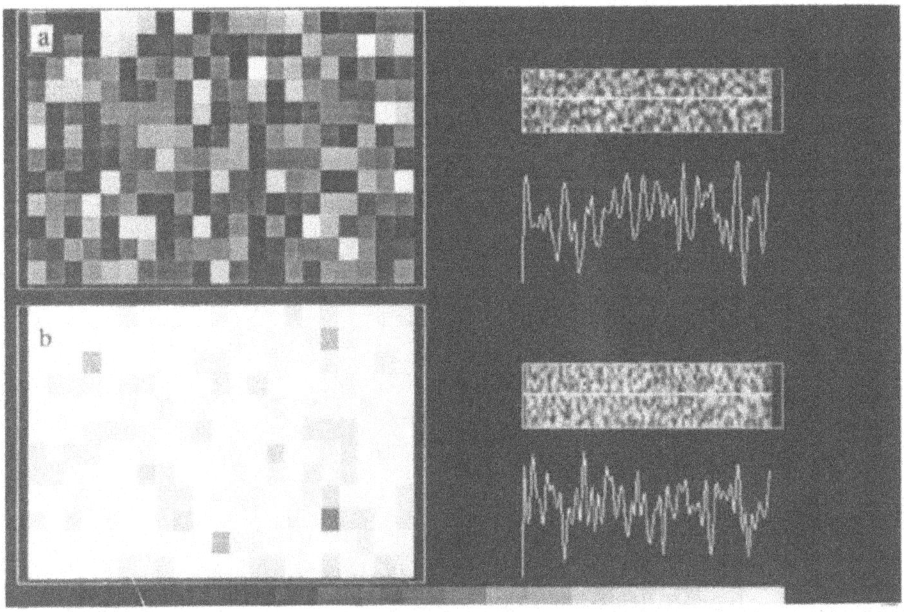

Figure 4. Spectral maps based on evaluation of the centroid's frequency via the AR modelling technique. Map a) represents a retinoblastoma having scatterers with diameters uniformly distributed between 20 and 100μm. Map b) represents a melanoma having scatterers with diameters uniformly distributed between 20 and 40 μm. Near the spectral maps the B-mode and the A-mode displays of the simulated pathologies are presented.

7. References

1. M.F.Insana,T.J.Hall *J. Acoust. Soc. Am.* **87**-1 (1990) 179.
2. A.T.Cheung,P.W.Buchen,C.Macaskill,D.E.Robinson,*J. Acoust. Soc. Am.* **86**-1 (1989) 407
3. F.L.Lizzi,M.Greenbaum,E.J.Feleppa,M.Elbaum, *J. Acoust. Soc. Am.* **73**-4 (1983) 1366.
4. F.L.Lizzi,M.Ostromogilisky,E.J.Feleppa,M.Rorke,M.M.Yaremko,*IEEE Trans. on UFFC* **33**-3 (1986) 319.
5. S.Kay,S.L.Marple, *IEEE Proc.* **69**-11 (1981) 1360.
6. M.Ueda, H.Ichikawa, *J. Acoust. Soc. Am.* **70**-6 (1981) 1768.
7. T.Wang,J.Saniie,X.Jin *IEEE Trans. on UFFC* **38**-2 (1991) 116.
8. E.A.Robinsons *IEEE Proc.* **70**-9 (1983) 885.
9. S.L.Marple *"Digital spectral anlysis with applications"* (Prentice Hall, Englewood Cliffs, N.J., 1987)

ACOUSTIC MICROSCOPY OF COMPOSITE MATERIALS

A. Kulik, J.-E. Bidaux, G. Gremaud
Ecole Polytechnique Fédérale de Lausanne
Institut de Génie Atomique
CH-1015 Lausanne, Switzerland

and

P. Bujard
Ciba-Geigy
Centre de recherche
CH-1701 Fribourg, Switzerland

ABSTRACT

Epoxy matrix composite material was studied using Scanning Acoustic Microscope working in GHz frequency range. Characteristic interference fringes were observed around silicate spheres. An interpretation is proposed, allowing one to calculate longitudinal wave velocity in the epoxy matrix and location of the subsurface spheres, from the spacing between fringes.

1. Introduction

Scanning Acoustic Microscopy (SAM) is often used for studying composite materials. Different acoustic properties give an excellent contrast and subsurface features can be also investigated. However SAM contrast has a interference nature. Different types of excited waves and different acoustic paths participate in image formation, leading in some cases to the difficulties in interpretation. We study appearance of the interference fringes in the expoxy matrix composite charged with small silicate spheres. The theory is proposed, allowing one to indirectly measure the diameter of the spheres and longitudinal wave velocity in epoxy matrix.

2. Model

Consider a sphere embedded in a acoustically different matrix and cutting or not the free surface at a distance **z** from its center (see Fig. 1). Interference fringes are produced which result from the interference of two rays: Ray A which strikes the surface normally and is reflected back to the lens and ray B which is refracted at the specimen surface and strikes the sphere normally being reflected back along its own path. Noteworthy, this problem is similar to that of interference fringes produced by a crack (Briggs, 1985).

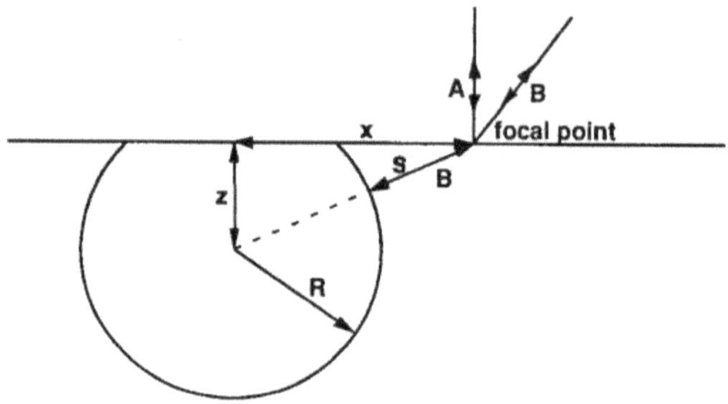

Fig.1. Schematic diagram of the rays generating interference fringes from a sphere embedded in a acoustically different matrix.

The phase difference between the two rays is:

$$\Delta\phi = 4\,\pi\, s\, \nu/c \tag{1}$$

where ν is the frequency, **c** the longitudinal acoustic waves velocity in the matrix and **s** the path length of ray B in the matrix.
s is given by a simple geometrical relation:

$$s = \sqrt{x^2 + z^2} - R \tag{2}$$

where **z** is the distance between the center of the sphere and the sample surface and **R** the radius of the sphere.
From the relation $\Delta\phi = 2\pi$ we can calculate the distance **Δx** between two consecutive fringes:

$$\Delta x = x_{i+1} - x_i = \sqrt{c/2\nu + \sqrt{x_i^2 + z^2} - z^2} - x_i \tag{3}$$

Δx is a decreasing function of $\mathbf{x_i}$. It tends to the limit $\Delta x_\infty = c/(2\nu)$ when $\mathbf{x_i}$ tends to the infinite which corresponds to half a wavelength of acoustic waves in the matrix. Therefore, from the distance between interference fringes far from the source we can, in principle, get direct information on the matrix sound velocity. In general, the exact position of the sphere with respect to the surface has to be taken into account.

3. Experimental results

The sample tested was an Araldite epoxy resin from Ciba-Geigy (density: 1.24) charged with spherical amorphous silicate (density: 2.2). The average diameter of the spheres was 30μm and the charge volume fraction 7.3%. The sample was mechanically polished and observed in a Leitz acoustic microscope. A typical micrograph is shown in Fig 2. Interference fringes are observed around the silicate spheres. These can be either subsurface spheres (not visible with an optical microscope) or truncated spheres (cut by the mechanical polishing). The most evident interference fringes are observed when the spheres are below but close to the surface. Only some of the truncated spheres are seen to produce interference fringes. This is in agreement with the model. It is clear that when the sphere center is above the surface no rays can be reflected back to the lens except those passing through the sphere.

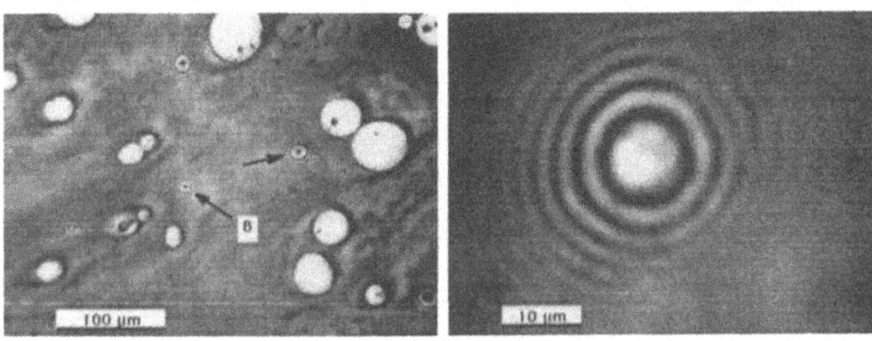

a) b)

Fig 2 Acoustic micrographs of interference fringes in Araldite epoxy resin charged with spherical amorphous silicate. Frequency: 0.8 GHz. Arrows indicate fringes generated by subsurface spheres.
a) image of a large area. B is the detail magnified in b)

The distance between consecutive fringes $\Delta x = x_{i+1} - x_i$ was measured as a function of the distance x_i to the center of the fringes (Fig. 3). Fringes are observed within a limited radius around the sphere. This radius is almost independent of the working frequency. Accordingly, the higher the frequency, the higher the number of the fringes that can be observed. This indicates low attenuation of acoustic waves in the resin. Accordingly, the limiting factor for observing high index fringes is most likely the opening angle of the lens. Note that the constant and small limit radius for observing the fringes makes it almost impossible to obtain quantitative data at low frequency ($\nu \leq 400$ MHz).

The full curves of Fig.3 correspond to best fits of the fringe spacing versus their distance from the center obtained using Eq.3. The adjustment parameters are the matrix sound velocity c and the distance z from the center of the sphere to the sample surface. As can be seen in Fig.3 good adjustments are obtained. Moreover, as shown in Table 1 the overall consistency of the velocity values is good, despite the fact that the fit was performed independently for each frequency,

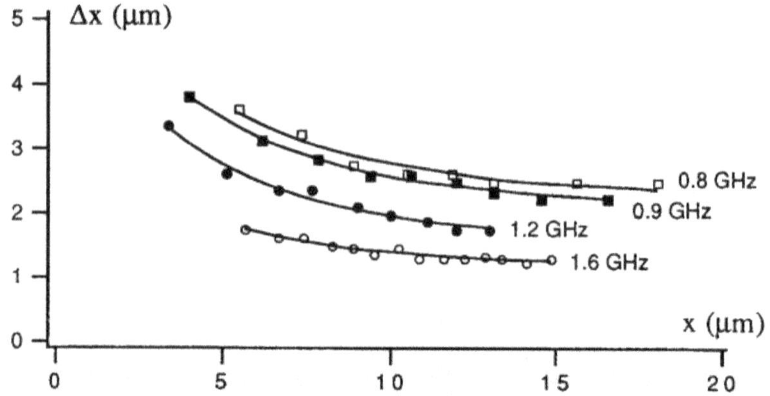

Fig 3 Fringe spacing Δx versus distance x to the center of the fringes measured for the subsurface amorphous silicate sphere of Fig 2.b.

Table 1

Frequency [GHz]	Sound velocity [m/s]	Distance [μm]
0.8	3345	9.6
0.9	3448	9.8
1.2	3227	11.0
1.6	3535	7.8

Neglecting the dispersion we can deduce a mean value for the sound velocity in Araldite epoxy resin in the GHz frequency range:

$$c = 3389 \text{ m/s}$$

This velocity is relatively high compared to the value measured by the pulse-echo method on the same specimen at 1 MHz:

$$c = 2480 \text{ m/s}$$

This discrepancy cannot be attributed to experimental scatter because measurements on other spheres (also truncated spheres) gave almost the same results. Besides, poor results have been obtained by simple fit of the experimental points for a fixed value of c equal to the value determined by pulsed-echo method. The discrepancy cannot either be explained by the different ratio between wave length and sample size in the two samples because in both cases the sample size was large with respect to the wave length. Now, it is possible that the reason for the large difference in velocity is wave dispersion. Actually, in our GHz data we have no indication of a frequency dependence of the sound velocity. However, a relaxation process occurring in the intermediate frequency range (100 MHz) cannot be ruled out.

4. Conclusions

Interference fringes are often observed in acoustic images. They contain the quantitative information. We were able to measure separately elastic properties of the epoxy matrix. Classical, bulk wave methods cannot be used in this case, since they allow one to measure only the mean value of the acoustic properties of the composite. Described method can be also helpful for interpretation of the SAM images.

5. References

[1] G. A. D. Briggs, *An Introduction to Scanning Acoustic Microscopy* (Oxford University Press, Oxford, 1985)

[2] J. M. R. Weaver, *Applications of The Scanning Reflection Acoustic Microscope to the Study of Materials Science, IEEE Trans. of Sonic and Ultrasonics* **SU-32** (1985) 302

[3] P. A. Tucker and R. G. Wilson, *Acoustic microscopy of polymers, J. Polym. Sci.* **18** (1980) 97

[4] K. Yamanaka and Y. Enomoto, *Observation of surface cracks with scanning acoustic microscope, J. Appl. Phys.* **53** (2) (1982) 846

[5] K. Yamanaka and Y. Enomoto, *Fringe pattern around surface crack observed with scanning acoustic microscopy, Electr. Lett.* **17/18** (1982) 2449

EFFICIENT MAGNETOSTRICTIVE EXCITATION OF BULK HYPERSONIC WAVES

A.S. Bugaev, V.B. Gorsky, Yu.V. Gulayev
Semiconductor Electronics Department of Moscow Institute of Physics and Technology, Dolgoprudny, Moscow, 141700, USSR

ABSTRACT

An efficient magnetostrictive excitation of bulk transverse acoustic wave (AW) in ferrite film at microwave frequencies is proposed. AW is excited in yttrium iron garnet film by dipol-exchange magnetostatic mode (MSM) and propagates perpendicular to the plane of the surface of the film.

High efficiency of th excitation is achieved by triple resonance dipol MSM-exchange MSM-AW interaction, i.e. under phase synchronism of exchange MSM and AW across the thickness of the film and under phase synchronism of exchange MSM and dipol MSM in the plane of the film. As a result AW is efficiently excited by microstrip microwave magnetic field.

The measured insertion losses of about 7 dB at 2-5,6 GHz are achieved. Theoretical estimations give the same value of insertion loss up to 20 GHz.

1. INTRODUCTION.

It is well know that magnetostriction can be used for acoustic wave (AW) excitation. Magnetostrictive AW excitation became possible in microwave frequency range after yttrium iron garnet (YIG) had been obtained. The characteristics feature of YIG is extremely low magnetic and elastic losses at microwave frequencies.

Natural oscillation of magnetization vector in bounded ferrite medium is know to be referred to as magnetostatic modes (MSM). This work presents theoretical and experimental results of AW excitation with the help of hybrid dipole-exchange MSM at 2-20GHz frequencies. AW was excited by MSM due to magnetostriction in YIG film and it propagated into gallium gadolinium garnet (GGG) substrate in the direction perpendicular to the plane of the film (Fig.1). AW wavelength was much less than the thickness of the film ($k_a d \ll 1$, k_a- AW wavenumber, d- film thickness).

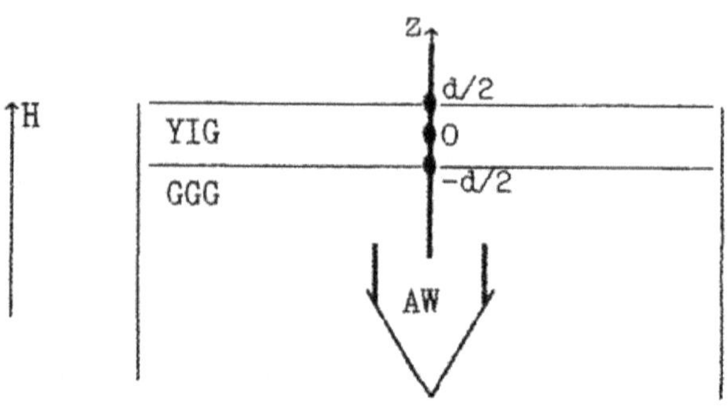

Fig.1. Geometry of the investigated structure. The arrow identifies the direction of AW propagation. H is dc magnetic field

The MSM spectrum in normally magnetized ferrite film is shown in Fig.2.[1] The spectrum consists of one dipole branch (the branch with the biggest incline angle) and a great amount of exchange branches different in rf magnetization distribution across the film thickness m(z). Dipole MSM is known to have quasi-homogeneous distribution of m(z) across the film thickness and its frequency is determined by dipole-dipole magnetic interaction energy. Exchange MSM has sinusoidal m(z) distribution and exchange MSM wavelength across the film thickness is so little that its frequency is determined by exchange interaction. It should be mentioned that exchange MSM's with q=0 are referred to as spin-wave resonance modes.

In the vicinity of dipole MSM branch and exchange MSM branches crossover the hybrid dipole-exchange MSM's are known to be excited.[2] The m(z) of hybrid MSM can be represented as the sum of dipole (subscript "d") and exchange (subscript "e") components.

$$m(z) = A_d \cos(k_d z) + A_e \cos(k_e z) \qquad (1)$$

A_d, A_e and k_d,k_e are amplitudes and wavenumbers of dipole and exchange components respectively; $k_d d \ll 1$, $k_e d \gg 1$. k_e roughly coincides with the wavenumber of exchange MSM interacting with dipole MSM. A_e/A_d ratio in m(z) depends on hybrid dipole-exchange MSM frequecy f & plane wavenumber q.

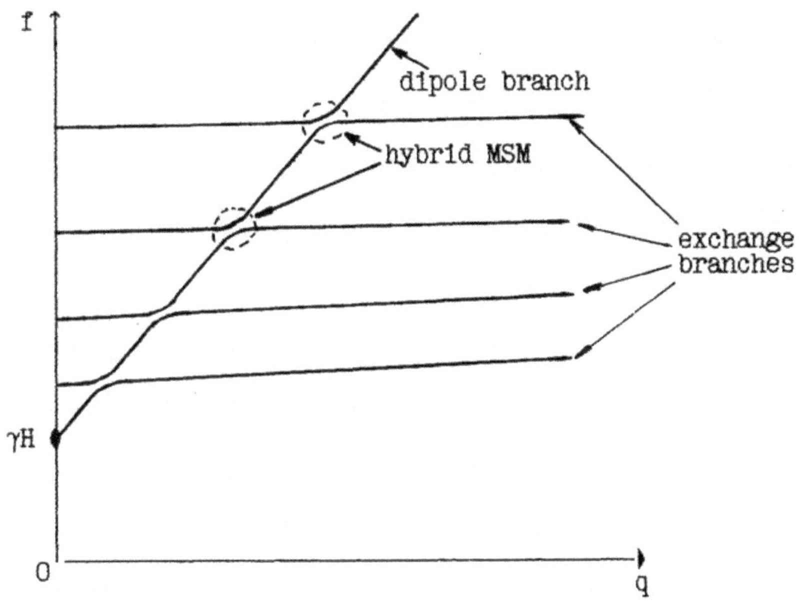

Fig.2. MSM spectrum in normally magnetized ferrite film. γ is gyromagnetic ratio. H is dc magnetic field q is wavenumber in the plane of the film.

2. THEORY AND EXPERIMENT.

Let's consider advantages and drawbacks of AW excitation with the help of dipole MSM, exchange MSM and hybrid dipole-exchange MSM. The AW excitation with MSM is schematically shown in Fig.3. Usually MSMs in ferrite films are excited by microwave magnetic field h created in ferrite film by microwave microstrip waveguide on the film surface. Dipole MSM (A_e=0 in Eq.1) is efficiently coupled with microstrip microwave magnetic field (since both h(z) and m(z) are quasi-homogeneous functions) but weakly excites AW (since $k_d \ll k_a$). Exchange MSM (A_d=0 in Eq.1) is efficiently coupled with AW under its phase synchronism across the film thickness ($k_e = k_a$) but it is weakly excited by microstrip microwave field (due to negligible value of h(z) & m(z) overlap integral). Thus AW excitation with the help of exchange MSM and dipole MSM is rather inefficient.

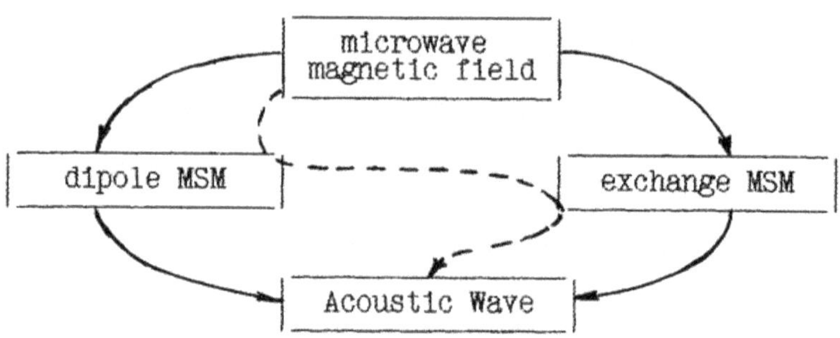

Fig.3.Schematic representation of main stages of the AW excitation by three types of magnetostatic modes: dipole MSM, exchange MSM, hybrid dipole-exchange MSM (dashed curve).

The efficiency proved to significantly increase if AW is excited by hybrid dipole-exchange MSM. In fact, dipole component of m(z) strongly couples hybrid MSM with microstrip microwave field; exchange component of m(z) efficiently excites AW (under phase synchronism condition ($k_e = k_a$) across the film thickness; while dipole MSM and exchange MSM strong coupling results from their phase synchronism in the plane of the film (hybrid MSM's are excited in dipole MSM & exchange MSM dispersion curves crossover (Fig.2)). Thus AW excitation by hybrid dipole-exchange MSM should be very efficient because under conditions of successive AW excitation: microwave magnetic field → dipole MSM → exchange MSM → AW the energy transformation is very efficient at all stages.

It should be noted that AW excitation efficiency is to depend on A_e/A_d ratio in hybrid MSM m(z) as this ratio influences MSM coupling with AW and microwave field.

Theoretical study was made for determining AW excitation efficiency dependence on frequency, ferrite film parameters, A_e/A_d ratio and for determining AW excitation optimal conditions.

We supposed magnetic spins pinning on the film surface so that m(d/2)=0 and accordingly $\delta u/\delta z(d/2)=0$, where u(z) is AW mechanical displacement. Thus u(z) can be represented in the following form

$$u(z) = \int_0^\infty u(k) \cos k(z-d/2)\, dk \qquad (2)$$

AW was considered to be excited by MSM with m(z) described by Eq.1. Considering equations of magnetization vector motion and elastic medium motion and taking into acount magnetostriction the following equation for AW amplitude U excited by hybrid MSM into the substrate was obtained:

$$U = \frac{\left[\dfrac{b_{44}M}{\pi c_{44}} d\right] \xi \varepsilon \left[1 - \varepsilon \dfrac{\sin \frac{k_e d}{2}}{\frac{k_e d}{2}}\right] \dfrac{M}{\Delta H} h}{1 + 2\varepsilon \dfrac{\sin \frac{k_e d}{2}}{\frac{k_e d}{2}} + \dfrac{\varepsilon^2}{2}\left[1 + \dfrac{\sin k_e d}{k_e d}\right]\left[1 + \eta \dfrac{M}{\Delta H} \dfrac{k_a d}{\pi} \xi^2\right]} \qquad (3)$$

where h is microwave magnetic field amplitude, M is saturation magnetization; b_{44}, C_{44} are matrix elements of magnetostriction and elastic tensors, $\eta = 4Mb_{44}^2/C_{44}^2$ is magnetostriction constant, ΔH is ferromagnetic resonance linewidth, ξ is determined by the following equation

$$\xi \cong \frac{\sin\left[\frac{k_e-k_a}{2}d\right]}{\left[\frac{k_e-k_a}{2}d\right]} \cos\left[\frac{k_e-k_a}{2}d\right] \qquad (4)$$

ε value denotes A_e/A_d ratio and is used as a calculation parameter.

Eq.3 is valid provided $k_d d \ll \varepsilon \ll k_e d$, $q/k_e \ll p$. From Eq.3 and Eq.4 it follows that the maximum efficiency of AW excitation is achieved on condition of: 1)coincidence of AW k_a and MSM exchange component k_e wavenumbers; 2)equality of the A_e/A_d ratio to its optimum value:

$$\left[A_e/A_d\right]_{opt} = \sqrt{2} \,/\, \sqrt{1 + \eta\, k_a d/\pi \;\; M/\Delta H} \qquad (5)$$

The optimal condition being satisfied, half the power of microwave oscillator is transferred into AW energy. Thus theoretically minimum losses to be achieved for transforming microwave oscillator energy into AW energy are 3dB. The maximum amplitude of AW to be excited by hybrid MSM is:

$$U_{max} = \frac{b_{44}M}{\pi\, c_{44}} d \; \frac{\sqrt{2}}{\sqrt{1 + \eta\, k_a d/\pi \;\; M/\Delta H}} \; \frac{M}{\Delta H} h \qquad (6)$$

Experimental (dotted line) and calculated by Eq.3 (solid line) $U(A_e/A_d)$ dependencies are shown in Fig.4. 10μm thick YIG film was used, ΔH=0.5 Oe, M=140 Oe, $\eta = 1.8*10^{-3}$. Microwave frequency was 4GHz. Points A_e/A_d=0 and $A_e/A_d \to \infty$ correspond to AW excitation with the help of dipole MSM and exchange MSM. AW was excited by hybrid MSM with k_e=k_a. Variation

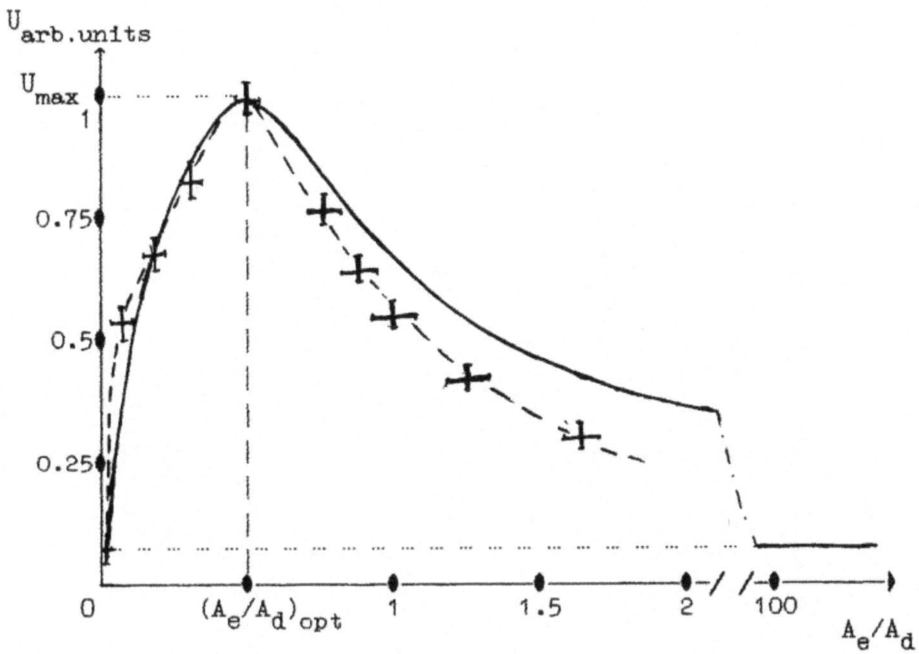

Fig.4. AW amplitude U dependence on A_e/A_d dipole & exchange component amplitude ratio in MSM exciting AW

of A_e/A_d ratio was achieved by changing q value (Fig.2). The impulse echo-method was used for determining AW excitation efficiency[3]. The microwave energy impulse excites MSM in YIG film. The same MSM was used for AW excitation and AW detection. The impulse length was less than the AW flight time across the film thickness.

According to theoretical predictions $U(A_e/A_d)$ experimental dependence is non-monotonous and there exists optimal $(A_e/A_d)_o$ value.

Under optimum condition insertion losses L for transforming microwave oscillator energy into AW energy were about 7dB. In 2 - 5,6 GHz frequency range insertion losses didn't exceed 8dB.

According to Eq.3 increasing the thickness of YIG films was to decrease the insertion losses L. It has been confirmed by reducing insertion losses when YIG film thickness has been increased from 4μm to 10μm. Taking into account that the AW power $P_a = \rho\omega^2 U^2$ the insertion losses are unlikely to increase at higher frequencies. Experimental insertion losses L frequency dependence shown in Fig.5 confirms this supposition. Thus AW can be excited at higher frequencies with the same losses of 7dB.

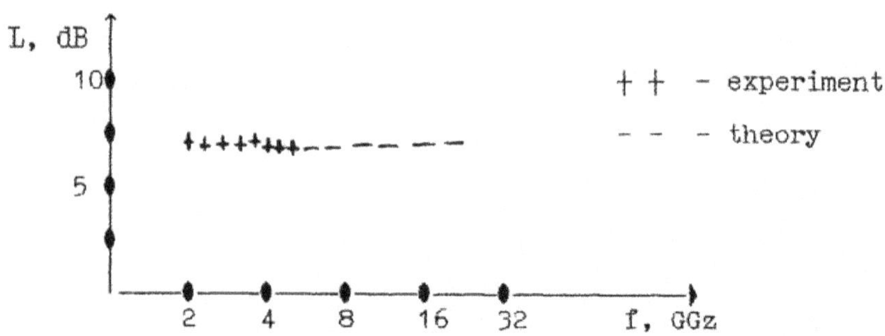

Fig.5. AW excitation losses L vs frequency.

The upper frequency fmax at which efficient AW excitation can be achieved using hybrid MSM is defined by the following equation:

$$f_{max} = V/2\pi \sqrt{q_{min} d / 4\alpha} \quad (7)$$

where α - exchange constant; V - AW group velocity; q_{min} - upper limit MSM wavenumber at which efficient MSM excitation can be obtained, using microstrip transducer. Indeed, MSM having too large wavenumber q in the plane of the film must be used to satisfy the optimum condition of AW excitation at higher frequencies by means of hybrid dipole-exchange MSM. MSM wavenumber q being increased, MSM excitation efficiency decreases, resulting in decreasing AW excitation efficiency. If we assume q_{min} to be 1.5 $10^2 cm^{-1}$ f_{max} is of the order of 20GHz for YIG film 40μm thick. Therefore efficient AW excitation at 20GHz frequency with 7dB insertion losses can be obtained using hybrid dipole-exchange MSM. Thus the efficiency of microwave AW excitation with hybrid MSM is much greater than that of other methods.

3. CONCLUSION.

Summing up, hybrid dipole-exchange magnetostatic modes can be used for acoustic wave excitation in microwave frequency range in selective microwave devices. The efficiency of acoustic wave excitation with the help of hybrid magnetostatic modes significantly exceeds efficiency obtained with exchange magnetostatic modes, dipole magnetostatic modes and efficiency of other acoustic wave excitation methods.

The acoustic wave excitation insertion losses at 2-5.6 GHz frequencies were 7dB. Calculation reveals that the same losses, i.e. 7dB can be obtained at 6-20 GHz frequencies.

REFERENCES.

1. T.Yukawa, K.Abe. *J.Appl. Phys.* V.45(1974). №7. 3146.
2. P.E.Zilberman, V.I.Kozlov, A.V.Pomyalov. *Sov.Phys.Solid state* 28(1986). 195.
3. R.C.LeCraw and R.L.Comstock, in: *Physical Acoustics: Principles and Methods* (ed. by W.P.Mason), Vol.3, Part B (Academic Press, New York, 1965) p.127.

NMR DETECTION OF ELASTIC WAVE EFFECTS

CZESLAW J. LEWA
Institute of Experimental Physics, University of Gdańsk, ul. Wita Stwosza 57, 80-952 Gdańsk, Poland

ABSTRACT

Suggestion is given of using an ultrasound field in order to periodically displace a sample to be analyzed in Nuclear Magnetic Resonance techniques, so as to detect viscoelastic properties through absorption spectra.

1. Introduction

It has been suggested previously [1-4] that elastic wave can be employed in the techniques of Magnetic Resonance Imaging (MRI) of the internal structure of large objects. The Nuclear Magnetic Resonance (NMR) response in the presence of a magnetic field gradient G has been analysed in terms of the imaging of viscoelastic properties [4]. This technique may find applications in medical diagnostics.

In the present report the NMR response in the longitudinal mechanical wave field is analysed having in view a detector for the elastic property imaging of large objects.

2. Elastic property imaging by MRI detection (EPMRI)

From the theory of NMR [5-7] the periodic change in the magnetic field intensity

$$B = B_0 + B_m \cos \Omega t \tag{1}$$

with angular frequency Ω results in the modulation of the Larmor precession frequencies of nuclear magnetic momenta. This is manifested in the NMR response by a spectrum consisting of lines with frequencies

$$\nu(0) \pm nV, \qquad n = 0, 1, 2... \tag{2}$$

where $\nu(0) = \gamma B$ is the Larmor precession frequency for a particular type of paramagnetic nuclei in an external field B, γ the gyromagnetic ratio divided by 2π, V the frequency of magnetic modulation.

The intensity of the n-th line is proportional to the square of the n-th component of Bessel's function, $J_n^2(\alpha)$, where α denotes the ratio $\alpha = \gamma B_m/V$.

As it has been demonstrated [2-4], for a time independent magnetic field with a constant G, the Larmor frequency can be modulated by varying the position of the observed sample element. This can be accomplished by a longitudinal mechanical wave with frequency V and acoustic pressure amplitude p_0. The displacement of the element

$$\xi = \xi_0 \cos(2\pi V - Kx) \tag{3}$$

whit $\xi_0 = \frac{1}{2} c_0 p_0 \Lambda$ the amplitude of the displacement, Λ and K the length and wavenumber of the mechanical wave respectively.

The modulation amplitude of the magnetic field intensity, B_m, affecting the nuclei contained in the displaced element, is

$$B_m = \vec{\xi_0} \cdot \vec{G} \tag{4}$$

(Fig.1)

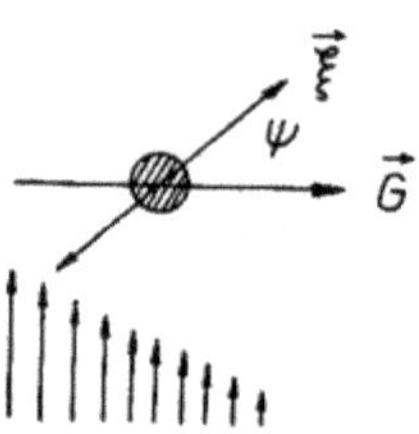

Fig.1 The displacement, $\vec{\xi}$, and the field gradient, $\vec{G}$, vectors.

The argument of Bessel's function can then be expressed by

$$\alpha = \frac{1}{V}\gamma \vec{\xi_0} \cdot \vec{G} = \frac{1}{2V}\gamma \beta \rho_0 \Lambda G \cos\psi \tag{5}$$

where ψ denotes the angle between the displacement $\vec{\xi}$ and the field gradient $\vec{G}$, and β is the coefficient of the adiabatic compressibility of the medium.

According to equation (5), the argument of Bessel's function depends on the properties of the medium, as well as on instrumental factors [4]. The existence of a simple relation between the properties of the medium and the instrumental factors, and the capability of determining the argument α of Bessel's functions by analyzing the NMR spectrum in the MRI method, make it possible to determine the distribution of the viscoelastic properties of samples by detecting the effects in the NMR response that are due to a controlled mechanical radiation distribution.

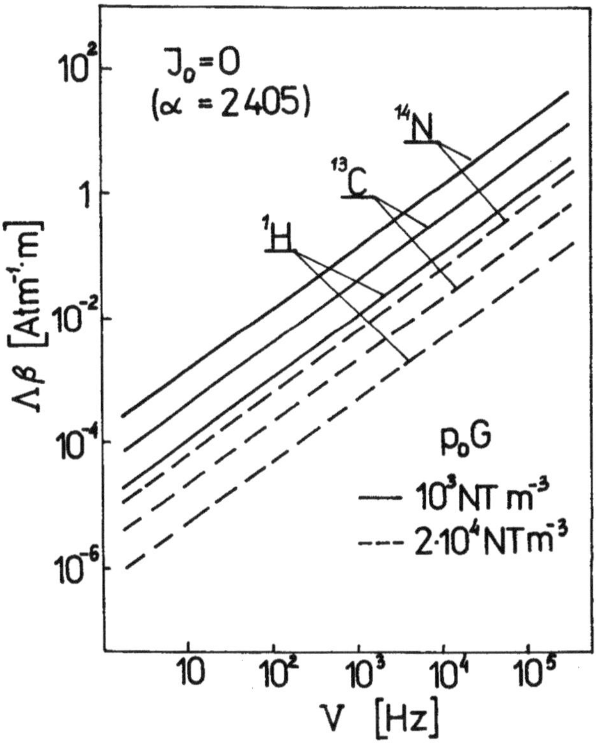

Fig.2 The product of the mechanical wavelength and adiabatic compressibility modulus, $\Lambda\beta$, as a function of frequency V for nuclei H, C, N and two values of p_0G, determined by setting the central line of the NMR spectrum to zero.

As an example, we report in Fig.2 the product of $\Lambda\beta$ as a function of mechanical wave frequency V for several types of nuclei and for two values of the product p_0G and $\psi = 0$, determined by setting the central line of the NMR spectrum equal to zero ($J_0^2 = 0; \alpha = 2.405$). If the viscoelastic properties of the medium are known, relation (5) can be used for mechanical wave field investigation. Using the MRI method, images of mechanical wave properties can be obtained.

3. References

1. C. G. Lewa, Bilateral Symposium DDR-Polen, Stolberg (1989)
2. C. G. Lewa, Spectroscopy Letters, 24, 55 (1991)
3. C. G. Lewa, Acustica - in press
4. C. G. Lewa, J. Phys. E.: Sci. Instrum. - in press
5. R. Karplus, Phys. Rev., 73, 1027 (1948)
6. I. Solomon and J. Ezratty, Phys. Rev., 127, 78 (1962)
7. J. Talpe, Theory of Experiments in paramagnetic Resonance, Pergamon Press, Oxford (1971)

Transducer Characterization with Steel Plate Targets

Christopher V. Kimball and Ralph M. D'Angelo

Schlumberger-Doll Research
Old Quarry Road
Ridgefield, CT 06877-4108

September 19, 1991

ABSTRACT

Characterization of a transducer in terms of a continuous plane wave spectrum is necessary to accurately predict pulse echo responses to plates. The kernel of the response, $\mathcal{W}(f,\theta)$, defined by Johnson and Devaney provides a convenient means for this characterization. The kernel for a Gaussian apodized disk transducer of radius, a, and standoff, S, is derived. A method of inverting responses from steel plates for (a, S) is described and is based on mean-squared error (MSE) fitting of model and measured waveforms. As implemented, the method is limited to frequencies where the transducer plane wave spectrum is concentrated below the critical angle of the water/steel interface; that is, where the transducer directivity is large, $|k|a > 20$. The method is applied to measured data from 10 steel plates of varying thickness. For the 7 plates with $|k|a > 20.0$, the normalized MSE errors between measured and model waveforms average -35.6 dB and the estimated transducer parameters agree with the actual parameters.

1. Introduction

Ultrasonic pulse-echo techniques often measure acoustic properties in geometries such as shown in Figure 1[1]. In this situation, a wideband ultrasonic pulse is reflected from the plate and the received waveform, such as shown in Figure 2, is processed to determine the impedance of the material behind the plate or other properties. The quality of these measurements depends significantly on characterization of the transducer. A theory for plate responses of transducers having a continuous plane wave spectrum is reviewed. The plane wave spectrum is described by the kernel of the response defined by Johnson and Devaney[2]. The kernel for a two-parameter transducer model is derived. These parameters correspond to the radius and standoff for a hypothetical Gaussian-apodized disk. Given the pulse-echo response of a plate, the model parameters are estimated by a MSE minimization between experimental data and the model. Laboratory data shows that these estimates closely match the parameters of the actual transducer.

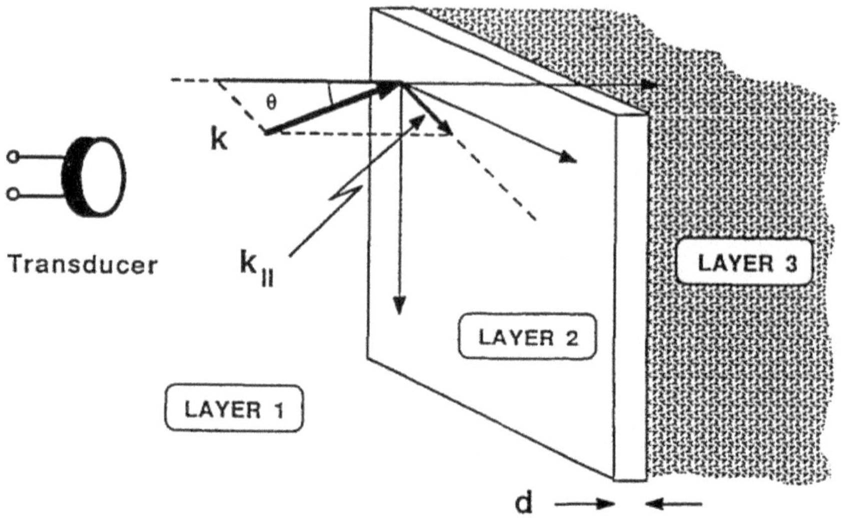

Figure 1: Pulse-echo measurement geometry.

2. Theory

2.1. *Plate Response Based on a Continuous Plane Wave Spectrum*

Johnson and Devaney[2] describe the Fourier transform $\mathcal{V}(f)$ of the pulse-echo response, $v(t)$, of a circularly symmetric, reciprocal transducer, oriented perpendicular to isotropic planar layers, $\mathcal{L}$, as

$$\mathcal{V}(f) = \int_0^{\frac{\pi}{2}} \mathcal{W}(f,\theta)\ \mathcal{R}(f,\theta \mid \mathcal{L})\ d\theta. \tag{1}$$

$\mathcal{R}(f,\theta \mid \mathcal{L})$ are the complex, planar reflection coefficients[3] for the layers, $\mathcal{L}$, at frequency f and incidence angle θ. $\mathcal{W}(f,\theta)$ is called the kernel and is given by

$$\mathcal{W}(f,\theta) = \frac{2\pi}{V_{fluid}}|k|^2 \sin\theta \cos^2\theta\ \ \mathcal{A}^2(f,\theta)\ \ e^{j2|k|S\cos\theta}\ \ \mathcal{T}(f) \tag{2}$$

where

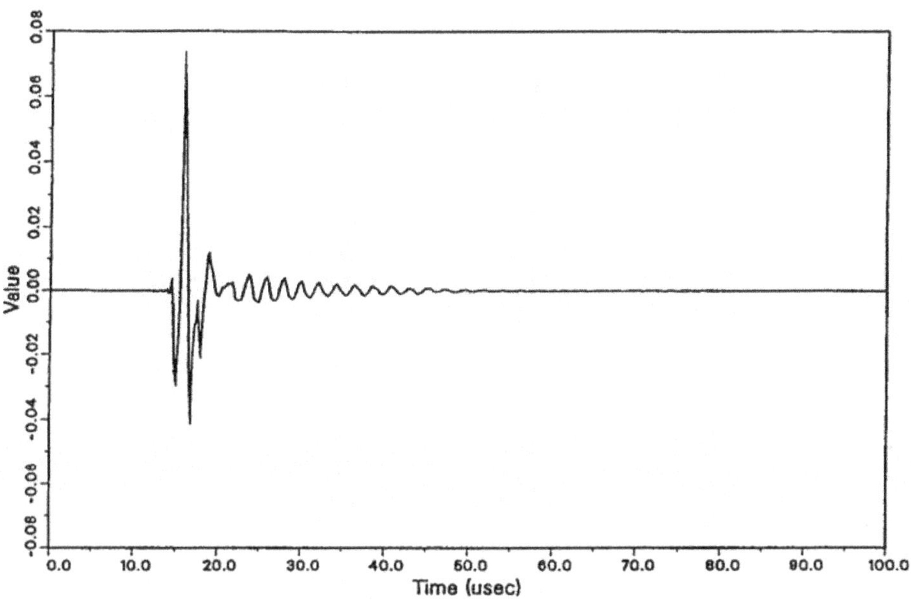

Figure 2: Measured time response to freeplate condition. d = 6.2 mm, S = 50.3 mm.

- V_{fluid}, $|k|$ are the velocity and wavenumber, respectively in the immersion fluid,
- $\mathcal{A}(f, \theta)$ is the output acoustic response function of the transducer,
- S is the distance from the transducer to the plate, called the standoff,
- $\mathcal{T}(f)$ is the spectrum of the electrical excitation.

This formulation is applicable for incidence angles, θ, up to 90°, but excludes secondary reflections and evanescent waves. $\mathcal{W}(f, \theta)$ can be considered the complex, two-way spectrum of the plane waves impinging on the plate.

Equation 1 might be expressed in terms of $k_{\parallel} = |k| \sin \theta$, the projection of the wavevector onto the plate. Incidence angle, θ, is used in this paper as the coordinate for two reasons. First, incidence angle is more intuitive to most readers than $k_{\parallel}$. Second, rapid means of calculating the planar reflection coefficients as a function of frequency for fixed incidence angle are available[4]. Because incidence angles considered here are small and $\sin x \sim x$ for small x, the differences between the two axes is approximately a scale factor.

If $\mathcal{R}(f, \theta \mid \mathcal{L})$ is known to correspond to the response of a plate of thickness d with water on both sides (the *freeplate* response), then the variations in the modal grooves of $\mathcal{R}(f, \theta \mid \mathcal{L})$ with θ near f_0 translate θ axis characteristics of $\mathcal{W}(f, \theta)$ into

$V(f)$. This is the physical basis for the characterization of the plane wave spectrum at a frequency f_0. Some readers may find a paradox in the fact that information for two dimensions (f, θ) can be extracted from a single (f or t) response; that is, without transducer movement. This is possible because of the special and known characteristics of $\mathcal{R}(f, \theta \mid \mathcal{L})$ for plates.

2.2. *Whitening*

The kernel, $\mathcal{W}(f, \theta)$, contains both the frequency and angular response of the transducer. Separation of $\mathcal{W}(f, \theta)$ into the product of the conventional frequency response multiplied by a whitened angular response is insightful. A convenient method of defining this separation is to consider the response to a very thick ("infinite") plate, the usual method of determining the frequency response of a transducer. Assuming that late reflections are truncated from the data, the planar reflection coefficients for a two layer, fluid/steel system can be used. These coefficients are independent of frequency and have no resonance grooves. A small θ dependence is present, however. Let $\mathcal{L}^\infty$ indicate this two layer situation. Then the spectrum of the response to this infinite plate, $\mathcal{V}^\infty(f)$, is given by

$$\mathcal{V}^\infty(f) = \int_0^{\frac{\pi}{2}} \mathcal{W}(f, \theta)\ \mathcal{R}(f, \theta \mid \mathcal{L}^\infty)\ d\theta. \tag{3}$$

Practically, $\mathcal{V}^\infty(f)$ is found from the measured response to a 140 mm thick steel block. The response is truncated in time to a 12 microsecond interval near the first reflection and Fourier transformed to yield $\mathcal{V}^\infty(f)$. In general, $\mathcal{V}^\infty(f)$ differs from the spectrum of the electrical excitation, $\mathcal{T}(f)$.

The kernel, $\mathcal{W}(f, \theta)$ can be factored into a product of the infinite plate response, $\mathcal{V}^\infty(f)$, and a whitened kernel, $\mathcal{W}^0(f, \theta)$. Define

$$\mathcal{W}^0(f, \theta) = \frac{\mathcal{W}(f, \theta)}{\mathcal{V}^\infty(f)} \tag{4}$$

$$= \frac{\sin\theta \cos^2\theta\ \ \mathcal{A}^2(f, \theta)\ \ e^{j2|k|S\cos\theta}}{\int_0^{\frac{\pi}{2}} \sin\theta \cos^2\theta \mathcal{A}^2(f, \theta) e^{j2|k|S\cos\theta} \mathcal{R}(f, \theta \mid \mathcal{L}^\infty)\ d\theta}. \tag{5}$$

Although $\mathcal{W}^0(f, \theta)$ has variations with frequency, variations common to both the infinite block and plate response such as $|k|^2, \mathcal{T}(f)$ are removed by whitening.

Suppose that $\mathcal{V}^\infty(f)$ has been determined for a transducer by measuring the response to an infinite plate. Then the response $V(f)$ to a plate can be whitened by dividing by $\mathcal{V}^\infty(f)$ to yield $\mathcal{V}^0(f)$:

$$\mathcal{V}^0(f) = \frac{\mathcal{V}(f)}{\mathcal{V}^\infty(f)} \tag{6}$$

$$= \int_0^{\frac{\pi}{2}} \mathcal{W}^0(f, \theta)\ \mathcal{R}(f, \theta \mid \mathcal{L})\ d\theta \tag{7}$$

from Equation 1. Thus if the infinite plate response of a transducer is known, a whitened response can be constructed which depends only on the whitened kernel.

Measured responses will always be whitened in subsequent estimation procedures so that the measurement of the kernel is for the whitened kernel $\mathcal{W}^0(f,\theta)$.

2.3. *Estimation by Fitting*

Estimation of $\mathcal{W}^0(f,\theta)$ is accomplished by fitting model waveforms to whitened data. Fitting repeatedly compares the model to the data, adjusting the model parameters until an error metric between the model and data is minimized. Then the model parameters are taken as estimates of the actual parameters for the data. Fitting has many recognized drawbacks. First, the fact that a certain parameter set minimizes the error doesn't necessarily indicate that the data has the same parameters. However, if the error metric is well-chosen, then an assertion can be made that the data is "equivalent" (in the sense of the error metric) to data that has the minimizing parameters. Second, local, rather than global minima of the error are the source of implementation problems. Ideally the error metric and model are selected so that only a single, global minimum exists for all plausible model parameters. The choice of the error metric and model given below have been made with these problems in mind.

2.3.1. *Error Metric*

The error metric is the conventional mean-squared error (MSE) metric. The signals are placed on a normalized basis before the MSE calculation to remove extraneous signal properties. This normalization, particularly the bandpass filtering, reduces the dimensionality of the signals and facilitates the fitting.

Since the resonance effects of interest are narrowband and have a center frequency, $f_0 = \frac{V_{plate}}{2\,d}$, which is easily estimated by apriori knowledge of thickness, the responses are compared in a band determined from the assumed plate thickness d. The filter $G(f,d)$ is given by

$$G(f,\ d) = \begin{cases} \dfrac{1}{\sqrt{2\pi}Bf_0}e^{-[(f-f_0)/(2Bf_0)]^2}, & \text{if } f > 0; \\ \dfrac{1}{\sqrt{2\pi}Bf_0}e^{-[(f+f_0)/(2Bf_0)]^2}, & \text{if } f < 0. \end{cases} \tag{8}$$

where Bf_0 is the σ-bandwidth of the filter.. This filter is applied to both waveforms. The gain of the signal or its delay do not effect the shape of the whitened kernel and consequently the filtered waveforms are normalized by setting the total energy to a constant $E^* = 1.56 \times 10^{-6}$,

$$\int_{-\infty}^{+\infty} |\mathrm{v}(t,d)|^2\, dt = E^* \tag{9}$$

and by shifting the maximum to $t^* = 31.25$ microseconds,

$$\max_t\{\ |\mathrm{v}(t,d)|\ \} = |\mathrm{v}(t^*,d)|. \tag{10}$$

These normalizations eliminate unnecessary factors and provide a standardized basis for viewing waveforms. All signals in this paper are normalized to satisfy equations 9 and 10 ; they are denoted in boldface.

Let $\mathbf{v}_1(t,d)$ and $\mathbf{v}_2(t,d)$ be two waveforms normalized as indicated above, then the error, $\epsilon(d)$, between $\mathbf{v}_1(t,d)$ and $\mathbf{v}_2(t,d)$ is defined as

$$\epsilon(d) = 10.0 \log \left[\min_{\pm\tau} \left\{ \int_{-\infty}^{+\infty} |\mathbf{v}_1(t,\ d) - \mathbf{v}_2(t+\tau,\ d)|^2 \, dt \right\} \right]. \tag{11}$$

This is the usual MSE or $\mathcal{L}_2$ metric with a minimization over time shifts, τ, to eliminate the effects small time misalignments. ϵ is expressed in decibels (dB). Large negative values of ϵ indicate good fits between $\mathbf{v}_1(t,d)$ and $\mathbf{v}_2(t,d)$; for independent noise inputs ϵ approaches +3 dB.

2.3.2. Models

The kernel of a measured response is characterized by that for a Gaussian-apodized disk. Because the ϵ is calculated in a narrowband, a computationally convenient bandpass approximation of $\mathcal{W}^0_{disk}(f,\theta)$ over frequencies near f_0 is developed.

2.4. Wideband Model

The acoustic output response for a Gaussian-apodized disk of radius $\mathbf{a}$ is given by :

$$\mathcal{A}_{disk}(f,\ \theta) = e^{-\frac{(|k|\ \sin\theta\ \mathbf{a})^2}{2}}. \tag{12}$$

The corresponding kernel for a transducer at standoff $\mathbf{S}$ from Equation 2 is

$$\mathcal{W}_{disk}(f,\ \theta) = \frac{2\pi}{V_{fluid}} |k|^2 \sin\theta \cos^2\theta \ e^{-(|k|\ \sin\theta\ \mathbf{a})^2} e^{j2|k|\mathbf{S}\cos\theta} \ \ \mathcal{T}(f) \tag{13}$$

Note that $\mathbf{a}$, and $\mathbf{S}$ independently determine the magnitude and phase, respectively, of the kernel. Thus the possibility of two different pairs ($\mathbf{a}$, $\mathbf{S}$) yielding the same kernel is eliminated. Further, the phase term is monotonically increasing with S. Both of these characteristics reduce the possibility of local minima in the fitting. The acoustic output response given by Equation 12 is real; minor adjustments to the phase in Equation 13 can be made by varying S. Hence S is called the "effective" standoff and may, in fitting, differ from the actual standoff, S_{actual}.

An alternative model for the kernel might be based on a circular piston[2]. The acoustic output response for a piston of radius a_{piston} is given by :

$$\mathcal{A}_{piston}(f,\theta) = 2.0 \ \frac{J_1(\ k\ \sin\theta\ a_{piston}\ )}{k\ \sin\theta\ a_{piston}} \tag{14}$$

where $J_1(x)$ is the Bessel function of the first kind and the factor of 2.0 allows matching of Equations 12 and 14 at $\theta = 0$. At first, this amplitude distribution appears more natural than that of the Gaussian apodized disk. Figure 3 plots $\mathcal{A}_{disk}$ and $\mathcal{A}_{piston}$ as a function of θ. For any fixed pair (f, θ), the value of $\mathcal{A}_{disk}$ is monotone-decreasing with a; that is, the magnitude of the plane wave always decreases as a increases. Because of its sidelobes, $\mathcal{A}_{piston}$ under the same conditions doesn't change monotonically with a_{piston}. Hence a possibility exists of local minima in a fitting of $\mathcal{A}_{piston}$ to the data which is avoided by the choice of an apodized disk.

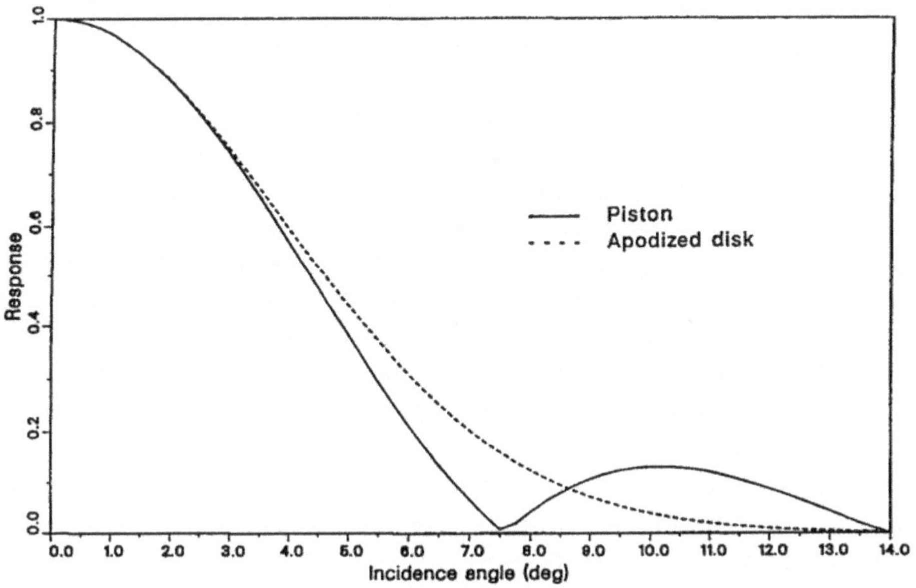

Figure 3: Comparison of one-way acoustic output response for a piston (a_{piston} = 14.3 mm) and a Gaussian apodized disk (a =7.2 mm).

Placing Equation 13 into Equation 7 gives the following result for the whitened, wideband model waveform

$$\mathcal{V}^{0}_{wideband}(f) = \int_{0}^{\frac{\pi}{2}} \mathcal{W}^{0}_{disk}(f, \theta)\ \mathcal{R}(f, \theta \mid \mathcal{L})\ d\theta \tag{15}$$

$\mathcal{W}^{0}_{disk}(f, \theta)$ is obtained from Equations 4 and 13. Numerical calculation of a wideband model waveform over a discrete grid of N_f frequency points by N_θ incidence angle points requires $N_f \times N_\theta$ evaluations of $\mathcal{W}^{0}_{disk}(f, \theta)$. If the calculation is done to fit the (a, S) pair to a specified layer structure, such as a freeplate condition, the $\mathcal{R}(f, \theta \mid \mathcal{L})$ matrix need be calculated only once, but $N_f \times N_\theta$ calculations are required at each iteration for $\mathcal{W}^{0}_{disk}(f, \theta)$. In this paper the frequency grid has 128 points and the default incidence angle grid has 49 points, so that 6272 calculations of $\mathcal{W}^{0}_{disk}(f, \theta)$ would be required for each model evaluation.

2.5. Bandpass Model

For the apodized disk, the frequency variations in plane wave magnitude are small over narrowbands. That is,

$$|\mathcal{W}^0_{disk}(f,\ \theta)| \sim |\mathcal{W}^0_{disk}(f_0,\ \theta)| \qquad \text{if f} \sim \text{f}_0. \tag{16}$$

Similarly, arg $\{\mathcal{W}^0_{disk}(f_0 + df,\ \theta)\}$ can be split into phase and group delay factors :

$$\arg\ \{\mathcal{W}^0_{disk}(f_0 + df,\ \theta)\} = [2k_0\text{S}(1-\cos\theta)] + [2\frac{2\pi\ df\ \text{S}}{V_{fluid}}(1-\cos\theta)]) \tag{17}$$

where $k_0 = \frac{2\pi f_0}{V_{fluid}}$. The first factor is the phase delay, the second factor is the group delay in the band centered at f_0. The phase and group delays have been defined so that both are zero at normal incidence, $\theta = 0$.

The group delay relative to normal incidence for a wave propagating at a 10 degree incidence angle from a source at a standoff of 50.3 mm is 0.52 microseconds. The duration of the impulse response of the narrowband filter of Equation 8 is greater than the period of the center frequency by a factor of 8. For d = 6.2 mm, the duration of the impulse response, $g(t, d)$ is 16.7 microseconds. Thus the group delay has a minor effect on the waveshape of the bandpass waveforms in the comparison. Let the group delay term be approximated by the delay corresponding to the actual standoff S_{actual}.

$$\arg\ \{\mathcal{W}^0_{disk}(f_0 + df,\ \theta)\} \sim [2k_0\text{S}(1-\cos\theta)] + [2\frac{2\pi\ df\ S_{actual}}{V_{fluid}}(1-\cos\theta)] \tag{18}$$

Then the bandpass approximation for the response of an apodized disk can be written as :

$$\mathcal{V}^0_{bp}(f,\ d) = \int_0^{\frac{\pi}{2}} \mathcal{W}^0_{bp}(\theta,\ d)\ \mathcal{R}_{bp}(f, \theta \mid \mathcal{L})\ d\theta \tag{19}$$

where

$$\mathcal{W}^0_{bp}(\theta,\ d) = C \sin\theta \cos^2\theta\ e^{-|k_0|^2 \sin^2\theta\ \text{a}^2} e^{j2k_0\text{S}(1-\cos\theta)} \tag{20}$$

$$\mathcal{R}_{bp}(f,\ \theta \mid \mathcal{L}) = \begin{cases} e^{j2\frac{2\pi\ (f-f_0)\ S_{actual}}{V_{fluid}}(1-\cos\theta)}\ \mathcal{R}(f, \theta \mid \mathcal{L}) & \text{if } f > 0; \\ e^{j2\frac{2\pi\ (f+f_0)\ S_{actual}}{V_{fluid}}(1-\cos\theta)}\ \mathcal{R}(f, \theta \mid \mathcal{L}) & \text{if } f < 0. \end{cases} \tag{21}$$

Here C is a complex, frequency-dependent scale factor. Since response energies are normalized and constant phase factors are removed in the MSE metric, C can be taken as 1 without effect.

Because the above approximation eliminates the frequency dependence of $\mathcal{W}^0_{disk}(f, \theta)$, fitting a new pair (a, S) to data requires calculation of only N_θ values (49) of $\mathcal{W}^0_{bp}(\theta, d)$ for the bandpass model, versus $N_f \times N_\theta$ (6272) for the wideband model. In net, fitting with the bandpass model is about 6 times faster than the wideband model. Only results based on the bandpass model will be presented.

2.5.1. Search Algorithm

Consider the estimation of the parameters (d, a, S) for a waveform $v(t)$ taken under freeplate conditions, that is, water on both sides of the plate. The outer loop is a search over thicknesses and is performed automatically; the result of a coarse-step search initializes the center of the range for a fine-step search. Gradient techniques are not used in the thickness search.

For each thickness d, the measured waveform is filtered by $G(f,d)$ and the $\mathcal{R}_{bp}(f,\ \theta \mid \mathcal{L})$ matrix is calculated. A constrained search routine from a standard package[5] then repeatedly constructs $v_{bp}(t,d) \leftrightarrow \mathcal{V}^0_{bp}(f,d)$, calculates ϵ, and adjusts (a, S) so as to reduce $\epsilon(d)$. Values of (a, S) from the previous thickness initialize the search. Typically 20 evaluations of the model are required at each thickness. Search results are examined manually to assure that the estimated values are not at the search constraints and that the time shift in the calculation of $\epsilon(d)$ is not at the limit.

3. Kernel Measurement

Laboratory data was acquired and processed to evaluate the estimation technique. The transducer was a lead metaniobate disk of diameter 28.5 mm (1.123 inch) of thickness 3.16 mm, with electrodes completely covering both circular surfaces. One side of the disk is welded to a cylindrical, high impedance, high loss backing of length 47.9 mm. This is an extremely simple transducer; the piston or Gaussian apodized disk models would be expected to predict its response very well.

In addition to an "infinite" plate of thickness 140 mm, 10 square plates with 305 mm sides ranging in thickness from 4.5 to 15.0 mm were used as reflectors in a water bath. To avoid differences in steel characteristics, the plates were cut from the same block and then machined to the desired thicknesses. The plates are smooth and parallel to at least 0.3 mm. For all measurements the actual standoff, S_{actual}, was carefully set to 50.3 mm. Care was taken to maintain the transducer oriented perpendicular to each plate. Waveforms were acquired with the transducer positioned over the center of the plates. The pulse-echo signal was generated and received with a Panametrics 5052UA pulser which excited the transducer with a -300 Volt impulse. The amplifier gain was 40 dB and did not require changing in the course of the experiments. The Panametrics 5052UA high pass filter was set at 30 kHz and an external low pass filter at 2.0 MHz preceded the 10 MHz sampling. To eliminate secondary reflections, the digitized waveforms were truncated approximately 60 microseconds after the peak of the initial pulse. Figure 2 indicates the response to the 6.2 mm plate.

The bandpass model was implemented over a discrete grid in (f, θ). The complex, bandpass signal was represented by 128 points centered on the demodulation frequency, f_0. The incidence angle grid was taken to be 0...14.4° [0.3°], approaching, but not exceeding, the critical angle of the water/steel interface, $\theta_c = 14.45°$.

The thickness search was first done for thicknesses within ±2.0% of the nominal thickness in steps of 0.2%. Initial values for a of 8.0 mm and S of 50.0 mm were used The best fitting thickness from the first search provided the center of a ±0.2% thickness range searched in a 0.02% step. At each thickness, the values of a were constrained to be in the range 0 .. 30 mm and the values of S were constrained to be in the range -50 ... 75 mm.

The kernel estimation technique was applied to the 10 measured waveforms. Although the model was intended to reduce the possibility of local minima in the MSE, local minima within 3 dB of the global minimum were found in the first thickness search for 5 of the 10 waveforms. All local minima were at least 1 dB greater than the global minimum. For example, Figure 4 depicts ϵ as a function of thickness for the waveform of Figure 2. A local minimum at 6.23 mm is observed in addition to the global minimum at 6.19 mm. The local minimum has an error value 1.6 dB greater than the global minimum. This demonstrates the necessity of a fine-grain, first thickness search and the significance of small differences in the MSE.

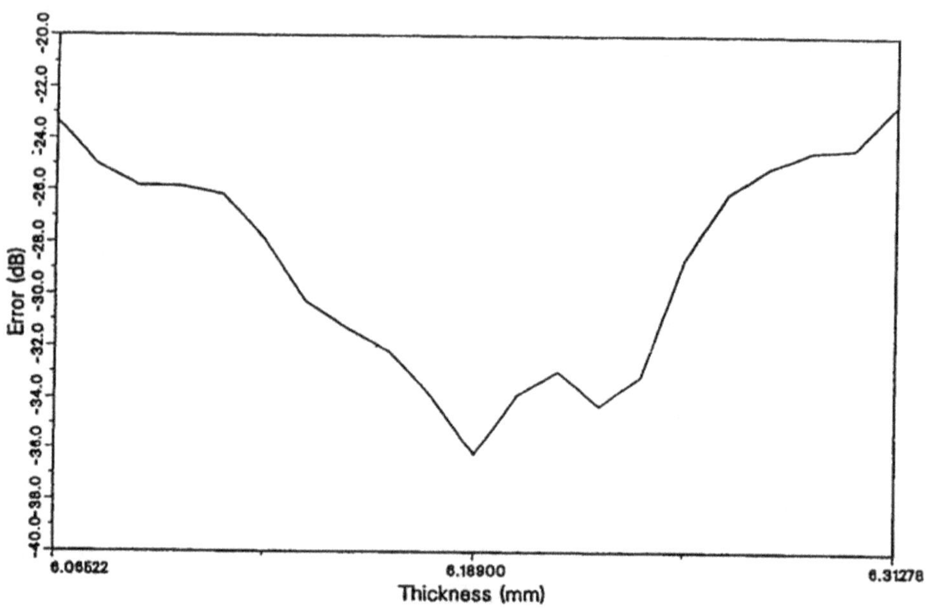

Figure 4: ϵ versus thickness for the fitting of the waveform of Figure 2.

The estimates for (a, S), and ϵ as a function of frequency are shown in Figures 6 and 5. The results for the three thickest samples appear anomalous; they shift greatly between 300 and 350 kHz. ϵ is also somewhat higher for these samples. The reason for these anomalies at low frequency is believed to be due to the limited range of incidence angles in the present implementation of the model. At these frequencies the directivity, $|k|a_{piston}$, of the transducer is less than 20.0 and significant excitation of plane waves at angles beyond θ_c occurs. Although the theory is not limited to incidence angles below θ_c, the present implementation for the planar reflection coefficients is limited to $\theta < \theta_c$. Consequently, only the data above 300 kHz (d > 9 mm) will be considered further and the restriction that $|k|a_{piston} > 20.0$ must be noted.

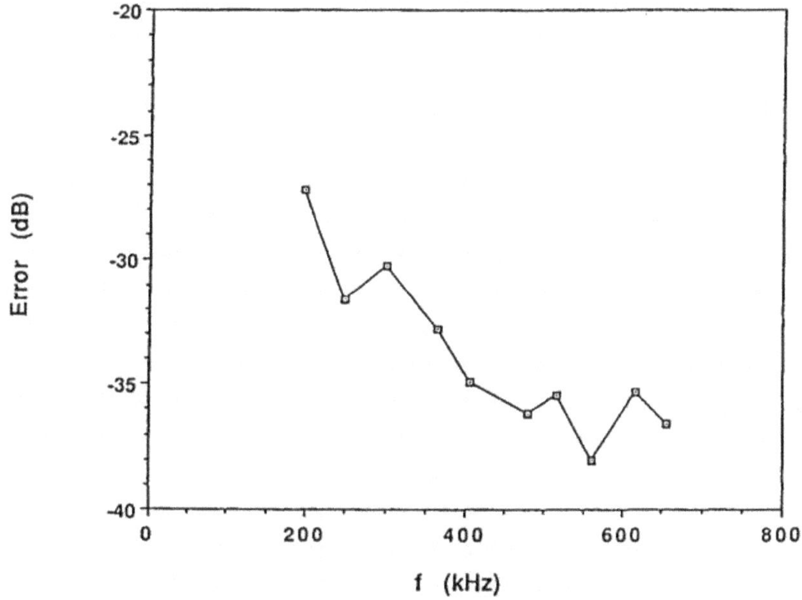

Figure 5: ϵ versus frequency for measured data.

If the apodized disk model were exact, the transducer parameters, (a, S), wouldn't vary as a function of frequency as evident in Figure 5. Two subsidiary studies where made to determine the source of these variations.

The possibility of an additive, waveform-to-waveform noise as a source of variance was considered. Ten waveforms were acquired at slightly different positions on the 6.2 mm plate. The error between 15 different waveform pairs was calculated. The average error for the 15 comparisons was -45.1 dB with a standard deviation of 2.8 dB, suggesting that the repeatability limit for the measurements is about -45.1 dB. This is below the errors of Figure 6 and hence the variations in the results are probably not due to additive noise.

Estimates of (a, S) were made for each of 10 waveforms taken at slightly different positions on the 6.2 mm plate. The standard deviations for (a, S) were 0.06 mm and 0.66 mm respectively. The standard deviation of the error was 0.5 dB. These standard deviations are smaller than those of Figure 6 and thus the fitting process is not the source of the variations.

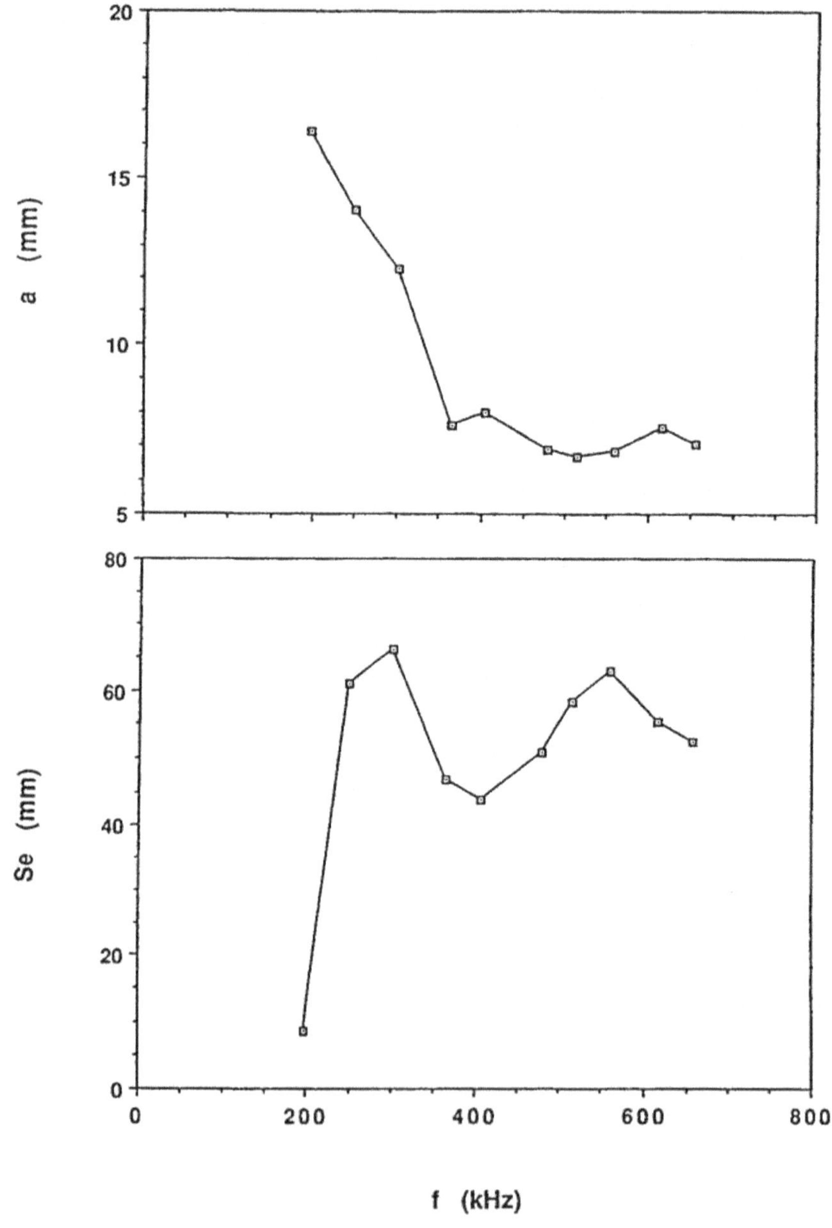

Figure 6: (a, S) versus frequency for measured data.

Thus the observed variations with frequency appear to be due to actual variations between the apodized disk model and the measured response. The variations are regarded here as a valid part of the transducer characterization. That is, the curves of (**a**, **S**) versus frequency in Figure 5 are more descriptive of the transducer than, for instance, the mean values computed over Figure 5.

Figure 3 compares the acoustic output response of a piston transducer having a radius equal to that of the laboratory transducer (a_{piston} = 14.3 mm) with that of a Gaussian apodized disk having a radius, a = 7.2 mm; the mean value of **a** in Figure 5. The agreement is very good and suggests that the mean estimated value for **a** does correspond to the transducer size. The mean estimated value for **S**, 53.0 mm, is close to the actual standoff, S_{actual} = 50.3 mm. Hence the technique provides a method of estimating transducer size and positioning from a single freeplate waveform.

4. Conclusions

The Gaussian apodized disk model characterizes the kernel, $\mathcal{W}(f,\theta)$, of a transducer in terms of its radius **a** and effective standoff, S. These parameters can be estimated from a plate response via a MSE fitting procedure.

For laboratory data taken with a disk transducer, the estimation yields fits to plate responses having MSEs of -35.6 dB for plates having resonant frequencies greater than 300 kHz, $|k|a_{piston} > 20.0$ (d < 9 mm). Results at or below 300 kHz are anomalous, probably due to the limited range of incidence angles in the present implementation of the model. The mean estimated transducer parameters agree with those of the actual transducer.

The proposed technique provides a convenient, single waveform method of characterizing the plane wave spectrum of a transducer and supplements present methods for the design and quality control of ultrasonic transducers.

5. References

[1] Havira, R. M., Ultrasonic techniques in oil well logging, *Proc. IEEE Ultrasonics Symp.*, (November 1986) 563–571.

[2] Johnson, R. K., and A. J. Devaney, Transducer effects in acoustic scattering measurements, *Applied Physics Letters*, **41** (1982) 622–624.

[3] Brekhovskikh, L. M., *Waves in Layered Media* (Academic Press, New York, 1980).

[4] Chang, H. W . Private communication, PREFPL Reflection Coefficient Routine (3 May 1989).

[5] IMSL Math/Library User's Manual : DBCONF, Volume 3, Chapter 8 (IMSL, Houston, Texas, 1987) p. 847.

ACOUSTIC SENSING AND PROBING

FOURTH COURSE OF THE

INTERNATIONAL SCHOOL ON PHYSICAL ACOUSTICS

Ettore Majorana Centre for Scientific Culture - Erice (Trapani), Italy

3-10 October 1991

LIST OF PARTICIPANTS

Laszlo Adler
Ohio State University
190 West 19th Avenue
Columbus Ohio 43210
USA

Adriano Alippi
Università di Roma 'La Sapienza'
Dipartimento di Energetica
Via A. Scarpa 4
00164 Roma, ITALY

Vladimir Anisimkin
Academy of Sciences
Institute of Radioengineering and Electronics
18 Marx Avenue
Moscow 103907
USSR

Walter Arnold
Universität Gebäude 37
Fraunhofer Institut IfzP
W-6600 Saarbrücken 11
GERMANY

Andrea Bettucci
CNR - Istituto di Acustica
Via Cassia 1216
00189 Roma, ITALY

John Blakey
University of Surrey
Physics Department
Guildford Surrey GU2 5XH
U.K.

Isabelle Bohnke
Lab. Genie Industriel et Biotechnologie
Ecole des Mines
Cours Fauriel 158
42023 St Etienne Cedex 2
FRANCE

Mack A. Breazeale
The University of Mississippi
The National Centre for Physical Acoustics
MS 38677
USA

Edmund H. Brown
University of Colorado
Boulder Colorado 80302
USA

Voichita Bucur
Centre National de Recherches Forestieres
Nancy - Amance
Champenoux 54280 Seichamps
FRANCE

Alexander Bugaev
Moscow Inst. Physics and Technology
Institutsky per 9
Dolgoprudny, Moscow region
141700 USSR

Cleofe Campos-Pozuelo
Centro de Fisica Aplicada
Instituto de Acustica
Serrano 144
28006 Madrid
SPAIN

Lorenzo Capineri
Dip. Ingegneria Elettronica
Via S. Marta 3
50100 Firenze, ITALY

Federico Cernuschi
ENEL/CRTN
Via Monfalcone 15
20132 Milano, ITALY

Robert C. Chivers
University of Surrey
Department of Physics
Guildford ENGLAND

Floriana Craciun
CNR - Istituto di Acustica
Via Cassia 1216
00189 Roma, ITALY

Bernard Cretin
C.N.R.S. - LPMO
32 Avenue de l'Observatoire
25000 Besançon
FRANCE

Carl Desmet
Dept. of Physics
Lab. Akoestek en warmtegeleiding
K.U. Leuven
Celestijnenlaan 200DB-3001
Leuven BELGIUM

Helge E. Engan
University of Trondheim
Norw. Institute of Technology
N-7034 Trondheim
NTH NORWAY

Ada Fort
Dip. Ingegneria Elettronica
Via S. Marta 3
50100 Firenze, ITALY

Italo Gabrielli
Università di Trieste
Dip. Fisica
Via Valerio 2
34127 Trieste, ITALY

Itziar Gonzales Gomez
Centro de Fisica Aplicada
Instituto de Acustica
Serrano 144
28006 Madrid
SPAIN

Tomas E. Gomez Alvarez-Arenas
Centro de Fisica Aplicada
Instituto de Acustica
Serrano 144
28006 Madrid
SPAIN

Yuri V. Gulyaev
Academy of Sciences
Institute Radioengin.& Electronics
18 Marx Avenue
Moscow 103907
USSR

Matt Haller
Stanford University
Edward L. Ginzton Laboratory
Stanford California 94305-4090
USA

Hugh W. Jones
Hugh W. Jones & Associated Ltd.
374 Viewmount Dr Allen Heights
Tantallon, Nova Scotia B0J 310
CANADA

Ernst R. Jorzik
CCR EURATOM
21020 Ispra (VA)
ITALY

Pierre Khuri-Yakub
Stanford University
Edward L. Ginzton Laboratory
Stanford California 94305-4090
USA

Siran Kikkarin
USSR Academy of Sciences
Institute Semiconductor Physics
Lavrentjev ave. 13
630090 Novosibirsk
USSR

Christopher V. Kimball
Schlumberger Doll Research
Old Quarry Road
Ridgenfield CT 06896
USA

Yuri A. Kosevich
Research Centre
All - Union Surface and Vacuum
Moscow 117334
USSR

Andreas Kulik
Institut de Genie Atomique
Dpt.de Physique
EPF-L Ecole Polytechn. Fed. Lausanne
PHB-Ecublens 1015 Lausanne
SWITZERLAND

Werner Lauterborn
Technical University Darmstadt
Institut fur Angewandte Physik
Schlossgartenstrasse 7
W-6100 Darmstadt
GERMANY

Oswald Leroy
Campus Kortrijk K.U. Leuven
E. Sabbelaan B - 8500
Kortrijk BELGIUM

Czeslaw Lewa
Institut Fizykiul.
Wita Stwosza 57
80-952 Gdansk
POLAND

Walter G. Mayer
Georgetown University
Department of Physics
Washington D.C. 20057
USA

Salvatore Mazzola
ITTP - C.N.R.
Mazara del Vallo, TP
ITALY

Francisco Montero de Espinosa
Centro de Fisica Aplicada
Instituto de Acustica
Serrano 14428006
Madrid SPAIN

Holgher Peehs
Universität Gebäude 37
Fraunhofer Institut IfzP
W-6600 Saarbrücken 11
GERMANY

Alberto Petri
CNR - Istituto di Acustica
Via Cassia 1216
00189 Roma, ITALY

Gabriele Pfannschmidt
SIEMENS AG HL QS AFA
2Balanstr. 73D -
8000 München 80
GERMANY

Gerard Quentin
Université Paris VII
Group de Physique des Solides
Tour 23 2 Place Jussieu
75251 Paris Cedex 05
FRANCE

David E. Raeside
2121 Grand View Avenue
Norman - Oklahoma 73072
USA

Rainer Reibold
Bundesanstalt Braunhschweig-Akustik
Physikalisch-Technische
Bundesallee 100
2300 Braunhschweig
GERMANY

Jacques Sapriel
Centre National d'Etudes des
Telecommunications
Bagneux
FRANCE

Natalia Shibanova
Moscow Inst. Physics and Technology
Institutsky par 9
Dolgoprudny Moscow region
141700 USSR

Antoni Sliwinski
University of Gdansk
Institut Fizyki
ul. Wita Stwosza 57
80-952 Gdansk
POLAND

Eugene Soczkiewicz
Silesian Technical University
Institute of Physics
ul. Krzywoustego 2
44-100 Gliwice
POLAND

Valerio Vignoli
Dip. Ingegneria Elettronica
Via S. Marta 3
50100 Firenze, ITALY

Igor Yakovkin
USSR Academy of Sciences
Institute Semiconductor Physics
Lavrentjev ave. 13
630090 Novosibirsk
USSR

Andrea Zanobini
Dip. Ingegneria Elettronica
Via S. Marta 3
50100 Firenze, ITALY

www.ingramcontent.com/pod-product-compliance
Lightning Source LLC
LaVergne TN
LVHW051053210926
848486LV00002B/101

* 9 7 8 9 8 1 0 2 0 9 1 2 4 *